Under Osman's Tree

Under Osman's Tree: The Ottoman Empire, Egypt, and Environmental History

Alan Mikhail

The University of Chicago Press :: Chicago and London

The University of Chicago Press, Chicago 60637
The University of Chicago Press, Ltd., London

Published 2017
Paperback edition 2019
Printed in the United States of America

28 27 26 25 24 23 22 21 20 19   1 2 3 4 5

ISBN-13: 978-0-226-42717-1 (cloth)
ISBN-13: 978-0-226-63888-1 (paper)
ISBN-13: 978-0-226-42720-1 (e-book)
DOI: 10.7208/chicago/9780226427201.001.0001

Library of Congress Cataloging-in-Publication Data

Names: Mikhail, Alan, 1979– author.
Title: Under Osman's tree : the Ottoman Empire, Egypt, and environmental history / Alan Mikhail.
Description: Chicago ; London : The University of Chicago Press, 2017. | Includes bibliographical references and index.
Identifiers: LCCN 2016035740 | ISBN 9780226427171 (cloth : alk. paper) | ISBN 9780226427201 (e-book)
Subjects: LCSH: Human ecology—Egypt. | Egypt—History—1517–1882. | Turkey—History—Ottoman Empire, 1288–1918. | Human ecology—History. | Human ecology—Middle East.
Classification: LCC GF13.3.E3 M55 2017 | DDC 333.70956/0903—dc23
LC record available at https://lccn.loc.gov/2016035740

♾ This paper meets the requirements of ANSI/NISO Z39.48-1992 (Permanence of Paper).

Contents

Preface: Osman's Tree xi

Introduction: The Global Environmental History of the Middle East 1

Part One: Water

1 Irrigation Works 19
2 History from Below 34
3 Silt and Empire 51

Part Two: Work

4 Rural Muscle 73
5 Expert Measures 93

Part Three: Animal

6 Animal Capital 111
7 Brute Force 131

Part Four: Elemental

8 Food and Wood 153
9 Plague Ecologies 169

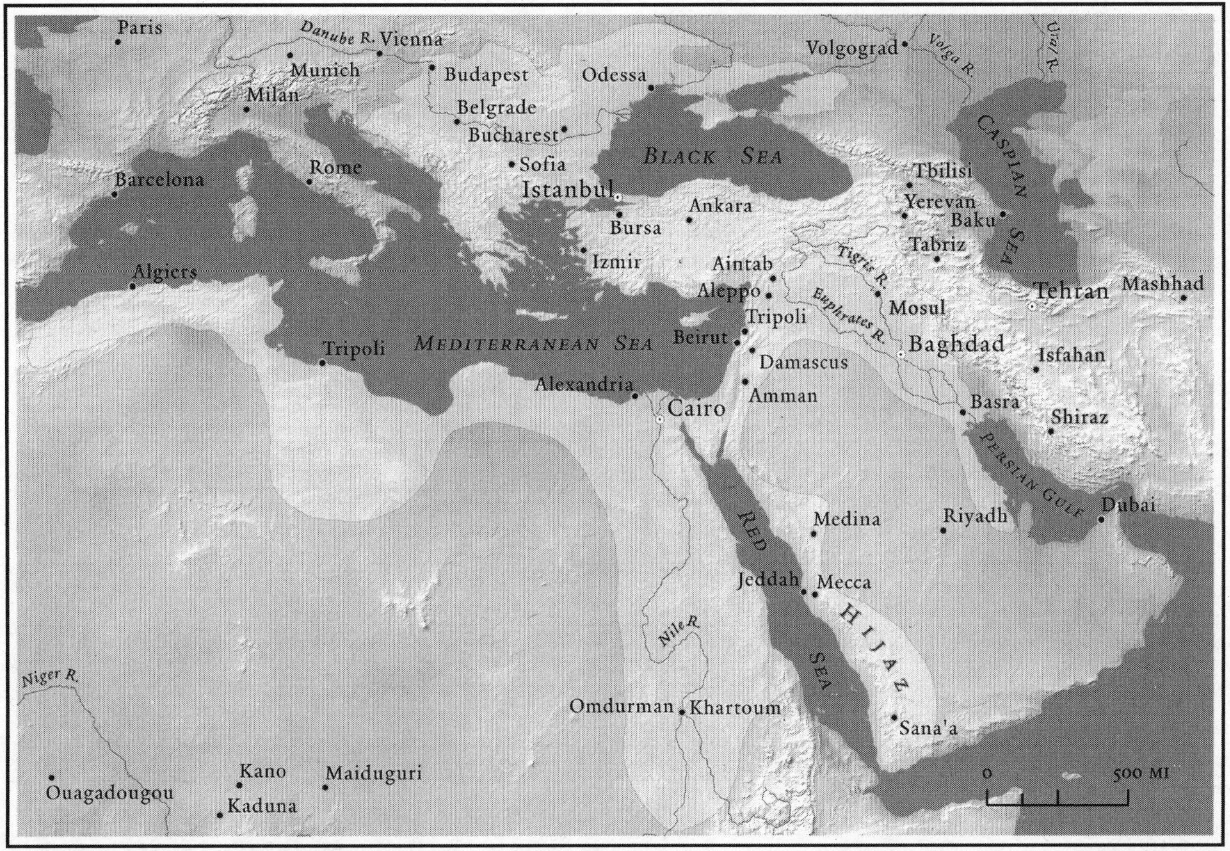

1. The Ottoman Empire, ca. 1650. Stacey Maples, 2014.

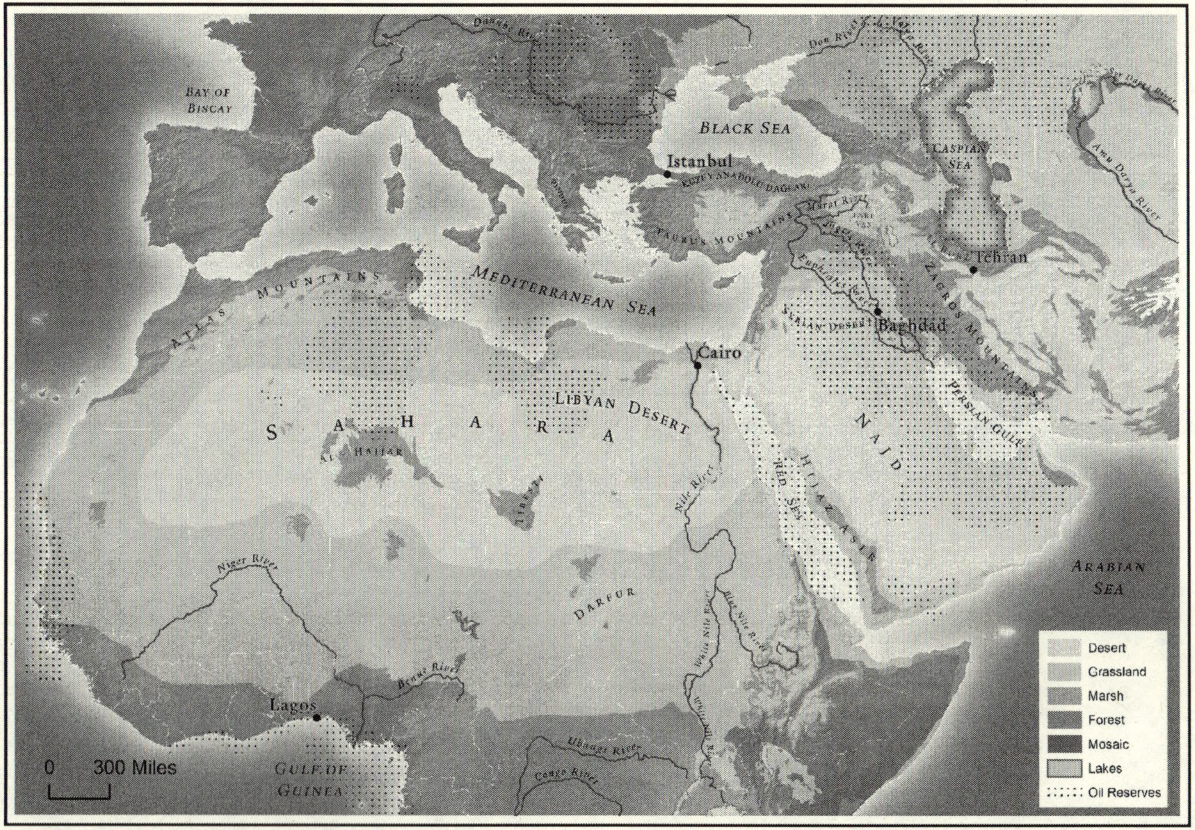

2. Natural resources in the Middle East and North Africa. Stacey Maples, 2012.

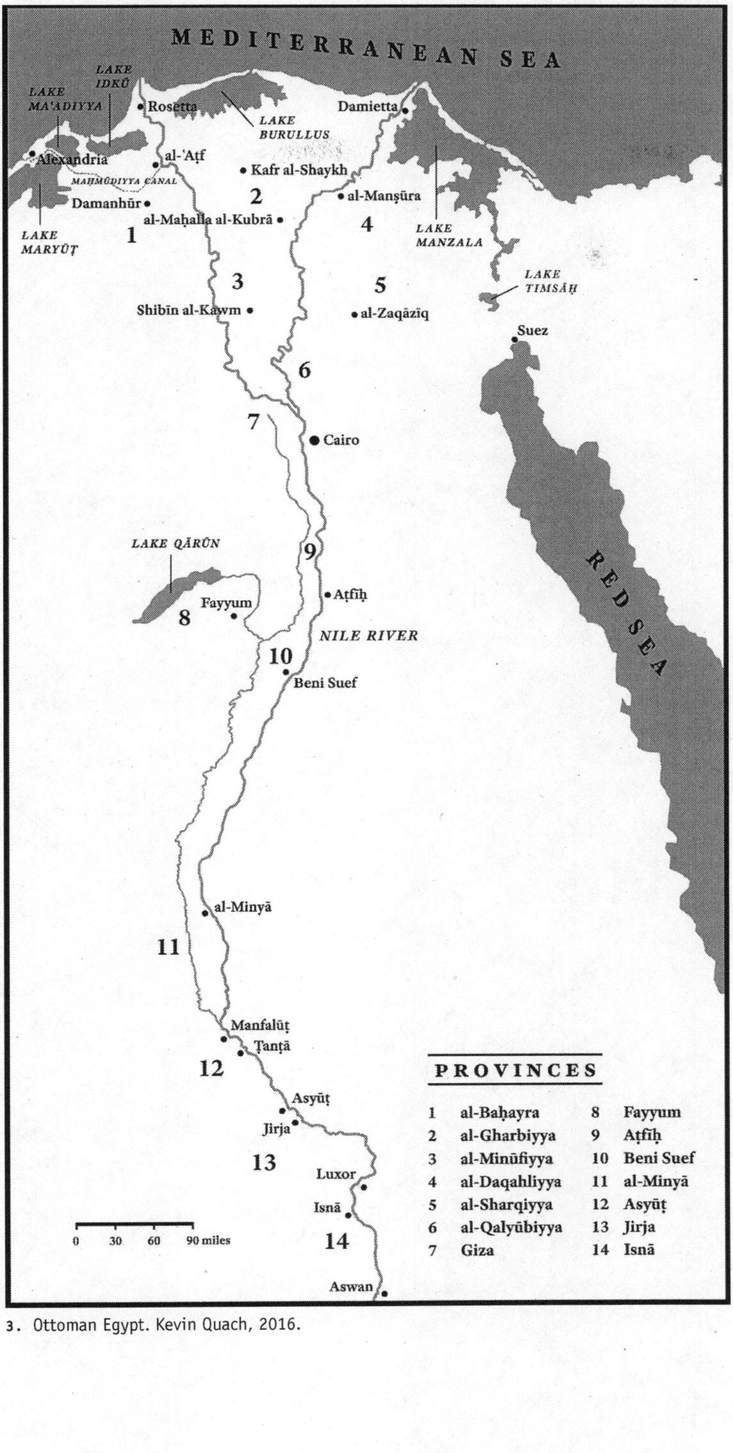

3. Ottoman Egypt. Kevin Quach, 2016.

Preface: Osman's Tree

The Ottoman Empire began under a tree. According to the most widely accepted foundation myth of the longest-lasting and most important political power in the Middle East over the last millennium, Osman, the purported founder of the empire, had a dream in which a moon entered his chest causing a tree to sprout from his navel.[1] This tree's shade then covered the world. At the foot of the mountains under its cover, streams flowed forth giving sustenance and life. Some drank from the streams, others watered their gardens, and still others built fountains. When Osman woke from his dream, he related it to the dervish in whose house he was staying, as it was from this man's breast that the moon in Osman's dream had originated. Upon hearing Osman's rendition of his vision, the holy man understood that Osman was to be the sovereign of a powerful state, and so he decided to marry his daughter Malhun to the rising leader.

This Ottoman origin story was put down on paper only at the end of the fifteenth century, nearly two centuries after Osman supposedly had his dream, at a time after the conquest of Constantinople when the empire was firmly established as a major political and military player in the Mediterranean. The dream has been analyzed in many different ways—as a story of the unity of religious and worldly power, the beginnings of the marriage politics of the Ottoman dynastic line, a declaration of territorial ambitions. It has, however, not been seen for what it most obviously is—a literal statement of the forging of political sovereignty with and through the power of natural resources. The main narrative arc of the story is an ecological one. Trees, water, lunar energy, mountains, gardens—these are the origins of imperial political power.

This book is an attempt to take seriously this connection between nature and power in Ottoman history. It brings the tools of environmental history to the study of the empire to show how analyzing the many relationships between peoples and environments offers a holistically fresh perspective on Middle Eastern history. Its focus is on the last five hundred years and on Egypt, the Middle East's most populous and historically most lucrative region. After the Ottomans conquered it in 1517, Egypt proved the most significant province of the empire, at the center of the entire Ottoman system of governance and economics. It was the most profitable province of the empire and its largest single supplier of foodstuffs. Egypt was crucial to Ottoman rule in the Mediterranean, Red Sea, and Indian Ocean; Ottoman functionaries in Istanbul maintained major economic interests in Egypt, as did Syrian, Venetian, and French merchants; and in the late eighteenth and early nineteenth centuries, some of the largest military threats to the empire's stability and durability came from Egypt.

In addition to its centrality to the Ottoman Empire, Egypt is also the perfect case study for understanding the role of ecology in politics and politics in ecology. Its reliance on the Nile, its rich agricultural history, and its geographic location between two seas and two continents have always put natural resources at the center of any project to govern Egypt. Still, while Egypt may have been the gift of the Nile, apart from such neat and tidy statements, the actual social, political, and economic history of Egyptians' relationships with the Nile and the environments it produced has yet to be fully understood. Using environmental history as an interpretative and methodological tool, the following chapters open up these many interactions between peoples and environments and between Egypt and the Ottoman Empire to understand the realities of living in the empire, on the land, and with the river. In forestless Egypt, in the shade Osman's tree provided from afar, humans, animals, the Nile, plants, silt, wind currents, and germs were locked together in relationships of reliance, conflict, and mutual constitution. This book tells part of this history.

: : :

Under Osman's Tree is both a synthetic treatment of and argument for Middle East environmental history. For the past decade, I and a small group of other Middle East historians have worked to try to bring an ecological perspective to Middle East history and to show environmental historians of other regions the utility of considering the Middle East (the introduction that follows will discuss this historiography in detail). *Under Osman's Tree* weaves together some of the major outlines of my thinking about Middle East environmental history into an accessible and sweeping methodological account. My hope is that it will serve as a kind of primer for Middle East environmental history.

Under Osman's Tree is informed by, updates, and extends my previous scholarship. Those familiar with my work will see some of the same themes in this book, supplemented, reworked, and recast. My first book, *Nature and Empire in Ottoman Egypt*, focused largely on the politics of irrigation in the Egyptian countryside and how the resources built around the Nile's watering of Ottoman Egypt affected imperial rule in Egypt and throughout the empire. A second book, *The Animal in Ottoman Egypt*, turned to changing relationships between humans and animals as a lens through which to analyze the political, social, economic, and environmental history of early modern and nineteenth-century Egypt. Irrigation and human-animal relations play a major role in this book too—not only because they are topics I know well, but, more important, because they are crucial for any understanding of the environmental history of the Middle East. In addition to water and animals, many other topics are covered in this book: climate, disease, labor, and economic transformation, to name just a few. This book thus aims to offer a capacious view of the current state of the field of Middle East environmental history and to point to some of the many possible fruitful ways to carry it forward.

Introduction: The Global Environmental History of the Middle East

One of the gaping holes in the global story of the environment has so far been the history of the Middle East. Likewise, one of the gaping holes in the study of the Middle East has so far been the history of its environment. This is a major problem for our understanding both of global environmental history and of the history of the Middle East as a region. From at least antiquity, the Middle East has been the crucial zone of connection between Europe, the Mediterranean, and Africa, on the one hand, and East, Central, South, and Southeast Asia, on the other.[1] It was an arena of contact where European and South Asian merchants traveled and conducted commerce and where each year religious pilgrims from Mali to Malaysia came not only to visit Mecca and Medina but also to buy goods in the many bazaars of cities throughout the Middle East. Pastoral nomads moved from northern India and Central Asia through Iran and into Anatolia.[2] Ships carrying goods, peoples, vermin, and ideas sailed from India, China, and Southeast Asia to ports on the Red Sea and Persian Gulf. Beginning in the late medieval and early modern periods, the world from the western Mediterranean to China was characterized by intense circulation, interconnection, and movement.[3] All the threads of these connections ran through the Middle East, with important environmental effects.[4]

Just as global trends in trade were deeply intertwined with what was happening in the Middle East, so too large-scale trends in climate, disease, and crop diffusion impacted and were impacted by the histories of the Middle East's many environments. For example, one of the results of the rise and spread of Islam in the seventh and eighth centuries was the movement of new kinds of crops and agricultural technologies across Eurasia. In what one scholar has termed the "Islamic green revolution," the unity provided by the new religion and the culture it produced for the first time brought disparate parts of the world together into a unified ecological contact zone.[5] In the medieval period, it was most likely through the Middle East that plague moved westward to the Mediterranean basin.[6] In another medieval example of environmental exchange, knowledge of irrigation technologies and waterworks was transferred from the Muslim world to Spain and Italy.[7] In the fifteenth and sixteenth centuries, coffee grown in the soils of Yemen, using techniques borrowed from East Africa, soon filled cups in Istanbul, Vienna, and Isfahan.[8] Trans-Saharan caravan networks ensured a steady circulation of animals, salt, paper, disease, and human slaves among the Middle East, North Africa, and West Africa.[9] And it was most likely through the Ottoman Empire that maize and other New World crops moved east.[10] Since the twentieth century, clearly the extraction, refinement, and distribution of Middle Eastern oil have been of enormous global environmental and geopolitical significance.[11]

The histories of various parts of the Middle East were also deeply affected by different instances of global climate change. Whether the Little Ice Age in the early modern period that greatly reduced agricultural yields in the Ottoman Empire, Icelandic volcanic eruptions at the end of the eighteenth century that affected Nile flood levels, El Niño–induced famines at the end of the nineteenth century in Anatolia and Iran, or today's global warming, the Middle East has clearly been deeply ensconced in the history of these instances of climatic alteration, and studying these cases can offer us much as we try to understand the global history of climate change.[12]

In other words, the Middle East was a region that served to diffuse agricultural and natural products, knowledge of environmental manipulation techniques, climatic forces and their effects, and disease agents across large swaths of Eurasia and the world. Therefore, to understand the history of these crops, diseases, commodities, weather patterns, and technologies, we must necessarily understand the Middle Eastern components of their global histories.[13] This is perhaps patently clear and unsurprising, but much of this history remains unknown and continues to be underresearched.

This book is an argument for Middle East environmental history. It explains and analyzes where the field came from, where it is, and where it is going. This introduction and the chapters that follow focus on the last five hundred years of Middle Eastern history, tackling major topics in environmental history: natural resource management, climate, human and animal labor, water control, disease, and the politics of nature. The book is centered on the Ottoman Empire, the Middle East's longest-lasting and most important political power since antiquity, and on Egypt, the Middle East's most populous and historically most lucrative region. Concentrating on the Ottoman Empire and Egypt allows for focused empirical analyses of the multiple relationships between peoples and environments and between one of the Middle East's most important polities and its most vital constituent part. From its first incorporation into the Ottoman Empire in the first half of the sixteenth century until its final takeover by British forces in 1882, Egypt was the most significant and lucrative province of the empire and its largest single supplier of foodstuffs. Egypt was crucial to Ottoman rule in the Mediterranean, Hijaz,[14] Red Sea, and Indian Ocean. In the late eighteenth and early nineteenth centuries, some of the largest military threats to the empire came from Egypt. Ottoman functionaries in Istanbul, moreover, maintained large economic interests in Egypt, as did Venetian merchants and others. And the province was a key staging ground for Ottoman expansionist efforts in Africa in the nineteenth century. Because Egypt was so vital to the whole of the Ottoman imperial system (and beyond), it consequently had greater potential than any other province to threaten and weaken the entire empire.

Although the geographic focus of this book is Ottoman Egypt, I strive throughout to explicate the global dimensions of this imperial province's history. Indeed, as already suggested, one of the arguments of this book is that environmental history is a methodology well suited to allowing us to see the global history of the Ottoman Empire. To understand Egypt and its place in the empire, we will travel throughout the imperium itself—to the forests of Anatolia, to the capital of Istanbul, to the grain markets of North Africa and Salonica, and to the Red Sea ports of the Arabian Peninsula. We will have to go far beyond the empire too—up the Nile to the Sudan and Ethiopia and even farther to places such as India and Iceland.

Using environmental history as an interpretative and methodological tool, this book thus offers a holistically fresh global perspective on Middle Eastern history; at the same time, it serves as a template for other possible environmental histories not only of the Middle East but elsewhere as well. In the rest of this introduction, I will first review the historiographical possibilities of writing Middle East environmental histories with an eye toward the question of the available sources for such histories.[15] I will then discuss three areas of research in Middle East environmental history that have received some attention from historians: climate, energy, and disease. This discussion will both introduce the broad outlines of some of the field's current literature and also lay the groundwork for the chapters to follow.

Middle East Environmental History: Seasons of Want or Plenty?

One of the most commonly cited reasons for the lack of environmental scholarship about the region is a paucity of sources. As much new work is showing, however, it is simply untrue that there are not enough sources to narrate environmental histories of the Middle East and North Africa. In fact, just the opposite is the case. Like China and South Asia, the Middle East offers a wealth of source material from antiquity to the present.[16] In the Nile valley, for instance, we have an unbroken documentary record of Egyptians' interactions with the river from at least as early as 3000 BCE to the present.[17] Similarly, archaeological and written sources from antiquity to the present exist on environmental management techniques in Anatolia, the Iranian Plateau, the Mesopotamian lands of Iraq, and elsewhere. Thus, the available published and unpublished sources for Middle Eastern history offer environmental historians nearly unparalleled empirical breadth and depth to be able to track environmental change, landscape manipulation, and human-environment interactions over millennia. Having the sources to operate over this vast time scale and at such a level of detail are luxuries not enjoyed by historians of most other regions of the globe.

Naturally, there are very few chronicles, histories, and archival collections specifically dedicated to Middle East environmental history. Almost all of the traditional primary sources for Middle East history, however, offer information about human engagements with nature. Middle East environmental history must thus be culled from close readings and collated from multiple source bases. Chronicles in Arabic, Turkish, and Persian—and before them Greek, Akkadian, ancient Egyptian, and old and middle Persian—document, for example, instances of agricultural hardship, extreme weather conditions, natural disasters, and cultivation techniques. As this book bears out, the archives of the Ottoman state, surely the most important repository for the history of the Middle East after 1500, are replete with all kinds of useful sources for Middle East environmental history—land surveys, records of infrastructural projects, legal disputes over rural property, and accounts of disease outbreaks are only the most obvious.[18] Colonial and postcolonial state archives as well offer sources on these and numerous other topics.[19] Because political, economic, and social power in the Middle East, as elsewhere, has for millennia derived primarily from control over natural resources, the productive cultivation of land, and the management of rural peoples, writers from various periods and places have given us a wealth of source material to use for Middle East environmental history.

Thus, in debates over the long-term effects of human manipulation of the natural world, the Middle East offers us one of the most complete and longest available documentary records of any region of the globe. And in conjunction with more traditional political and economic histories, there is also a sizable geographical, palynological, biological, and geoarchaeological literature that gives us a historic and cultural geographic picture of ecological systems and their operations.[20] This literature can be roughly divided into two major categories: descriptions of the continuities of various delicate equilibriums of human settlements and environments, and studies of human groups inevitably doomed to collapse because of inherent contradictions in their ecological systems. Both of these modes of narrating the environmental past are more concerned with painting a picture of ecological systems and their function than they are with either how humans were integrally connected to environments or how the many contingent relationships between humans and nature changed over time.

In the first story of continuity, the picture is one of dynamic stasis—of human communities constantly working and struggling to survive in the face of enormous environmental challenges. These histories of societies constantly running only to stay in place generally devote very little attention to the effects of human agency on ecological systems. Likewise, the second narrative of inevitable and eventual decline also gives primacy of place to ecological limits while sidelining historical specificity and contingency. If certain human communities were predetermined to decline because of a fundamental flaw or contradiction inherent in their spot on this earth, then what, other than documenting this downward spiral, is left for the historian to do?

Two examples from this geographic literature are Karl Butzer's work on Egypt and Peter Christensen's study of Iran.[21] Both accounts sketch a picture of the ecological limits within which human communities were made to live and survive. Butzer's study of Egypt fits the first paradigm of continuity, while Christensen's work analyzes an example of decline and eventual demise on the Iranian plains. In both of these cases and in others as well, the descriptions are of systems in overall equilibrium on the time scale of generations despite massive fluctuations from year to year. It is no coincidence that both examples—Egypt and the Iranian plains—are areas reliant on the yearly floods of massively complex river systems. Floods could be good or bad from year to year, but overall, through the development of various technologies of environmental manipulation, chiefly irrigation and agriculture, communities were able to ride out these vagaries to achieve some semblance of ecological balance that either continued for millennia, in the case of Egypt, or eventually broke down, as in the case of Iran. Other examples of this kind of description of ecological systems that prevailed throughout the Middle East before 1500 include J.M. Wagstaff's general study, Carlos Cordova's work on landscape change in Jordan, Russell Meiggs's and J.V. Thirgood's examinations of wood supplies around the Mediterranean, and Robert McC. Adams's study of southern Iraq.[22]

Some of these and other historians have also successfully used the wealth of available evidence to examine how and with what techniques humans in extremely ecologically fragile environments managed to survive for millennia and what perturbations to these environments made them no longer viable.[23] In this vein, Christensen successfully shows how various enclave communities on the Iranian Plateau were able to manage their extreme environmental vulnerability for centuries through the resilient and careful use of irrigation technologies, agriculture, and nomadic pastoralism. Similarly, J.R. McNeill paints a picture of Mediterranean mountain environments in Turkey and Morocco that makes clear just how fragile these environments were over time.[24] With a long documentary record at his disposal, he is able to show that from year to year the fortunes of any one environment might rise and fall drastically, yet, when seen over the course of centuries, these environments were actually quite resilient and able to deal with intense change. Thus, it is only because of the presence of a comparatively long record that we are able to see how these environments functioned over time. Without this record, we might mistakenly judge a year of acute food shortages to be the norm in the mountains.

The presence of such a long record of human interactions with the environment in the Middle East also affords environmental historians ample opportunity to offer new perspectives on some long-debated issues in the field of environmental history. The following section, for example, addresses some of the implications of Middle East environmental history for current understandings of the history of climate change.[25] One of the more obvious and perhaps the most discussed historiographical intervention of the Middle East in the field of environmental history revolves around Karl Wittfogel's thesis of Oriental despotism.[26] Several of his examples concerning the emergence of despotic forms of authoritarian government come from the Middle East—Mesopotamia and the Nile valley. However, it is precisely these examples that refute his argument. An examination of the documentary record Wittfogel fails to consult shows that it was the very people he claims were exploited by sultans, shahs, and emperors who were the ones actually in control of the day-to-day function and maintenance of the large-scale irrigation networks that are so crucial to his idea.[27] This politics of irrigation is the subject of part I.

Environmental historians interested in the history of natural disasters and their effects on human communities will likewise find much material in the Middle East to test their hypotheses. We have records and accounts of earthquakes and other events from antiquity to the present.[28] The history of public works is also similarly usefully addressed by considering the record of such projects in the Middle East. As with elsewhere in the world (India and the American Southeast, for instance), the middle of the twentieth century was a period of intense dam building throughout the Middle East—at Aswan, in Iran, and in Turkey.[29] What was the connection between these major public works projects and earlier instances of this kind of environmental work (the Suez Canal, the Ottoman Don-Volga canal-building project, or Safavid plans to build a dam to divert the course of the Karun River toward the new capital at Isfahan) or current ones (the Red to Dead pipeline project in the Jordan valley or Turkey's dam projects in its southeast, for example)?[30] Again, because we have a record of such projects and social, economic, cultural, and political commentaries about them, they offer a rich source base for thinking through the environmental history of public works both in general and in the Middle East in particular.

Alongside this potential for *longue durée* environmental analyses of the Middle East, recent scholarship has also shown some of the ways the region's copious source materials can be used to construct an environmental story that elucidates the history of much more concentrated periods of time.[31] Most of these studies rely on masses of archival source materials that, when used judiciously with physical data, allow us to understand how humans have interpreted and dealt with environmental change and fluctuation. Unlike some older works, these newer studies put human engagement with nature much more squarely at the center of their analyses, recovering some of the histories physical data alone cannot capture. This is an important distinction between the two time scales involved in the different uses of sources for the history of the Middle Eastern environment. In some of the older geographic literature, human agency is often limited to a constant timeless fight against the inevitability of environmental circumstances—a kind of environmental determinism, in other words.[32] More recent environmental histories, by contrast, focus more on the back-and-forth, push-and-pull relationship between humans and nature.

These newer works therefore exemplify a kind of environmental history of the Middle East that gives center stage to a dialectical understanding of humans and environments. According to William Cronon's well-known formulation, "Environment may initially shape the range of choices available to a people at a given moment, but then culture reshapes environment in responding to those choices. The reshaped environment presents a new set of possibilities for cultural reproduction, thus setting up a new cycle of mutual determination."[33] Cronon continues that environmental history therefore seeks "to locate a nature which is within rather than without history, for only by so doing can we find human communities which are inside rather than outside of nature."[34] This book, along with other work in Middle East environmental history, seeks to locate the kind of nature that is within rather than outside of Middle Eastern history.

Climate

One of the richest areas of current research in the environmental history of the Middle East is the study of climate change.[35] Much of this work challenges various propositions put forth in the literature on global climate change.[36] One of these is the idea that the Middle Ages were a period of global warming. Evidence for this worldwide weather trend comes mostly from European and Chinese sources. Using Persian and Arabic sources from medieval Iran, however, Richard W. Bulliet has shown that in fact the opposite seems to have been occurring in Iran in this period.[37] That is, an intense period of cold—what he calls the "Big Chill"—gripped Iran in the eleventh and early twelfth centuries, leading to reductions in cotton cultivation levels. The Big Chill facilitated Turkic migrations west into Iran and eventually Anatolia and perhaps even contributed to the emergence of Shiʿism as the dominant form of religiosity in Iran. This period of cooling decimated large landholdings and much of the economic and social base of Iran, which likely made the Iranian Plateau vulnerable to Mongol and other incursions. After this period of cold, the Middle East and parts of Central Asia enjoyed a few centuries of temperate weather patterns that contributed to the emergence of the great empires of the early modern period—the Ottomans, Safavids, Mughals, Uzbeks, and Mamluks.

In the sixteenth and seventeenth centuries, however, another more well-known period of cooling set into the Middle East and elsewhere. The Little Ice Age contributed to various economic and political troubles for the Ottoman Empire, forcing a realignment of resources and sovereignty, and was also a contributing factor to the general crisis of the seventeenth century.[38] A consideration of this climatic event puts the historiographically vexed issue of the decline, decentralization, or realignment of the Ottoman Empire in the seventeenth and eighteenth centuries in a wholly new light in which ecological factors stand alongside political, economic, and social dynamics as causes for change in the early modern period. One of the important and most discussed facets of this period in the Ottoman Empire is the Celali peasant revolts at the end of the sixteenth and the beginning of the seventeenth centuries.[39] As Sam White's research shows, there is now ample evidence linking these rural rebellions to ecological pressures precipitated by the Little Ice Age.[40] The manifestations of this climatic change in central Anatolia—drought, famine, plagues of livestock and people, and eventually increasing rates of mortality among Ottoman peasants—led to the violence and banditry that finally coalesced into an organized rebellion against the Ottoman state. Thus, the causes of the Celali revolts are not only, or perhaps even most importantly, to be found in the realms of the economic and the political, but rather must also be seen as results of ecological forces. This middle period of Ottoman history that so many historians have interpreted as the crucial hinge between the rise and sixteenth-century efflorescence of the empire and its eventual efforts at reform in the nineteenth century is thus usefully thought of as a time when the empire was attempting to manage a set of new realities that included massive climatic fluctuations and their attendant effects on agricultural production, human and animal populations, and flood levels.

Likewise, the last two decades of the eighteenth century were also a period of weather fluctuation, albeit less dramatic, but no less crucial. The eruption in 1783 and 1784 of the Laki fissure in Iceland began a climate event of global proportions.[41] As chapter 10 discusses, monsoon rainfall levels decreased over the Indian Ocean, temperatures plunged around the Mediterranean, flood levels of various riverine systems in the Middle East were dramatically reduced, and a twenty-year period of famine, drought, disease, and hardship began. These climatic changes were no doubt contributing factors to the political turmoil of the end of the eighteenth century that resulted in, among other things, various European incursions into the Middle East and eventually the period of Ottoman reform known as the Tanzimat.[42] In the latter half of the nineteenth century, a similar period of climatically induced famine and food shortages contributed to discontent among provincial notables in the Ottoman Empire and to calls for political reform in the empire.[43]

In short, there is plenty of good reason to conceive of the last millennium of Middle Eastern history along the lines of climate and weather patterns rather than political states or military confrontations. Indeed, it seems that periods of cooling or climatic fluctuation set off by events like the Laki eruption in Iceland often had much more impact on the vast majority of rural peoples of the Middle East than did the political powers ostensibly ruling over them. The point is not to say that climate single-handedly determined the political and economic fortunes of human communities.[44] Clearly not. The point is rather to accept that historical change resulted from a complex mesh of factors that included climatic fluctuations and extreme weather events, as well as, of course, political, economic, and social transformations.

Energy

The Middle East also has much to contribute to understandings of the global history of energy. Accepting that there has only been one major shift in energy regimes over the course of human history—from solar energy to fossil fuels—the Middle East, with the largest known petroleum supplies in the world, clearly played a central role in this story.[45] Understanding how humans cultivated and then exploited their energy supplies must be at the heart of any history of human communities.[46] Since the latter half of the nineteenth century, much of the world's energy supplies have come from the Middle East, and the region must therefore be included in any history of energy since this period. The Middle East, however, was clearly also very important to the global story of energy long before oil was ever discovered in the region. As in most of the preindustrial world, energy in the Middle East before the nineteenth century was characterized by massive reliance on human and animal power.[47] Water and wind power never made up a large portion of the energy resources utilized by people from Morocco to Iran. This was a result not of a lack of running-water sources, but rather the availability of abundant quantities of cheap animal resources. Waterwheels were therefore pushed by oxen; goods were transported by donkey rather than by river; and fields were tilled by animal plow. Thus, the history of animal and human power in the Middle East, as discussed in parts II and III, helps us to understand this most vital of preindustrial environmental energy stories as it played out elsewhere in the world too.

The more recent global transition to a fossil fuel economy therefore represents the replacement of animals with oil as the region's most abundant and cheapest source of energy.[48] This energy transition is one that has largely occurred throughout the world, increasing the geopolitical importance of Middle Eastern oil reserves. Putting petroleum utilization and consumption in the larger historical context of transitions between energy regimes shows us that oil and the supposed need to seek out alternative sources of energy in the Middle East are the latest iterations of a much longer story of low-cost energy.[49] Because animal power was (and in some places still is) so effective and cheap, there was no incentive to find and use other sources of energy, like water or wind. Thus, animals remained the primary means of transport and power in the Middle East and Central Asia while Europeans began experimenting with water-powered mills. Oil, like animals before it, therefore, represents a new phase of energy regime in the Middle East, and as long as petroleum remains abundant and cheap, there will be little incentive for Middle Eastern oil states to seek out other forms of energy.[50] By making energy outputs, whether caloric or carbon organic, our unit of analysis, then, the politics of oil in the Middle East no longer comes to be seen as only a problem of modern industrial economies, but rather as a much older problem of energy and society.

Disease

One of the most productive areas of research on the global import of the environmental history of the Middle East has been and continues to be the epidemiology and etiology of various diseases. Because so many disease epidemics in the Middle East, whether plague, cholera, or something else, deeply impacted populations and almost always traveled between the region and elsewhere, we have ample documentary evidence of these outbreaks, which historians have used to great effect.[51] Some of this work is beginning to put disease in a wider ecological context as one of various other environmental phenomena.[52] This is partly a result of the realization of how disease functioned in Middle Eastern environments and of how people in those environments conceived of disease. In other words, people who experienced plague and other diseases in the Middle East considered them to be one in a series of environmental forces that included famine, rain, drought, flood, and wind.

For example, in the case of Egypt, plague was just one aspect of a regular biophysical pathology of the environment.[53] As we will see in chapter 9, the disease came and went as one stage of a cycle that included famine, flood, drought, price inflation, wind, and revolt. Climatic factors, the level of the Nile's inundation, and relative populations of rats, fleas, and humans all contributed to the incidence and severity of plague for any given year in Egypt. As such, like the annual Nile flood, increases in the prices of grain, famine, or other hardships, plague was an accepted and expected environmental reality. Egyptians understood it to be one pathological element of the Egyptian environment. Part of the reason for this acceptance of plague as part of the Egyptian environment was its regularity and frequency. A new plague epidemic visited Egypt an average of every nine years for the entire period from 1347 to 1894,[54] and plague was reported in Egypt in 193 of these 547 years.[55] Plague in Egypt owed the regularity of its outbreaks to Egypt's strategic location as a center of trade and to the many thousands of people—and rats and fleas—that came to Egypt every year from elsewhere.[56] The two primary entry points of the disease were through the main trading areas of Egypt: its Mediterranean ports and the southern route from the Sudan.[57] Thus, although plague was not epidemiologically endemic to Egypt, it was a historically endemic phenomenon that functioned in conjunction with other environmental and natural forces. The high incidence of plague in Egypt suggests that the disease and its epidemics were regular and expected occurrences in the lives of most Egyptians and that historians should pay closer attention to the role of the disease in the shaping of Egyptian and Middle Eastern history.

Another example of the centrality of the Middle East to the global environmental story of disease is the role of the annual Islamic pilgrimage (*ḥajj*) in the spread of various epidemic diseases. As perhaps the largest single annual gathering of people anywhere in the early modern and modern worlds, the *ḥajj* had important epidemiological implications. For a few days every year, hundreds of vessels, thousands of people, scores of caravan animals, and millions of germs from all over the world gathered and jostled together on the western coast of the Arabian Peninsula. Upon returning home, many people found that they had contracted some form of epidemic disease. In the nineteenth century, this was usually cholera. All over the world, but especially in South Asia, North America, and Europe, in the decades after the 1830s, cholera struck major cities with great force, leading to the implementation of various kinds of protective measures.[58]

The pilgrimage was one of the main avenues through which cholera moved between South Asia and the Mediterranean and between areas of Europe and the Middle East.[59] Both before and after the opening of the Suez Canal in 1869, Egypt was one of the main transit points for *ḥajj* pilgrims—and hence disease—traveling to and from the Mediterranean. Egyptian ports thus became especially important as choke points to prevent the movement of cholera to and from the Hijaz during the pilgrimage season. In 1833, for example, two thousand pilgrims from various parts of the Ottoman Empire and several hundred Russian Tatars from around the Black Sea were quarantined in a lazaretto in Alexandria on their way to Mecca.[60] Once it was determined that they were not infected with cholera, plague, or any other disease, they were allowed to join the caravan to the Hijaz. The Alexandria lazaretto could accommodate up to twenty-five hundred people; the government of Ottoman Egypt provided those in quarantine with food, clean quarters, and various other provisions. The British consul general who visited the Alexandria lazaretto in 1833 made special note of the facility's cleanliness, convenience, and comfort.

Such quarantine efforts notwithstanding, cholera epidemics originating in the yearly pilgrimage would at various other points in the nineteenth century overwhelm those stations charged with controlling the disease's diffusion.[61] For example, in 1881, there was a particularly virulent outbreak of cholera in the Hijaz.[62] In response, the quarantine board of Alexandria ordered all pilgrims returning to Egypt from the Hijaz to spend at least twenty-eight days in one of three quarantine stations established in the city. These three facilities, however, were soon overwhelmed by the enormous number of pilgrims who came to the city. Overcrowding, a lack of food and water, and a decision by the quarantine board to extend the mandatory detention period made an already bad situation even worse. A contained outbreak of cholera thus became a full-blown epidemic throughout these camps and in surrounding areas. Many of the pilgrims being held in these facilities soon began to riot, and some even succeeded in burning down one of the lazarettos before escaping altogether. These and other experiences of diseases associated with the pilgrimage led various governments at the end of the nineteenth century to institute medical regulations for pilgrims going on and coming from the yearly pilgrimage.[63] These measures were attempts to manage the centrality of the *ḥajj* to the global ecology and etiology of disease.

Conclusion

Given the Middle East's enormous cultural traditions and long-standing diversity; its centrality to the global traffic of economics, politics, and ecology; and the vast quantities of source materials available from various regions and from all periods, it should be no surprise that the Middle East has much to teach us about various facets of global environmental history and that environmental history opens up new research avenues for Middle Eastern history. *Under Osman's Tree* is an attempt to bring these two fields together to understand the history of the last half millennium of the Middle East in new ways.

Part I explains the nature of the politics forged through water management, arguing that in Ottoman Egypt irrigation did not produce Oriental despotism, but instead afforded local communities an amount of autonomy and power. Part II shows how a massive polity such as the Ottoman Empire, centered as it was thousands of miles away in Istanbul, functioned on the ground in Egypt through peasant labor, knowledge, and expertise. Just as the empire used peasants, so too peasants used the empire. Part III inserts animals into the story of early modern empires and Middle Eastern history. And part IV analyzes the roles of other nonhuman things—grain, wood, pathogens, and SO_2—in the shaping of Middle Eastern and environmental history. With its specific focus on Ottoman Egypt, *Under Osman's Tree* explores only a small slice of Middle East environmental history. As a marker of the current state of this growing field, the hope is that this book will serve as both a guide for how to move forward and a jumping-off point for new research.

Some possibilities for future work include the environmental impacts of war, gender and the environment, pollution, science and nature, colonialism and the environment, the extraction of resources other than oil, literature and the environment, and the role of environmentalism and other kinds of environmental thought in the Middle East.[64] Additionally, although some contemporary scholars are rethinking the role and history of the environment in Islam (a brand of Islamic environmental ethics), the general subject of Islam and nature remains a ripe area for further research.[65] As Islam was and is a global cultural idiom, both what various Islamic texts from different places and times say about nature, environmental manipulation, the nonhuman living world, and so forth and the history of how Muslims have actually put into real-world effect their religious ideas about nature should be a major area of scholarly interest.[66] The Middle East's extreme demographic heterogeneity and rich cultural geography also afford environmental historians the opportunity to examine how different cultural groups have thought about, cultivated, and lived in the same natural spaces and environments. In southeastern Turkey, northern Syria, Iraq, and Iran, for example, communities of Turks, Kurds, Arabs, Armenians, Iranians, and others share the same rivers, mountains, and plains.[67] Is there a difference in access to environmental resources or in environmental manipulation techniques between groups? What is unique or similar about how each of these groups thinks about and interacts with their shared environments? More generally, both in the past and today, what are the different ways distinct cultural and religious groups engage nature?

The presence of the world's longest river along with other important watersheds; the meeting point of three continents and five seas; the global significance of the region's oil; the Suez Canal and other crucial past and present public works; the historic importance of cotton, coffee, and other agricultural commodities from the region; the presence of a vast diversity of landscapes, cultures, religions, ecologies, and borderlands; and countless other reasons all make the environment crucial to any study of the Middle East and the Middle East crucial to any study of the global environment.

Part One: Water

1 Irrigation Works

Understanding water management is crucial to any understanding of the political, social, and economic history of Egypt. Water management during the period of Ottoman rule in Egypt (1517–1882)—one of the longest single periods of Egyptian history—was characterized by the local control of water resources by rural communities.[1] Peasants, local notables, and others in the countryside participated in a collaborative and collective political system with Ottoman authorities to harness the Nile's water resources in the most effective, sustainable, and equitable manner for the largest number of water users.[2] Those users included rural cultivators in Egypt, consumers of food all over the empire (from Tunis to Salonica to Istanbul to Aleppo), and the entire imperial bureaucracy, which largely ran on the tax revenues produced by Egypt's irrigated soils.[3] To irrigate Ottoman Egypt, peasant communities relied on the administration and money of the empire, and the empire relied on the knowledge, labor, and experience of Egypt's rural cultivators.[4] Like most other regions of the Ottoman Empire, Egypt could only be irrigated through this cooperative arrangement of imperially coordinated localism that delegated authority to thousands of local communities throughout the province. Analyzing various aspects of this water management history, this chapter argues that peasants were the most vital actors in the politics and economics of irrigation in Ottoman Egypt and that they were recognized as such, indeed encouraged in these roles, by the empire's administrative structure. Rather than looking down from the sultan's throne in Istanbul, this chapter stands in the mud of Egypt's canals to look out at Ottoman governance.

Irrigation Surveyed

Perhaps the single most important effort by Ottoman authorities to delineate and understand the complexities of irrigation in the Egyptian countryside was the imperial irrigation survey undertaken after the conquest of the province in 1517. This gargantuan bureaucratic enterprise was an imperial recognition and legal instantiation of local Egyptian peasant knowledge as the key determinant and driver of water management in the countryside. By beginning with a state-centric effort like the imperial survey, I want to make the point that even a centralized administrative tool of this kind was built on the knowledge, expertise, and work of local peasant communities. It could only ever be this way. Irrigation was radically different from place to place. Knowing the water situation of one village in no way helped one understand the situation in neighboring villages. Particular, specific, on-the-ground local knowledge was needed to manage water in Egypt. The Ottomans therefore dispatched imperial representatives to the countryside to get this knowledge, interviewing locals about canals, sluice gates, embankments, and other waterworks in their villages. The information these rural cultivators provided then became the basis for organizing Ottoman water management in Egypt, the empire's most lucrative province. The administrative strictures and structures built on the empire's irrigation survey determined the outcomes of legal disputes over water in the countryside and shaped peasants' interaction with one another and with the imperial bureaucracy. In the end, it was peasants working through the Ottoman system, not the Ottoman state itself, who controlled irrigation in rural Egypt.

The Ottomans' irrigation survey of Egypt, first compiled in 1539 or 1540, as best we can tell from internal and paleographic evidence, was known as al-Jusūr al-Sulṭāniyya.[5] The survey aimed to delineate and map the many communities created by water and irrigation in the Egyptian countryside. Determining which villages used which waterways would reveal aqueous connections between communities and, perhaps more important, show who was collectively responsible for maintaining irrigation works, since those who shared water also shared the responsibility for maintaining the irrigation works carrying that water.[6] This collection of information—a kind of liquid map of the countryside—could then be put toward the ultimate goal of water management in Ottoman Egypt: that all the province's villages and agricultural lands be agriculturally productive and that Egypt enjoy a state of full and total irrigation (*ḥuṣūl al-riyy al-kāmil al-shāmil*).[7]

To achieve the empire's aims of mapping the irrigation network of rural Egypt, determining responsibility for the upkeep of irrigation works, and maintaining the agricultural productivity of Egypt, this survey sought answers to the following questions: Did a particular canal serve the common interests and needs of a large group of peasants, their lands, or a *waqf* (a pious endowment whose revenues funded a particular purpose or group in perpetuity), or, conversely, did the canal only serve the more narrow interests of a particular set of people? Did the repair of a canal lead to equality among peasants and serve the common good (*al-maṣlaḥa al-ʿāmma*)?[8] This irrigation report was also meant to ascertain which canals had been abandoned and were no longer maintained in any regular way. These were canals that had dried up, become filled in with earth, or were no longer clearly visible. Surveyors were to determine why these canals had been abandoned and for how long. Moreover, given the history of these canals and their current poor condition, if resources were put toward their repair, could they ever return to full use? Was there some inherent ecological problem with the land, water, geography, or topography of the particular area of a canal or with a canal's water flow? Was it, in short, better to accept the environmental reality of the canal and to leave it abandoned, or to attempt to revitalize it? And who would be responsible for the reconstruction of an abandoned canal? Was it the charge of the ruling Ottoman *divan* (imperial council), the holders of *waqf*s, or those who used the canal to irrigate their lands? Likewise, what were the costs of previous repairs to a particular irrigation work? Above all, did human technical knowledge and experience allow for the construction and maintenance of canals within the limitations of a particular natural environment? The overall aim was to match lands to canals and people—to link agricultural areas to sources of water and to identify the people responsible for both. Thus, the survey had to establish, among other things, which villages were irrigated from waterways and which were irrigated from lakes or other bodies of water. It was also tasked with determining which waterways were free from obstruction and which were clogged. Moreover, was it beneficial to open the clogged waterways or better to leave them closed?

How to answer all of these highly technical and complicated questions? The Ottomans' surveyors leaned on the knowledge of the communities that lived along Egypt's canals. These Ottoman state employees walked along the banks of every canal in Egypt with locals. Together they measured the length, width, and height of these canals; decided which needed dredging, cleaning, and general maintenance (*al-jarāfa*); and specified how these canals served as borders of property and dividers of rural space. The more the Ottoman administration could learn from locals about the irrigation situation on the ground in rural Egypt—village by village and canal by canal (*baladan baladan wa jisran jisran*), as stated in the survey—the better it could manage Egypt's water and maximize the province's agricultural productivity and thus profitability.[9] Creating a bureaucratic map of the thousands of communities of water in the Egyptian countryside demanded and deserved this kind of work. Moreover, this was a task that could only be accomplished through the participation of rural peoples in the Ottoman system—through the sharing of their knowledge of local irrigation environments; through their labor; and through the Ottoman imperial administration's recognition of the power, utility, and autonomy of this local expertise. Egyptian peasants understood that the Ottoman state could improve their particular individual situations by providing them with resources for irrigation repairs. Likewise, the imperial administration understood that Egyptian peasants helped the empire achieve its overall goals of stability and economic solvency. This was a mutually beneficial working relationship.

In the section of al-Jusūr al-Sulṭāniyya dedicated to the subprovince of al-Minūfiyya in the northwest delta, the role of local communities in shaping water management policies emerges quite clearly.[10] The subprovincial district leader (*al-kāshif*) of al-Minūfiyya, a man named Muḥammad ibn Baghdād, was tasked with gathering information about the subprovince's canals and irrigation features. Himself a local who had risen within the empire's rural governance structure, Muḥammad relied on his local connections in undertaking the survey. Led by rural cultivators, Muḥammad and his assistants walked up and down the subprovince's waterways—again, canal by canal (*jisran jisran*)—inspecting and measuring each of them thoroughly and attempting to answer all of the state's questions. Throughout the text of this and other similar cases, the voices of locals are invoked for their expertise and know-how. This particular elder said this about a local canal. This group of men understood that when the annual flood came, this embankment often broke. In this particular case from al-Minūfiyya, the elders of the districts (*mashā'ikh al-nawāḥī*) were singled out for the information they gave the surveyors. Moreover, the case emphasized that these were the men most immediately in charge of overseeing irrigation and water management in the area. After the survey had been completed and recorded, they were responsible for ensuring that all would be managed properly and efficiently going forward.

These men—*mashā'ikh al-nawāḥī*, alternatively known in Arabic as *ahl al-wuqūf* or *ahl al-khibra*—were locals who had likely spent most of their lives in al-Minūfiyya; they were familiar with the canals, dams, and the normal course of the flood in their area; they understood the soils and topographies of their villages' fields.[11] Because the environment of the countryside, and irrigation within it, could only be understood through sustained and intimate local interaction and knowledge, the Ottoman bureaucracy of Egypt had to rely on Egyptian peasants like these men to maintain the province's rural irrigation network. The local knowledge, experience, and expertise of these rural cultivators ultimately determined how the Ottoman Empire managed irrigation works throughout Egypt. With administrative tools like al-Jusūr al-Sulṭāniyya, the on-the-ground directives and collective local knowledge of these Egyptians were codified as all of this information literally entered the bureaucracy of the Ottoman state to influence and sustain its rule.

Once the initial highly detailed survey of al-Minūfiyya had been completed, two copies were made—one for the high Ottoman divan in Cairo, the most important imperial body in the province, and the other for the local court in al-Minūfiyya.[12] The latter copy was to serve as a kind of local handbook for all matters related to irrigation in the subprovince. As a baseline statement of irrigation in al-Minūfiyya at the start of the Ottoman period, it was to direct how newly appointed imperial officials were to manage the subprovince's irrigation. Because so much of the survey represented a codification of local understandings and practices of water management, the document instantiated local knowledge as the law of water management in al-Minūfiyya. From then on, moreover, if there were changes to the subprovince's irrigation situation that required a reworking of various aspects of the survey and its legal purview, it was the same class of local men of knowledge and expertise who would be consulted to amend the imperial record. This again indicates the great local autonomy Egyptian peasants enjoyed with respect to the management and use of the countryside's irrigation infrastructure and water resources.

Irrigation Negotiated

Al-Jusūr al-Sulṭāniyya was in many ways a snapshot: a picture of the countryside at a particular moment meant to establish a sort of year-zero for Ottoman Egypt's irrigation. Water levels, however, rose and fell, canal beds silted, embankments broke. One could never step in the same canal twice. Egypt's water network and rural environment were in constant flux, and so the survey was in many ways always already incomplete. Still, Egypt's continually changing environment had to be managed—it affected property holdings, cultivation levels, tax revenues, people's livelihoods, and indeed their very lives. Thus, turning from the sort of static picture captured on the pages of the survey al-Jusūr al-Sulṭāniyya to some of the thousands of cases of negotiation, conflict, and challenge that emerged around water management in rural Ottoman Egypt's constantly changing, ever-evolving environment affords us a deeper understanding of how local peasant control and autonomy over irrigation worked on the ground—in the mud.

Peasants were the water management eyes and ears of the countryside. They flagged irrigation works in need of repair, undertook the work of repairing them, and then oversaw their proper function. In 1771, for example, a group of Egyptian rural cultivators in the northeastern subprovince of al-Daqahliyya initiated a set of repairs to a canal.[13] Peasants, village elders, and the poor from the village of al-Manzala came to the court of al-Manṣūra to report that certain sections of the canal of al-Baḥr al-Ṣaghīr had silted up with dirt, sand, and debris, restricting the waterway's flow and hence preventing irrigation water from reaching their villages.[14] Moreover, the presence of many ditches along the length of the canal drew water out of this main conduit, adversely affecting the strength of its overall downstream flow.[15] The Egyptian peasants who came to the court thus suggested that areas of buildup in al-Baḥr al-Ṣaghīr needed to be dredged and that many of the ditches and small canals along its length should be dammed or otherwise sealed to improve the waterway's overall flow.[16] These rural cultivators also directed that each village along the length of the canal should be held responsible for cleaning and dredging the section of the waterway nearest it and for the reinforcement of the canal's embankments in each of those areas. This series of actions would improve access to water resources for all the villages along the length of the canal and also those elsewhere that drew their water supplies from it. Citing the many advantages of these peasants' suggestions, court officials forwarded their request to the Divan of Egypt, standard procedure to gain permission for the repair of a

4. Typical canal in contemporary Egypt, suggestive of what a canal may have looked like in Ottoman Egypt. Photograph by the author, 2007.

large subprovincial canal of this sort. The divan sent a *firman* (imperial order) back to the court of al-Manṣūra granting imperial permission to begin executing the peasants' plans.

Following the bureaucratic trail in this case shows how Egyptian peasants directed the Ottoman bureaucracy in the management of the complicated and ever-changing rural environment of irrigation in the Egyptian countryside. Set in motion by a group of peasants and villagers in their local Ottoman court in the subprovincial city of al-Manṣūra, the bureaucratic flow of this case took the peasants' ideas from the court to the Divan of Egypt in Cairo for approval and sanction.[17] Bureaucratically, this was an extremely unproblematic case since all seemed to agree that the peasants' ideas were practical and exigent. The Divan of Egypt's function was thus merely to approve the peasants' requests and to register them in the administrative record of the empire.[18] As part of the Ottoman solution to the complexities of the administration of irrigation and environment in rural Egypt, we see here (and in thousands of other cases) that the empire deferred to the knowledge, expertise, and initiative of Egyptian peasants. They suggested the most efficient and expedient means of managing the water resources and maintenance of the canal of al-Baḥr al-Ṣaghīr.[19] They lived on it and understood that if peasants in each of the villages along the canal managed the small portion of the waterway that served their village, then the entire length of this main artery would be well maintained to serve all those—in Egypt and elsewhere—who relied upon Egypt's imperially consequential irrigation infrastructure.[20]

At the very end of the seventeenth century in Asyūṭ in the south of Egypt, another group of peasants used the institution of their local court to improve their particular situation by pushing the hand of the Ottoman state to undertake canal repairs.[21] These locals went to the court to request that the state dredge several canals near their villages, since the flood season was near and the canals were not prepared to properly channel water to their fields. Evincing in court that the absence of proper irrigation hurt both their villages' capacities for food production and potential to generate tax revenue, these rural cultivators had a clear sense that they could leverage the imperial implications of their local irrigation situation to their own advantage. Irrigation in rural Egypt mattered to the Ottoman state because of food and taxes; improperly irrigated fields in Asyūṭ thus hurt the empire far beyond just its interests in Egypt itself. Peasants knew this and used this fact to improve their lives and livelihoods. In response to the petition in this case, the Divan of Egypt essentially did as it was told. An order was sent to all the relevant local officials—the head of the subprovince, the judge in the court of Asyūṭ, and the commanders of the seven military blocs stationed there—instructing them to inspect all of the area's canals and to dredge them as needed.[22] They were further directed to do all they could to avoid any long periods of time during which agricultural land remained improperly watered. Harnessing the looming threats of a lack of irrigation water, and therefore the inability to cultivate food, these peasants in Asyūṭ were able to leverage the Ottoman bureaucracy to help them repair and maintain their community's vital irrigation features.

Egyptian peasants were able to force the hand not just of the Divan of Egypt, the highest imperial body in the province, but sometimes also of the sultan in Istanbul himself. That is, peasant requests for repair work in very small Egyptian villages would often bubble all the way up through the empire's administrative structure to the palace, showing just how vital these locals' concerns were for the function of the entire imperial system. For example, in 1713 a petition from a group of peasants in the southern city of Manfalūṭ reached the court of Sultan Ahmet III.[23] They wrote to say that the supports of a weir on a canal in their village of Waḥīshāt had been broken by the force of the water's incessant pounding and were now in a dangerous state of disrepair. If the supports gave way entirely, water would rush through the canal uncontrollably, and the many villages and people that relied on the waterway would lose their primary source of water, causing agricultural lands to remain parched and dry and food supplies and tax income to suffer. This threatening situation occasioned great concern for the Ottoman sultan and his retinue. They understood the downstream implications of the potential massive loss of agricultural resources and revenues that would follow from the ruin of an area of high cultivation like Manfalūṭ. Indeed, the very fact that this petition was not handled by the court of Manfalūṭ, nor the Divan of Egypt in Cairo, but instead reached the palace in Istanbul is itself evidence of the gravity of the situation.

Faced with these impending losses of revenue, land, and life, the sultan issued a firman instructing the provincial governor of Egypt to hire, as before, a group of local experts (in this Ottoman Turkish case, the terms are *ehl-i hibre* and *erbab-i vukuf*) to inspect the situation.[24] Thus, in response to a request for a repair job initiated by peasants, a different group of peasants who possessed rural technical expertise took the lead in inspecting and repairing the irrigation works in question. After going to the site of the deteriorating weir and completing their measurements, these men reported back to the palace, through the provincial governor, that the area of the weir in need of repair measured 9,110 square *zirāʿ* and would cost 18,130 *para* to fix.[25] After some deliberation, and with the stated goal of properly reinforcing the weir to prevent future damage, imperial funds were made available to undertake the repairs. Throughout the text of the firman issued to release these funds, the authority of local experts was continually invoked to evidence the exigency of the repair work and to justify its costs. Both peasant autonomy and local control of the imperial management of irrigation in rural Egypt were clear in this case. Rural cultivators from the rather small southern Egyptian city of Manfalūṭ began and ended this case. They pushed the hand of the Ottoman sultan himself to fix a broken irrigation work that they needed to irrigate their lands. The bureaucratic distance between Manfalūṭ and Istanbul was thus much shorter than the thousands of kilometers between them. From the perspective of Manfalūṭ's peasants, the Ottoman state represented a way to fix a weir.

The direct link between rural Egyptians and the Ottoman state comes through clearly in cases involving disputes between peasants and tax farmers (*multazim*s) or other local notables in the countryside.[26] These cases offer a more detailed accounting of Ottoman Egypt's rural social structure, making the point yet again that peasants, over all other social groups in the countryside, were the final arbitrators of the function and management of Egypt's rural irrigation network. One such case from the court of al-Manṣūra in 1703 pitted a group of peasants against two tax farmers, Ibrahim Ağa and Bektaş Çelebi.[27] The dispute revolved around a set of four canals in need of repair that provided water to the peasants' villages. The canals ran through the lands of the two tax farmers, and during the flood season they regularly overflowed their banks, flooding the tax farmers' fields and causing great destruction and hardship for these men. Thus, the tax farmers came to the court to request that the Ottoman state repair the canals to prevent them from flooding their fields. They suggested decreasing the volume of water that was allowed to flow through the canals, building much larger embankments along the waterways, or, most radically, covering them over altogether. These men drew their water from sources other than the four canals in this case and were hence unconcerned about the downstream effects their actions would have on the many peasants who relied on the canals. For their part, the downstream peasants in this case were obviously very concerned by these tax farmers' efforts. Thus, their main impetus in coming to court was to try to prevent the tax farmers from affecting their water supply.

As in other cases, once the basic facts had been presented, the court ordered an inspection of the irrigation works in question, a process that involved conversations with locals along the waterways and in the affected areas.[28] The inspection report returned the following: the canal farthest downstream was badly damaged; the canal directly upstream from it was completely destroyed; a third near these two had been reconstructed in the previous year by peasants in the area and was functioning properly; and the last canal inspected was upstream from these other three, much larger, and had recently been rebuilt over the remains of an older canal. This fourth and final canal was the most important since it fed the others and was the main source of both irrigation and drinking water for all the users of the canal system.

With this information in hand, the court could now adjudicate between the competing desires of the peasants who wanted to preserve their sources of water and the tax farmers who wanted to protect their lands from flooding. The court sided with the peasants. Its decision to repair the canals stated that the peasants in this case should be the final arbitrators of the canal network's fate since they were the ones who most directly relied on it and, hence, would be most affected by any changes made to it. As part of the inspection process, the court consulted these rural cultivators for information about the canal system, later returning for further testimony as the case progressed. The peasants unsurprisingly wanted the canals fully repaired so as to have the maximum possible amount of water for their fields. Bound by the court's decision, the tax farmers on the losing side of this case resigned themselves to the reconstruction of the four canals and the overflow that would continue to affect their fields. The Ottoman bureaucracy of course recognized that the interests of downstream peasants aligned with its own interests, as securing adequate supplies of irrigation water would ultimately help to maximize the agricultural and economic output of the Egyptian countryside. Rural cultivators were the primary motor of the countryside's economic productivity, and were thus given the control they demanded—often trumping the needs and desires of others in the countryside—to manage rural Egypt's irrigation system.

In some cases, peasants had to take quite drastic measures against tax farmers and other local actors to achieve their desired ends. One of these measures was the wholesale abandonment of villages.[29] Without peasant labor, the countryside, even with copious amounts of water and perfectly functioning canals, was effectively moribund. Peasants made the countryside run. They used this reality to their advantage. A court case from May 1688 illustrates how a group of peasants was able to use the collective abandonment of a village to reverse an especially threatening situation. Earlier that year, their village of Famm Ẓāfir was transferred to the tax farm (*iltizām*) of a rural elite named Aḥmad Jalabī ibn al-Marḥūm al-Amīr Muṣṭafā Bey.[30] Located on the banks of the major canal of al-Baḥr al-Ṣaghīr, Famm Ẓāfir had been administered by the village of al-Kashūfiyya from time immemorial (*min qadīm al-zamān*) and had generally enjoyed prosperity, proper irrigation, and ample cultivation. When control of this village was handed over to Aḥmad Jalabī, however, things quickly took a turn for the worse. Prompted by his selfish efforts to extract more and more revenue from the village, fighting and many other kinds of evils (*shurūr*) took hold of the people, visiting great harm, disrepair, and hardship on the social, economic, familial, and public works structures of the rural community. Irrigation, sowing, and planting all stopped, as did the village's payment of taxes to the state. Things came to a head in the year this case was brought to court as an enormous and deadly fight broke out in the village between some of its peasants and Aḥmad Jalabī and his men. During the course of this battle, numerous fields in the village were destroyed and could not be planted. As a result, many peasants decided to leave Famm Ẓāfir for good out of fear that its problems would never be solved.

During the upheaval, the people of Famm Ẓāfir stopped dredging the imperial canal that ran through their lands, a waterway known as Jisr al-Ẓafar.[31] This purposeful political move on their part meant that little or no water reached many of the villages downstream from Famm Ẓāfir. Moreover, because the canal had not been properly dredged, water collected behind silted-up areas of the waterway, overflowing the channel's banks, and flooding parts of the village of al-Manzala. Thus, problems of administration in Famm Ẓāfir led to the village's peasants' strategic withholding of their labor, which in turn resulted in a lack of proper irrigation both in Famm Ẓāfir and in other nearby villages. From an imperial Ottoman perspective, the destruction wrought in this village and its effects on surrounding communities meant great losses to the empire's budget. The potentially dire economic consequences of this lack of irrigation in Famm Ẓāfir created a pressingly urgent state of affairs.

The court of al-Manṣūra moved quickly to try to deal with this situation. The court dispatched the tax farmer of the village of Tahākān, a village neighboring Famm Ẓāfir, to investigate.[32] He asked those still living in Famm Ẓāfir why the community had fallen into ruin and destruction and why so many peasants had fled their lands. These peasants informed him that since time immemorial, their village had fallen under the administration of the village of al-Kashūfiyya and, moreover, that no individual—not a subprovincial governor, a powerful tax farmer, or one of the heads of the seven military blocs—had the right to take the village away from al-Kashūfiyya for his own personal gain. The peasants went on to say that when Aḥmad Jalabī assumed control of the village, he caused all sorts of civil strife and infighting to break out among the villagers. These remaining peasants threatened to abandon the village as well and never again to maintain its canal or cultivate its fields as long as it remained under the sway of Aḥmad Jalabī. As if this were not a strong enough threat, the villagers also vowed to prevent any other peasants—those loyal to Aḥmad Jalabī or those from other villages—from planting the fields of Famm Ẓāfir. The entire village of Famm Ẓāfir was, in other words, under threat of permanent desertion unless returned to the control of the village of al-Kashūfiyya.

If all was restored to its former state, however, the peasants of Famm Ẓāfir promised to return to their village to plant its fields, restore its built structures, and maintain and protect (*ḥifẓ wa ḥarāsa*) its canal in accordance with previous laws.[33] Understanding the magnitude of these peasants' threats and its inability to act in the face of them, the court acquiesced to the peasants' demands, recommending that Famm Ẓāfir be reattached to the administration of al-Kashūfiyya. Thus, the peasants of Famm Ẓāfir successfully leveraged their powers as cultivators and sources of revenue to expel an unjust and corrupt ruler and to return their village to the administrative control of al-Kashūfiyya. By abandoning their village and preventing it from functioning as a productive agricultural unit (by essentially breaking the factory's machines), these peasants were able to force rural elites and the Ottoman administration of Egypt itself to agree to their demands to remove Aḥmad Jalabī.

Like peasants throughout the Egyptian countryside, the people of Famm Ẓāfir were the most important actors controlling irrigation in rural Ottoman Egypt. The case of Famm Ẓāfir was in some ways an extreme one, given the extraordinary lengths to which the peasants of the village went to flex their muscle. Some of them voted with their feet by leaving the village.[34] Those who stayed purposefully allowed their village's canal to fall into disrepair, thereby hurting numerous other villages and forcing them to become invested in the hardships visited upon the peasants of Famm Ẓāfir by Aḥmad Jalabī. The people of Famm Ẓāfir ultimately won. In yet another recognition of their power as the first and final arbitrators of irrigation in rural Egypt, the Ottoman state once again sided with peasants and their demands and desires against the claims and interests of others.

Conclusion

That the Ottoman Empire devolved authority over irrigation works to local communities in rural Egypt was not a function of the state's largesse but rather a recognition of the fact that this was the most expedient, efficient, and effective means of managing so complex a system of irrigation as existed in Ottoman Egypt. Everywhere in the empire, local peasants were the ones who knew their canals, embankments, and fields best. They were hence best suited to determine how they should be maintained. As an early modern polity dependent on managing the natural worlds of agriculture and irrigation, the Ottoman Empire was clearly subservient to the courses and movements of the environments under its suzerainty.[35] This was the primary reason local populations were so centrally involved in the management of water and irrigation works in rural Ottoman Egypt.[36] While the empire could of course never fully control the environmental forces within its realm—could never churn its bureaucracy quickly enough to keep up with the flow of water in Egypt's canals—the presence of Egyptian peasants on the ground watching over rural irrigation features and informing the state of their status meant that the Ottoman imperial administration could much more efficiently and effectively rule over Egypt's irrigated ecosystems.[37] At the same time, Egyptian peasants on their own could not repair large irrigation works in their communities without the resources of the Ottoman state—resources fundamentally based, of course, on their agricultural labor. In short, neither peasants nor the Ottoman imperial administration could repair irrigation works in the Egyptian countryside alone; it was only together that they could undertake this vital and mutually beneficial work.

This reality was a consequence, on the one hand, of the constant demands of a changing natural environment that regularly overwhelmed local communities and affected the resources on which they relied and, on the other hand, of an Ottoman imperial system that privileged local expertise and historical specificity over any notion of empire-wide homogeneity or cultural or social uniformity. The relationship between the Ottoman capital in Istanbul and peasants in rural Egypt was predicated on mutual and collaborative political, infrastructural, and economic benefit. From an imperial perspective, the Ottoman state pooled the revenues generated by Egypt's rural cultivators and used some of these resources to finance the maintenance of the province's irrigation system.[38] In return, peasants ensured that the system functioned properly and productively. From a rural perspective, peasants used the resources of the Ottoman state to fix urgent local problems with their irrigation works. The price of this was some of the food they grew and their labor. This was the Ottoman solution to Egyptian irrigation.

I have argued in this chapter that in contrast to descriptions of rural Ottoman Egypt that stress the oppressive nature of Ottoman rule in the countryside, it was really Egyptian peasants, calf-deep in Egypt's mud, who controlled Ottoman bureaucrats—not the other way around—in the determination and execution of repairs to irrigation works.[39] The imperially coordinated localism that characterized the management of irrigation in rural Egypt invested local peasant actors with the freedom and wherewithal to initiate and control repairs to Ottoman Egypt's irrigation network from beginning to end. Peasants managed the empire's irrigation.

2

History from Below

The Nile delta and the Mediterranean have been pushing against each another for the past ten million years since the river first began carrying dirt to the sea.[1] For most of the last seventy-five hundred years, though, the delta has enjoyed the upper hand. As the fifth-century BCE Greek traveler and historian Herodotus sailed toward Egypt's northern coast, he wrote of how "as you approach it and are still within one day's run from the land, and you drop a sounding line, you will bring up mud, though you are in eleven fathoms' depth."[2] For Herodotus, the presence of all this mud so far out at sea was evidence enough of how the delta had been steadily made over the course of thousands of years by the accumulation of sand and dirt carried by the Nile.[3] In his words, "The Delta, according to the Egyptians themselves (and I certainly agree), is alluvial silt and, one might say, a contribution of the day before yesterday."[4] As evidence that the delta was indeed a product of "the day before yesterday," Herodotus noted the region's lack of any of the ancient ruins responsible for bringing so many visitors such as himself to Egypt.

Another of these travelers who wrote of the Nile delta and its steady creation over millennia was an American named John Antes, who lived in Egypt in the eighteenth century, the period of most immediate concern to us in this chapter. Of the delta he wrote, "The large quantities of mussel and oyster beds, with other productions of the sea, which are to be found under ground in various places, even not far from Grand Cairo, made me sometimes think, that most probably the whole Delta was originally nothing but a shallow bay of the sea, of unequal depth. . . . Near Rosetta there seems to be striking proof that the country is still encreasing by the sediments of the river; by every appearance it seems that Rosetta was formerly situated close to the sea."[5] In addition to these fossilized remnants of the sea, Antes also observed how the yearly flood moved impressive amounts of dirt to increase the area of the delta. "When I thus noticed what large pieces of ground were yearly carried away, and of course removed towards the sea, and considered that this must have been the case from the first existence of the river, it seemed to me a very strong argument . . . that perhaps the greatest part, if not the whole of the Delta has been thus produced, and must still be encreasing by an encroachment upon the sea."[6] And like Herodotus many years before him, Antes also observed "that no monuments of very great antiquity are to be found in these low places, but only on some few elevated spots, and even these few do not seem to be so old as those found in the upper parts of the country."[7] Thus, the period from Herodotus's visit to Egypt twenty-five hundred years ago to Antes's observations at the end of the eighteenth century saw the steady expansion of the delta into the Mediterranean. This reign of Egypt's northern coast over the sea, however, would soon begin to come to an end a few decades after Antes wrote his account.

Indeed, on the basis of much more contemporary accounts, there is clear evidence that the multiple millennial domination of the delta's dirt over the Mediterranean has slowly been coming to an end since about 1800.[8] Like other deltas around the world, Egypt's is steadily retreating. Some parts of the coastline are being eroded at a rate of 125 to 170 meters per year.[9] This is primarily a function of the detrimental effects of two hundred years of gigantic public works projects meant to manipulate the Nile's waters for what were stated at the time to be exigent political and economic needs. From the efforts of the early nineteenth-century Ottoman provincial governor Mehmet ʿAli to irrigate more of the delta to President Gamal ʿAbd al-Nasser's Aswan High Dam hydroelectric project in the middle of the twentieth century, the ecology of the lower Nile was changed more rapidly and more fundamentally in the past two hundred years than ever before. One consequence of these changes to the river has been that the delta no longer receives the full impact of the yearly flood and the rich silt it contains (over 125 million tons of sediment a year).[10] These gifts of the Nile now pile up behind the Aswan High Dam. Thus, unlike Herodotus, someone sailing toward the Nile delta today would not find much mud in the sea at even a very short distance from the shore. Among the other regrettable environmental consequences of these public works projects—each of course with its own significant human costs as well—are salinization, coastal erosion, massive increases in the usage of chemical fertilizers, and extreme water loss due to evaporation from Lake Nasser behind the High Dam. These grand environmental stories of the creation of the Nile delta and its current erosion are clear enough to anyone interested in Egypt and have received much attention from geologists, historians, environmental activists, hydrologists, and others.

Instead of these well-known millennial tales of creation and destruction, this chapter focuses in on some of the thousands of smaller-scale daily interactions between Egyptians and the Nile delta's silt in the seventeenth and eighteenth centuries—a few decades before the nineteenth and twentieth centuries' massive projects of river manipulation.[11] This history of human interaction with water and dirt in Ottoman Egypt (as elsewhere) is at its heart a story about the outlines of society—literally—and about how ideas of the rural landscape were formed and maintained through and by water. The yearly flood and the massive amount of sediment it brought ensured that the shape of Egypt in the Ottoman period and before was in constant flux. Water ebbed and flowed, embankments broke, canals were dredged, silt settled, water evaporated, and dams collapsed. These and other environmental and infrastructural realities of life in rural Ottoman Egypt meant that peasants and the imperial bureaucracy they were a part of had to adapt to a constantly changing physical landscape. The history of how peasants and the Ottoman state dealt with this environment in flux reveals how and why they conceived of, negotiated with, and tried to harness the dirt of their countryside.

The annual meeting of water with dirt was a process that fundamentally shaped Egyptian peasant and Ottoman imperial experiences and understandings of the rural environment. As is made clear in a seventeenth-century satirical Arabic literary account of the countryside, proximity to the Nile and the particular ways its floodwaters settled in land created a hierarchy of rural spaces.[12] At the bottom of this hierarchy were swampy lands (*bilād al-malaq*) on the margins of the Nile watershed. These areas received water but not enough to properly irrigate agricultural fields to grow food. Next were villages very close to the river with highly sophisticated irrigation networks that fed water to very productive areas of cultivation. And at the top of this social ladder were Egypt's large towns and cities. That Egyptian peasants and the imperial bureaucracy equated relative levels of irrigation to social status, political and economic import, and cultural sophistication clearly indicates the centrality of water to the development of understandings of rural society and environment in Ottoman Egypt.

At the same time, an examination of both local and Ottoman imperial conceptions of the environment further suggests something of the cooperative nature of water management introduced in the previous chapter. Controlling, sharing, and using water both necessitated and fostered cooperation and compromise among all parties. As we will see below, many conceptions of how best to manage water were built on this cooperative ideal, one understood and cultivated by *both* the imperial bureaucracy and Egyptian peasants. This common acceptance of the cooperative nature of water utilization goes a long way in explaining the remarkable similarities of many of the shared imperial and local views of the rural Egyptian environment. As I show below, however, this is not to say that peasant and imperial interests and actions were always in lockstep. Taken together, these cooperative and contested negotiations over environmental resource management help delineate a set of environmental ideals and ideas at play in the early modern Ottoman-Egyptian countryside that included notions of community, responsibility, precedent, and resource allocation.

Two aspects of life in rural Ottoman Egypt bring the imperial bureaucracy's and peasants' engagements with dirt and water into the starkest of reliefs—canal dredging and the changing shape of alluvial islands in the Nile and its tributaries. Canals had to be dredged regularly throughout Egypt to keep irrigated water flowing in the countryside, and likewise, the vicissitudes of the Nile ensured that alluvial islands were constantly getting larger and smaller, reforging their spatial form, and connecting and disconnecting from canal banks. Both of these processes, among others, altered the physical landscape of Egypt, changed political and social relationships between peasant groups and between peasants and the Ottoman state, affected rural labor practices, challenged the abilities of rural Islamic courts to adjudicate complex disputes, and reshaped economic interests.[13] In more specific terms, this chapter shows how canal dredging involved the establishment of legal precedents for the responsibility of maintaining properly functioning waterways and other irrigation works and how alluvial islands contributed to notions about what constituted evidence for the continuous use and cultivation of property. These ideals of community, precedent, sharing, and the establishment of responsibility were all integral facets of both imperial and local principles of how to manage the rural Ottoman-Egyptian environment.

Ideals in the Dirt

The water of Egypt's vastly complex irrigation network was forever in motion. This perpetual aqueous movement ensured that land as well was constantly appearing, disappearing, and changing shape. It also ensured that the Ottoman state had to address this continually shifting terrain through various bureaucratic and legal mechanisms. It did this largely by upholding in its network of legal courts certain notions of how communities shared and used water resources.

One of the most important factors determining how water flow shaped the banks of canals and the borders of land was the amount and character of silt on the bottoms of the beds of Egypt's waterways. A highly silted-up canal could, for example, force water to flow with more force and in considerably different directions than it usually did. This could in turn erode canal embankments or completely overtake them. Alternatively, such a canal could stop flowing altogether. To attempt to gain a semblance of control over how water changed the shape of the rural Egyptian environment, the Ottoman administration of the countryside in the early modern period relied on dredging as one of the most crucial elements in its management of rural spaces. Because cleaning canal beds greatly impacted the local environments of all villages sharing a particular waterway, the bureaucratic organization of dredging points to conceptions of ecological community and responsibility and to how these notions were established, maintained, and manipulated to manage Egypt's rural irrigation network.

Dredging is one of the most common issues in the archival record of irrigation in Ottoman Egypt.[14] Certain canals were notoriously susceptible to large buildups of silt and were therefore constantly in need of attention. One of these waterways was a canal branching off of a large central canal known as al-Baḥr al-Ṣaghīr in al-Manzala in the subprovince of al-Manṣūra in the northeast delta.[15] This auxiliary canal was dredged in every one of the twenty years between 1684 and 1704 to remove what were termed the many small "islands and steps" (*cezireler ü atebeler*) of underwater buildup that had formed on the canal bottom.[16] The regular opening and closing of smaller canals feeding off of this main canal allowed silt and debris to settle on its bed, resulting in its nearly perennial state of siltation.

The dirt obstructing the canal's flow created all sorts of problems for peasants living in the forty villages along its length. In an effort to deal with these problems in a more permanent way so as to avoid the need to dredge the canal every year, the people (*ahali*, *ahālī* in Arabic) of these villages served by the waterway came to the court of al-Manṣūra in 1704 with village notables and local engineers from the region to discuss what could be done about this situation. Not surprisingly, it was agreed by all that, in the words of this case, the canal should be cleaned of all the dirt clogging it up so as to repair it with maximum strength and sturdiness to ensure that funds and effort were not continually expended on its dredging and maintenance. Functionaries of the court were thus dispatched along with peasants living on the canal to determine the costs of such a repair. They returned to the court to report that this work would total 50,000 *para*.[17] The twenty-three upstream villages near the mouth of the smaller canal at al-Baḥr al-Ṣaghīr were each to contribute 1,000 *para* to this repair effort, and the other seventeen downstream villages were to contribute 600 *para* each (the remaining difference of 16,800 *para* was to be made up by Ottoman state funds).[18]

This adjudication of the canal's dredging reflected an understanding by the Ottoman administration and by Egyptian peasants of how water flow and canal siltation affected communities differently based on their location along a waterway. Because downstream siltation was largely a by-product of the opening and closing of upstream canals and of the water consumption of upstream villages, these villages had to pay more for the dredging of the canal. Undergirding this and other similar court settlements was an understanding shared by all the parties involved that the collective usage of a canal tied them together into a community of water utilization and consumption. This reflected a conception of environmental resource management in which actions in any one part of this irrigated ecosystem were seen to affect and implicate *all* canal users. This principle of irrigation was a basic tenet of the shared rural environment of Ottoman Egypt and maintained in almost every dispute involving water in the countryside.[19] Clearly not all actions on a canal were equal. The water usage of upstream villages greatly impacted the quantity and character of the water and silt that reached downstream villages. The opposite was, however, obviously not true. Thus, in dealing with the sediment carried by water, it was always important to remember in which direction water flowed.

One of the best examples of the implementation of this principle in Egyptian irrigation was the management of the flow and dredging of the Ashrafiyya Canal, which coursed through the subprovince of al-Baḥayra in the northwest delta to connect Alexandria, Egypt's second city, to the

Rosetta (western) branch of the Nile. As with the previous canal, the Ashrafiyya's flow was extremely weak owing both to its lack of incline from its mouth toward Alexandria and to the constant breaking of the canal's embankments by peasants seeking to siphon water off to their fields.[20] Unlike in other cases of communities of water in Egypt, however, with the Ashrafiyya it was obvious (at least from the perspective of the Ottoman state and Alexandria's residents) which of the populations served by the waterway was most important—those at the canal's terminus in Alexandria. The imperial bureaucracy therefore expended a great deal of energy attempting to prevent villages along the length of the canal from breaking into the waterway, since this removal of water from the canal made its problems of siltation all the worse.[21] The canal was shallow, had a weak current, was surrounded by very loose soil, and consistently lost water. All of these factors, combined with the desire to have copious amounts of fresh and clean drinking water reach the people of Alexandria, contributed to near constant dredging efforts.

In the middle of the eighteenth century, for example, there were several major dredging and cleaning initiatives designed to improve the canal's flow. In the summer of 1751, peasants from the village of Minyyat Ḥiṭṭiyya were charged by the Ottoman state to dredge the bed of the canal and to reinforce embankments near their village to prevent soil and debris from falling into the waterway.[22] Almost exactly a year later (in 1752), other villages along the canal were instructed to carry out similar infrastructural work on the Ashrafiyya.[23] The canal was divided into three sections to make its dredging and cleaning more orderly and efficient. In each section, one village was put in charge of overseeing work on that part of the canal. In this period as well, a large waterwheel was constructed at the mouth of the canal in the village of al-Raḥmāniyya to attempt to quicken its flow.[24]

Despite these and other similar efforts, however, siltation and the collection of debris, rocks, and sand in the canal remained constant problems throughout the seventeenth and eighteenth centuries. In a firman sent from Istanbul to the *vali* (provincial governor) of Egypt Mustafa Paşa in February 1738, for instance, the palace complained that the canal had been badly neglected over the previous few years and was currently so clogged up with sand and dirt that water was barely reaching Alexandria.[25] When the canal was clean and properly functioning, the normal flood height of sixteen cubits was more than sufficient to fill Alexandria's 210 cisterns with water for the city's residents, which were estimated here to be sixty thousand or seventy thousand.[26] In this year, however, the canal was in such a bad state that even the waters of the exceptionally high flood mark of twenty-two cubits did not reach Alexandria. Thus, the imperial divan ordered the immediate dredging and cleaning of the Ashrafiyya. Nevertheless, a few decades later, in the spring of 1763, Alexandrians living near the canal filed a petition with the Ottoman bureaucracy complaining that the waterway's flow was again being restricted by all the silt, thorny branches, and garbage that were collecting in it.[27] These petitioners continued on to say that this shortage of flowing water exposed them to great difficulty and hardship, and they thus implored the state to clean and dredge the canal as soon as possible. Realizing the urgency of this situation, the imperial capital issued a firman to its vali in Cairo to immediately undertake this work.

In these cases about the dredging of the Ashrafiyya and other canals, the Ottoman state and Egyptian peasants sought to preserve the flow of canals so that their water would serve as many people as possible all along the waterways. At play in these cases were attempts to balance the needs and desires of the upstream against the demands and necessities of the downstream—a fundamental aspect of the imagined ideal of how to manage the consumption of water by multiple parties. Community welfare was always privileged over individual rights. This was one of the basic principles guiding the management of water and dirt in Ottoman Egypt. In establishing how to dredge canals and who was to be charged with this work, several other conceptions of environmental management were also at play. Foremost among them were notions of how proximity and shared usage determined responsibilities for the maintenance of irrigation works and the dredging of canal beds. Those in the immediate vicinity of an irrigation work who directly benefited from its presence and proper function were responsible for its upkeep; likewise, those who shared the water of a canal were also to share in the work of dredging and maintaining that canal. Telling examples of these ideals in action in the Egyptian countryside were instances of the dredging of canals shared between two or more villages on opposite sides of a waterway.

In June 1724, three villages—two from the subprovince of al-Daqahliyya (Kafr Ghannām and al-Jazīra Bākhir) and one from al-Sharqiyya (al-Hajārsa)—came to the court of al-Manṣūra (the subprovincial seat of al-Daqahliyya) to report on the successful dredging of a shared canal that served as the border between the two subprovinces.[28] The *hakim* (subprovincial governor) of each of these two subprovinces was responsible for dredging and cleaning the canal every year from the water's edge at the border of his subprovince to the middle of the canal.[29] In 1724, the court of al-Daqahliyya sent its representatives to villages near the canal to ask local village elders (*mashā'ikh*) whether or not their half of the canal had indeed been properly dredged. These local notables reported to the court's functionaries that the hakim of al-Daqahliyya had indeed carried out his required charge efficiently and properly. His men (*rijāl*) had worked for thirteen days to clean the canal and to reinforce its embankments, and its waters were now flowing quickly and without obstruction. The judge in this case then reminded these local village elders that the responsibilities to keep the canal properly functioning were now completely in their hands. The imperial bureaucracy had, in other words, carried out a major dredging operation on a canal and was now handing off its fate to villagers living around it. Moreover, the judge added that should these local notables fail to maintain the canal's proper function, they would have to pay for this failure with their lives.[30]

Water and silt were thus literally matters of life and death. With this dramatically unambiguous threat of execution, the Ottoman bureaucracy clearly signaled that irrigation and dredging were two of the most important aspects of its rule in the Egyptian countryside. Moreover, it was also making a strong declaration in this case, as in others like it, that ensuring the steady supply of irrigated water to dozens or scores of villages and thousands of peasants was of greater concern to the state than preserving the life of one or a few peasants. This ideal of resource management and access was, needless to say, one that only *some* Egyptian peasants benefited from and hence consented to and willingly implemented. Thus, unlike previous cases, this was an instance in which various Egyptian peasants and the Ottoman bureaucracy obviously held very different environmental ideals. This disjuncture of outlook and priority further evidences that the Ottoman Empire consistently privileged the interests of the rural whole over the life of the individual.

Moreover, these examples of the sharing of a canal among multiple villages also illustrate how authority over dredging and irrigation was conceptualized, organized, and delegated in Ottoman Egypt. Proximity was again key to the empire's understanding of environmental management. Like the peasants and their village heads in the above case, those who directly benefited from a properly functioning canal were responsible for keeping it flowing. This authority invested in the local control of irrigation works was meant to serve as a preemptive measure against massive infrastructural destruction and the repairs it would surely necessitate. Steady maintenance of canals by those directly served by them would ensure the overall health of Egypt's irrigation network. The Ottoman administration of Egypt calculated that if all peasants took control of their immediate surroundings, then the irrigation system would work together as a whole. The line of authority traced in this present case makes this abundantly clear: from the Ottoman state through its imperial institution of the court, to subprovincial hakims, to representatives of the court, to local peasant elders, and finally to those peasants who actually carried out the canal's repairs. Cases like this one, then, in many ways represented the ideal function and most efficient execution of the empire's conception of irrigation management.

In other cases, these privileged conceptions of proximity, shared authority, and communal welfare were again challenged by peasants attempting to gain some degree of advantage over their neighboring peasants through the manipulation of a canal's water or silt or both. As before, some of the most common and instructive of these disputes over water and dredging were those between upstream and downstream villages. In February 1682, the heads of two neighboring villages on the same bank of a shared canal came to the court of al-Manṣūra.[31] The head of the downstream village Nūb Ṭarīf complained to the court that the peasants of the upstream village of Ṭummāy had failed to dredge and clean the section of the canal that ran past their village.[32] As a result, the canal's embankments were crumbling and silt mounds were beginning to peek through the water surface. Thus, an insufficient amount of water was reaching Nūb Ṭarīf, causing many of its fields and those of other villages near it to become parched and dry. The representative of this village thus asked the judge to send state officials to inspect the situation so that they could see for themselves that the peasants of Ṭummāy had failed to clean and dredge the canal as was their duty "from times of old" (*min qadīm al-zamān*)—as the oft-repeated phrase went—and so that they could force these locals to fix the waterway.

The peasants of Ṭummāy in this instance either through choice, incompetence, or irresponsibility let their canal silt up, which in turn prevented water from reaching their downstream neighbors. This case gives no indication as to why the canal's dredging was ignored in contravention of the empire's idealized goal of equitable environmental resource management. Perhaps the people of Ṭummāy had acquired some other source of water that made the older canal no longer relevant for their own irrigation purposes. Perhaps there was some crisis in the village that took peasants' attention away from the canal. Alternatively, perhaps a dispute between the two villages caused the people of Ṭummāy to use their more advantageous upstream position as a weapon against Nūb Ṭarīf. Whatever the case may have been, the actions of the court instigated by the head of Nūb Ṭarīf upheld the ideal of proximity in determining responsibility for canal maintenance. Moreover, in a statement asserting the power of community above all else, the physical presence of the canal in Ṭummāy did not give its residents ultimate authority over the canal's usage, consumption, and management rights. The downstream village of Nūb Ṭarīf indeed exercised its authority as an invested member in a community of water to force peasants in another village to dredge a canal they all shared.

The constantly changing shape of rural Egypt caused by the movement of water and sand, siltation, the collapse of dams and embankments, and the various actions of users of canals and other irrigation works ensured that dredging was a central aspect of Ottoman rule in Egypt. As a site of regular cooperation and contestation between the empire and peasants and between different peasant communities, dredging indeed served as a crucial indicator of how various challenges to Ottoman imperial ideals of the Egyptian environment were handled. Within this realm of negotiation over how to deal with silt, water, and their multiple and often unpredictable effects on rural life, ethics of community, proximity, and the sharing of responsibility were both developed and maintained to govern canal dredging and cleaning. This was accomplished through various legal institutions, infrastructural formations, and social principles that operated to uphold and reinforce these ideals and to protect the overall health and productivity of agriculture in the countryside. Nile sediment thus not only made Egypt the most lucrative agricultural and financial province of the Ottoman Empire, but it also shaped much of Ottoman rule in Egypt and much of both imperial and local social, ecological, and political understandings of the countryside and of how rural physical spaces were to be governed.

Dots on the River

Some of the best examples of how Ottoman imperial officials and local Egyptians understood the rural environment are the multiple ways in which the Nile and the dirt it carried altered the topography of the Egyptian countryside through the changing shape, size, and connectivity of islands in the Nile and its tributaries. The rise and fall of the Nile meant that islands were constantly getting bigger and smaller and often connecting to and disconnecting from the mainland. This geographical flux challenged conceptions of terrestrial fixity, on which was built the whole edifice of ideas about responsibility, sharing, and communal governance. Even perhaps Egypt's most famous island—al-Rauḍa, whose

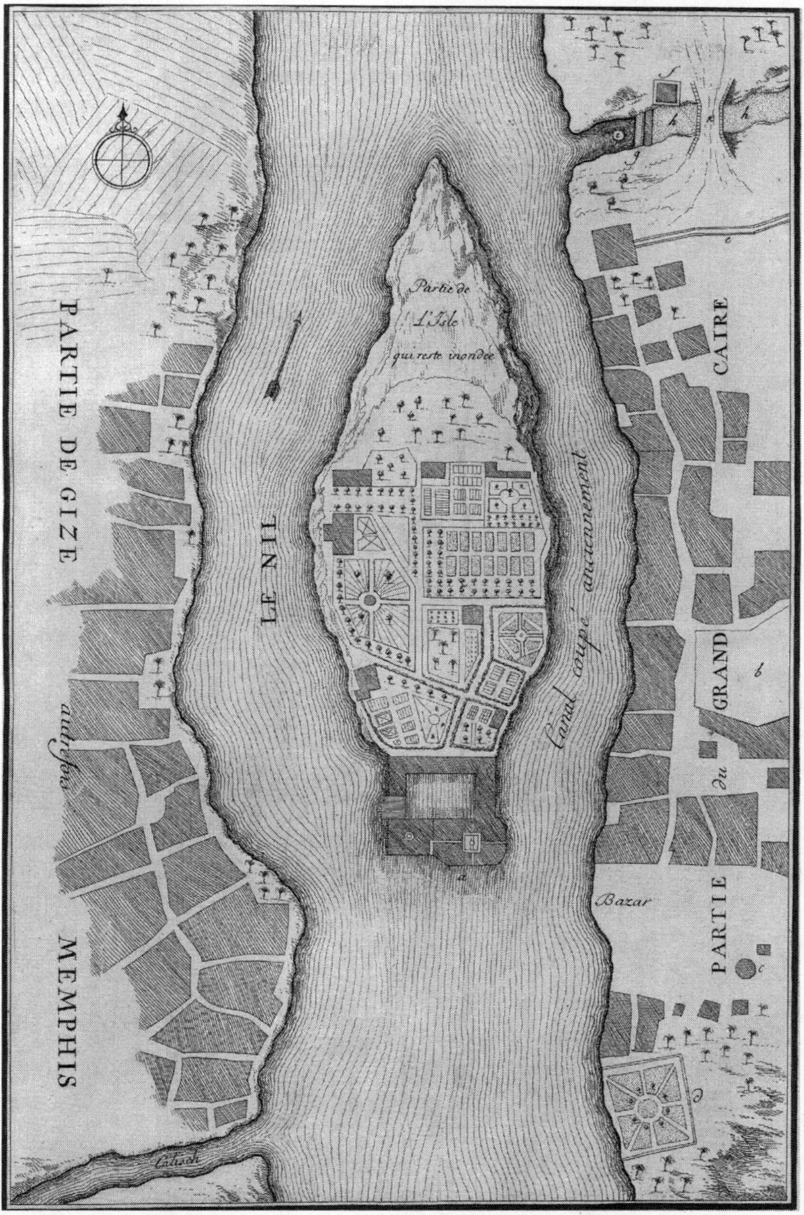

5. Aerial view of al-Rauḍa in the eighteenth century. Frederik Ludvig Norden, *Voyage d'Égypte et de Nubie, par Frederic Louis Norden, ouvrage enrichi de cartes & de figures dessinées sur les lieux, par l'auteur même*, 2 vols. (Copenhagen: Imprimerie de la Maison Royale, 1755), vol. 2, plate 24.

prominence derived chiefly from the Nilometer (*miqyās al-Nīl*) on its southern tip, the device used to measure the official level of the annual flood for purposes of taxation and agricultural production—often found itself connected to Cairo's coastline due to the river's recession.[33] Such was the case in the spring of 1792 when the eastern portion of the Nile between al-Rauḍa and Cairo began to dry.[34] Huge piles of sand formed all along the channel and even farther north, serving essentially to connect al-Rauḍa to the eastern bank.[35] Water levels in this season were so low that even to the west of the island, in the much wider portion of the river separating al-Rauḍa from Giza, land had begun to emerge from underneath the river, turning this western section into a pathetic trickling brook (*salsūl jadwal*) in which little children played and through which only the smallest boats could pass.[36]

With the river's rapidly changing water levels and its slower and faster currents, the amount of land available for cultivation on islands like al-Rauḍa was constantly fluctuating. As we saw with the example of dredging, because of erosion and other environmental changes, some areas of land along the banks of the Nile and its branches were exposed for parts of the year and submerged in others. Thus, some dirt formerly covered by water often became permanent agricultural land as a result of changes—humanly induced or otherwise—to the direction and flow of the river. This emergence and disappearance of cultivatable earth meant that there was a constant need to reimagine and redetermine the legal, social, and agricultural status of these new and old lands. The working out of resolutions to these administrative imperatives shows again how the bureaucratic mechanisms of the Ottoman Empire in Egypt sought to maintain precedent as the empire's overriding conceptual framework for understanding and managing the persistently changing rural environment. And furthermore, it also reveals how Egyptian peasants attempted to harness these changes for their own local advantage.

In a dispute over an island from the late eighteenth century, a group of tax farmers (*multazim*s) from the subprovince of al-Daqahliyya came to the court of al-Manṣūra to testify on behalf of the rights of the village of Ṭalkhā al-Gharbī to an island that had recently been seized by peasants from the neighboring village of Qūjindīma.[37] Whereas this island had historically been separated from the mainland by water on all sides, it had recently attached to the shoreline near the two villages due to the receding or ṣhallowing of the river. When the former island connected to the mainland, the people of Qūjindīma moved quickly to take control of it by crossing over the newly formed land bridge. Although the peasants of Qūjindīma freely admitted to seizing the island, they refused to give up their claims to it, and thus the present quarrel between the two villages came to pass. All the tax farmers and other locals who came to court and everyone else mentioned in this case—other than the peasants of Qūjindīma, of course—agreed that the island, no matter that it was now connected to the mainland, belonged to the peasants of Ṭalkhā since they had historically cultivated it. And thus on the basis of this testimony, the judge in this case ruled in favor of the peasants of Ṭalkhā. Unlike other disputes over islands in which the presence of irrigation works and other built features evidenced a history of cultivation, in this case no such structures existed, and hence the assertion of rights over the island rested purely on the testimony of various parties as to which group of peasants had historically farmed the island.[38]

For their part, through their takeover of the island, the peasants of Qūjindīma clearly prescribed a reimagination of its status. Not only had its new connection to the mainland fundamentally altered a particular local geography, but it also necessitated a completely different bureaucratic and legal configuration. Given the new land bridge, where were the new borders of the former island's cultivated space? To whom did the cultivation rights of the new land bridge belong? Did it belong to the people of Ṭalkhā since the land fell between two areas they farmed? Or was its status as yet undefined since it was a new piece of land that had never before had an assertion of rights applied to it? For the imperial court, the answers to these and other similar questions were reached through a determination of legal precedent—the precedent of the historic cultivation of the island by the people of Ṭalkhā. To this end, the court sought out the testimony of peasants from neighboring villages to help establish who had previously cultivated the island. Thus, in this instance of its conceptualization of the rural Egyptian environment, the Ottoman state sought to reinstitute a system of resource management that had obtained before the island had attached to the mainland.

At the same time, the peasants of Qūjindīma wanted to establish their own new superseding precedent of cultivation on the island by moving quickly to take it over once it was connected to the shore. Theirs was thus an effort to create new facts on new ground. They attempted to entrench a new physical reality so as to precipitate a reimagining of the island as their possession. In this case, however, the court upheld that the recent connection of the island to the shore did not change who enjoyed cultivation rights to the island. Thus, even in the face of literally shifting geographies and challenges to established modes of governing the Egyptian countryside, here, as elsewhere in the Ottoman Empire, precedent continued to reign supreme.

What is not made explicitly clear in this case is the legal status of the new land that emerged from underneath the water. The judge made no determination in this regard. Thus, here again it seems that there was resistance to reimagining and marking this land as having a legal status other than the one that was previously maintained (namely, that it had no legal status). Perhaps establishing rights to cultivate this new piece of earth was considered unnecessary since all knew that land near the Nile and its canals regularly appeared and disappeared as the river rose and fell. It was only a matter of time, therefore, before this land was again flooded over, making the complexities of its legal status rather irrelevant. As we saw previously, however, in cases involving the dredging of canals between villages on opposite sides of a waterway, there was an imaginary line established down the center of canals splitting the responsibilities of cleaning and dredging them into equal parts. Was such a notion at play in this case that made the establishment of rights to the new land unnecessary since it was clearly that of the nearest village? These questions are at their base inquiries into how silt and dirt were imagined, understood, and legally assessed in rural Ottoman Egypt. Despite the lack of explicit answers to some of these questions, we should nevertheless not assume that these issues were somehow not debated or disputed in Ottoman Egypt. Clearly they were, as evidenced by the numerous court cases initiated to deal with Egypt's ever-shifting rural terrain.

Another of these cases shows how conceptions of precedent and a notion of shared usage similar to that discussed previously came to govern the management of peasant actions precipitated by physical changes to alluvial islands. In the spring of 1792, a series of cases came to the court of al-Baḥayra about an island in a branch of the Nile between the villages of Nitmā and Kafr al-Gharīb.[39] From times of old (*min qadīm al-zamān*), these two villages had equally shared the island in sowing and cultivation.[40] Beginning a few years earlier, however, the tax farmer of the village of Nitmā started preventing the peasants of Kafr al-Gharīb from coming to cultivate their fields on the island. Moreover, because of the strength of the Nile's crashing waters, the river "ate" (*akala*) most of the section of the island that had historically been cultivated by the people of Kafr al-Gharīb. Thus, needless to say, this situation caused great hardship and concern for the peasants of Kafr al-Gharīb, since not only had the total area of their section of the island's agricultural land decreased because of the Nile's encroachment, but the whole of the lands that did remain were taken over by the peasants of Nitmā. In the face of this difficult state of affairs, the imperial firman issued in this case directed the people of Kafr al-Gharīb to take back half of the reduced total area of the island so that, as before, both villages cultivated equal halves of this bounded piece of land. Indeed, these cases cite the imperative to return to the precedent from times of old that the two villages were to split the island's area equally.[41] As such, these cases also end with an admonition, if even a formulaic one, to both villages to never again act against this principle of equal usage—one of the basic conceptual ideals shaping how the rural landscape was to be managed.

In this example, as in most, the set of strictures that determined how disputes over a shared island were to be resolved was based on precedent. Precedent, then, was the "law" as it came to be practiced in rural Ottoman courts. Over and over in these cases, judges, litigants, and witnesses underscore their ideas of environmental resource management by invoking the way things have always been done, again "from times of old" (*min qadīm al-zamān*), as a justification for why they should remain that way. Thus, in the face of a rural topography that was constantly changing, the Ottoman bureaucracy of Egypt and the province's peasants sought to preserve a social order based on their understandings of the past shape of the Egyptian countryside. Principally through its courts, the Ottoman imperial bureaucracy strove to make the countryside line up with its ideals of the history and function of the rural Egyptian world. And through their recourse to these courts, groups of Egyptian peasants sought to preserve and then further their history of cultivation and water usage rights.

To put it differently, the Ottoman state attempted to prevent environmental change from dictating its imperial rule. Although the empire could do very little to slow down or stop the environmental impacts of erosion, siltation, or flood, it could attempt to prevent these natural forces from changing its management of the Egyptian countryside.[42] This is perhaps an obvious point—that existing political powers want to preserve the status quo of their rule. For their part, Egyptian peasants were likewise so invested in this imperial system of natural resource management that they too had a stake in its preservation. Nevertheless, neither Ottoman imperial power nor Egyptian peasant knowledge of local environments could prevent islands from becoming larger and smaller or from connecting to and disconnecting from the mainland. Thus, to preserve a sustainable rural social order given these constants of change, the state—through cooperation and contestation with Egyptian peasants—upheld and defended precedent, cultivated ideals of sharing and community, and financed and supported irrigation repair work in an effort to fashion the countryside in accord with its own imperial environmental imagination and ethics.

Conclusion

When analyzing the environmental history of political entities like the Ottoman Empire, it is essential to define the specific spatial demarcations within which such polities existed. In the case of Ottoman Egypt, and also Ottoman Iraq and elsewhere in the empire, the physical outlines and shapes of these provinces were constantly changing.[43] As the Nile's waters pushed silt along or allowed it to settle, some lands were submerged and others appeared. Canal embankments were eroded away, and islands often changed size or connected to the shore. Through a consideration of how precedent was defended and of the legal and social mechanisms used to establish responsibility over irrigation works, this chapter outlines part of the environmental worldview of Ottoman state bureaucrats and Egyptian peasants as it emerged in their dealings with the rural environment's forever shifting terrain. In contemporary parlance, we would say that the Ottoman Empire's "environmental policy" in dealing with the dirt and water of early modern Egypt was to return irrigation works to their former states of functionality through dredging or some other means of environmental manipulation and to uphold previous local cultivation rights no matter the changes made to the countryside's geography. In sum, the Ottoman Empire's goal was to maintain the physical parameters of the rural environment in accord both with local Egyptian imagined conceptions of its past and with perceived notions of its most efficient function in the present moment.

The twin examples of dredging and islands thus highlight two very different aspects of Ottoman rule in Egypt as it relates to local communities—respectively, peasant responsibilities of maintenance and peasant rights of cultivation. Whereas the former was an imposition on village communities, the latter was a privilege. Both, however, were ultimately derived from and sustained by notions of what proximity, shared usage, community, and precedent represented for the users of canals and the farmers of lands fed by these waterways. Such environmental ideals and the imaginations that underlay them—though often submerged under water and buried in the dirt—are fundamental to any understanding not only of the history of rural Ottoman Egypt but also of the Ottoman Empire as a whole and of riparian communities more generally.

3

Silt and Empire

This chapter examines the history of irrigation in and around the Egyptian town of Fayyum to show how a shared reliance on natural resource management bound together extremely rural regions of the Ottoman Empire such as Fayyum with centers of power in Istanbul and Cairo. Unlike most cities and regions in Egypt, Fayyum has not received major attention from historians of the Ottoman period.[1] This is rather surprising given the importance of Fayyum as a population center, the many references to it in various narrative accounts from the ancient period on, and its large agricultural-productive capacities. It was said at the turn of the fifteenth century, for example, that *each* village in the subprovince of Fayyum produced enough foodstuffs to provision the *whole* of Egypt for one day.[2] The most obvious reason for this lack of historical work on Ottoman Fayyum is the absence of Islamic court records from the region—both from Fayyum and from the town of Beni Suef, to which Fayyum had been sporadically administratively attached since at least the late Mamluk period.[3] Court records have long been a mainstay of historians of Ottoman Egypt and other Ottoman provinces. Yet these records for Fayyum and Beni Suef have either not survived or have not yet been made available to researchers in the Egyptian National Archives (Dār al-Wathāʾiq al-Qawmiyya).[4] Without these rich sources of social and economic history, we are left to seek out other documentary evidence for the history of Fayyum in the Ottoman period.

As part of the effort to reconstruct the history of Ottoman Fayyum from sources other than court records, this chapter uses the rich archival collections of both the Prime Ministry's Ottoman Archive (Başbakanlık Osmanlı Arşivi) and the Topkapı Palace Museum Archive (Topkapı Sarayı Müzesi Arşivi) to uncover the environmental history of irrigation in the region during the first half of the eighteenth century. This period has not received much consideration from historians but is critical to any understanding of the transition from the early modern to the modern eras in Egypt and the Ottoman Empire.[5] Although numerous historians have employed these predominantly Ottoman Turkish archival documents to write political and economic histories of Ottoman Egypt —the empire's most agriculturally productive and financially lucrative province—none has dealt with the history of Fayyum specifically or the history of irrigation or environment in Egypt more generally.[6]

As we have seen already, irrigation necessitated a very direct relationship between those in the immediate vicinity of a dam or canal who witnessed and experienced firsthand any infrastructural disrepair and those "downstream" who relied on the proper function of irrigation works for their food and revenues. This cooperative relationship forged by irrigation meant that Egyptian peasant cultivators and leaders, although not the only local Egyptian and imperial actors involved in the management of water in Ottoman Fayyum, were certainly the most important because they were the closest to irrigation works on the ground. There were several different local groups who brought instances of disrepair to the attention of the Ottoman imperial bureaucracy. Perhaps the most commonly referenced group in the archival record are *ahālī*. This somewhat vague term refers to "the people" of a village. Here I take it to mean those peasant cultivators who worked the land, were in the most direct contact with irrigation works, and hence were the ones most immediately affected by a deteriorating canal or dam. The documents cited in this chapter draw a clear distinction between this undifferentiated group of peasants and other local Fayyumis. Those who held higher rank in the rural community included *mültezimīn*, engineers, *kāşifler* (district administrators; s. *kāşif*), village elders, and a group we will meet below known as *ehl-i vukūf*. The *ahālī* of Fayyum most usually brought their complaints to the Ottoman state in consultation with and through the mediation of these other rural actors.[7]

Although the military-administrative Mamluk grandees of the Ottoman period wielded great power and authority in other subprovinces of Egypt, the primary sources used in this chapter suggest they held little political influence in Fayyum during the first half of the eighteenth century. The example of Fayyum therefore serves as a useful historiographical counterweight to much of the scholarship on Ottoman Egypt, which has tended to argue for the continued importance of Mamluk political factions throughout the centuries of Ottoman rule in the province.[8] This limited Mamluk involvement in Fayyum goes a long way toward explaining, for example, why the "civil war" of 1711 and the political upheavals of the 1740s (both times during which irrigation repairs were undertaken in Fayyum) had seemingly very little impact on the region's history of water management.[9]

As in other parts of the empire, irrigation provides a particularly good lens through which to view the history of Ottoman Fayyum and the region's many relationships to the rest of the Ottoman Empire because it was an intensely local process, differing according to each particular village environment, canal, sluice gate, and embankment.[10] Water had to be managed and controlled by individuals on the ground with intimate experience in and knowledge of their local environments. At the same time, irrigation was a process of wide imperial concern. In an agriculturally rich region such as Fayyum—however distant it was from the imperial capital—irrigation water was essential to produce food that would be transported to feed individuals all over the Ottoman Empire. The water that irrigated Fayyum was converted into calories to feed stomachs in Istanbul, Cairo, Dubrovnik, Mecca, and Medina.

Beyond simply pointing to a new potential source base for historians of Ottoman Fayyum, Ottoman Egypt, and Ottoman environmental history, this chapter shows how the history of a region like Fayyum during the Ottoman period can be understood *only* by considering it in relation to other parts of the empire and to the imperial bureaucracy as a whole. Thus, this study contributes to a critique of the model of center and periphery that has long been used to explain relations between Istanbul and the empire's numerous provinces.[11] Critiques of this model have benefited greatly from considerations of the historiography of other early modern and modern empires, in which simple center-periphery descriptions no longer maintain much traction. It would, for example, now seem highly dubious to tell the story of modern Britain or early modern Spain without recognizing the complex and mutually determinative relationships between component parts of the British and Spanish Empires.[12]

Briefly put, the model of center and periphery asserts that the imperial capital drew to itself cash, military manpower, and grain and other raw materials from the rest of the empire with little interest in the cultural or social production, economic expansion, or history of these regions. Accordingly, the economic, human, and material output of the provinces was extracted and pulled solely toward the capital along vectors that weakened the farther they emanated out from the center. This model does not consider interactions between various "peripheries" (or various "centers") that often bypassed Istanbul altogether. In this schema, Ottoman provinces come to be seen as piecemeal collections of sealed agricultural or economic units with little interaction with locales other than Istanbul. Karen Barkey usefully summarizes this system using a "hub-and-spoke" metaphor: "One can say that most of the different segments of the polity remain largely unconnected among themselves. That is why an imperial system is best represented in terms of the hub-and-spoke network structure, where the rim is absent."[13]

Challenging this historiographical representation, I show that Fayyum, like most Ottoman regions, did not operate in a vacuum with little connection to other parts of Egypt or the empire. On the contrary, Fayyum existed as just one part of an Ottoman imperial system of natural resource management. The region's relationship with Istanbul was not simply one of unidirectional resource extraction toward the imperial capital. Fayyumi locals, both peasants and elites, were in a consistently reciprocal, though of course unequal, relationship of push and pull with Istanbul. In the cases that follow, we will see how these locals came to inform the center's management of rural irrigation works. Simply put, power and knowledge moved in both directions along the spokes.

Likewise, a rim was most certainly present in the Ottoman Empire. Grains grown in Fayyum's rich soils—soils fed by irrigated water that originated in monsoon rains over the Ethiopian highlands—were moved across the Mediterranean and Red Seas while timber supplies from Anatolia, Greater Syria, and the Black Sea littoral were brought to Fayyum and elsewhere in Egypt to construct irrigation works and other infrastructure.[14] Ottoman Egypt, in other words, was a world in which natural resources and products were in constant motion into, out of, and through the province. The exchange of resources, knowledge, and authority along the spoke between Istanbul and Fayyum (a spoke that usually ran through Cairo) ensured the proper irrigation of the Egyptian countryside. This water in turn made agricultural land productive, forging the multiple "rim" relationships Fayyum maintained during the early modern period through its export of food. The bureaucratic energy and coordination that kept this system functioning, productive, and sustainable was a constant negotiation between Egyptian peasant knowledge of local environments, Ottoman imperial administrative acumen, and the agency of natural processes such as rainfall, siltation, and plant growth.

Thus, the perpetual back-and-forth between Fayyumis and the imperial capital and the many "rim" relationships created by water in this one rather rural region of Egypt help us to conceptualize the Ottoman Empire more usefully and, to my mind, with greater historical accuracy than can be achieved by using a model of center and periphery. Hubs and spokes paint a picture of discrete points and fixed lines of connection, with many empty spaces between the spokes and little suggestion as to how points at the ends of different spokes, or for that matter the spokes themselves, fit together or affect one another. The Ottoman Empire was surely much more complicated than this. Perhaps there were multiple "rims" on the wheel at different distances from the center, perhaps certain spokes only reached certain rims and never touched others, or perhaps most spokes crossed and curved in very intricate ways. Or perhaps we should discard this metaphor entirely in favor of something like a web or a Penrose or Venn diagram. All of this granted, my goal is not to advocate for one schematic metaphor over another. Any and all such diagramming impulses would ultimately fall short of capturing the historical details, nuances, contingencies, contradictions, and multivalent relationships that made up the Ottoman Empire. In considering the role of Egypt—and of a rather marginal locale like Fayyum—in the empire, I hope to be able to capture and understand some of this complexity.

Something Like a Rim

Fayyum's closest and most intense "rim" connection during the Ottoman period, as under earlier regimes, was with the Hijaz. This periphery-periphery relationship was built in large part on the movement of Fayyum's surplus grain to the region to feed yearly pilgrims to Mecca and Medina. Food, or the transference of caloric energy, was one of the primary factors in the maintenance of links between Fayyum—and the whole of Egypt, the proverbial breadbasket of the Ottoman Empire—and other parts of the Mediterranean as well as the Red Sea littoral.[15] Grains grown in Egypt went to fill stomachs not just in the Hijaz[16] but also in Tunis,[17] Yemen,[18] Aleppo,[19] Morocco,[20] Izmir,[21] western Tripoli,[22] eastern Tripoli,[23] Crete,[24] Salonica,[25] Algeria,[26] and Istanbul.[27] Apart from provisions shipped to the Ottoman capital, most of this food moved out of Egypt independently of the direct administrative oversight of Istanbul.

This massive network of commodity movement relied on steady supplies of irrigation water and on Egypt's established trade links with its Mediterranean and Red Sea neighbors and with regions of sub-Saharan Africa and South Asia.[28] Indeed, Egypt was in many ways the fulcrum of world trade in the late medieval and early modern periods before the realignment brought about by European penetration of the New World.[29] Throughout the early modern period, most Indian Ocean trading networks had a terminus on the Egyptian Red Sea coast; similarly, European trading routes that served to import coveted goods from East Asia and India into Europe loaded these commodities onto ships in Alexandria, Rosetta, and elsewhere on the Egyptian Mediterranean coast.[30] Egypt always had one eye on the Mediterranean and the other on the Indian Ocean. Indeed, exemplifying just how far early modern Egyptian trading connections went, the late sixteenth-century Cairene merchant Isma'il Abu Taqiyya, as Nelly Hanna has skillfully shown, maintained investments and commercial relationships from Venice to Goa and from Aleppo to Kano (in what is today Nigeria).[31]

Though perhaps not explicitly couched in terms meant to challenge a center-periphery model of early modern empire, some work has already considered Egypt's intraimperial links to other Ottoman provinces and its transregional connections to areas outside the empire (both sorts of relationships functioning in ways that often bypassed the bureaucratic purview of Istanbul). Some historians have, for example, taken the Nile as their unit of analysis and, hence, have necessarily delved into Egypt's connections to Ethiopia, the Sudan, and other regions of Africa.[32] Others have considered how the cultivation and trade of a particular commodity such as coffee enmeshed Egypt in larger networks of commerce and consumption.[33] Similarly, studies of plague, cholera, and other diseases in the Ottoman Middle East show how disease traveled from and to Egypt and elsewhere with little regard for political boundaries or the strictures of the imperial bureaucracy.[34] Likewise, Peter Gran, Khaled el-Rouayheb, and others have documented the intimate intellectual connections between scholars in Egypt and their counterparts in Greater Syria, North Africa, and elsewhere that, again, did not run through Istanbul.[35] In short, Egypt cultivated and maintained multiple "rim" relationships with other provinces of the Ottoman Empire and with areas far beyond the empire's borders.

It was within this context of an early modern global economy that Fayyum's connections to the Hijaz, Istanbul, and elsewhere functioned.

In the eighteenth century almost all annual imports to Jedda, the primary port of the Hijaz, came from Egypt on thirty to fifty rickety three hundred– to six hundred–ton ships.[36] About half of this merchandise consisted of European goods transported from Alexandria to Suez through Cairo then on to Jedda; the other half were Egyptian foodstuffs from Fayyum, the Nile delta, and elsewhere earmarked for the provisioning of Mecca and Medina. It is unfortunately very difficult to estimate what percentage of the foodstuffs shipped to the Hijaz in the early eighteenth century originated in Fayyum's soils.[37] What is clear, though, is that the majority of Fayyum's surplus grain went to the Hijaz (often via Bulaq). This had long been the case and remained so even in the fifteenth and sixteenth centuries, when the delta emerged as Egypt's most lucrative area of agricultural cultivation, and Suez became the main port of export to the Hijaz. Fayyum is geographically much closer to Suez than to Qusayr, Aydab, or any of the other Upper Egyptian ports that saw reductions in the volume of their exports beginning in the fifteenth century.[38] Fayyum's geographical proximity to Suez ensured that the overall export volume of grains from the region to the Hijaz was not detrimentally affected by the waning importance of the southern ports. Fayyum also maintained its status as a major exporter of grain to the Hijaz because of the many *awqāf* (pious endowments, plural of *waqf*) earmarked for the holy cities in the region and throughout Upper Egypt more generally.[39] The chief function of these Upper Egyptian endowments attached to the Awqāf al-Ḥaramayn was to provide grains for both the pilgrimage caravan and the people of Mecca and Medina.[40]

In bringing together the following stories of historical actors that include Egyptian peasants, Ottoman bureaucrats, silt, the sultan, religious pilgrims, and water, I hope to elucidate the connections of the bureaucratized eco-community that bound a rural region like Fayyum to Istanbul, the Hijaz, and other areas of the empire in the context of the globalized economy of the eighteenth century. In the process, I hope to demonstrate the limitations of some long-cherished overarching narratives of the nature of early modern imperial rule. After a brief discussion of the geographic and environmental circumstances of Ottoman Fayyum, I will consider three instances of major irrigation repair jobs carried out in Fayyum during the first half of the eighteenth century. I will use these cases to show how food connected Fayyum to other parts of the Ottoman Empire and how the imperial bureaucracy sought to maintain these productive connections by ensuring reliable supplies of irrigation water. I will then explain the role of local Egyptian Fayyumis, both peasants and village elites, in this imperial system of natural

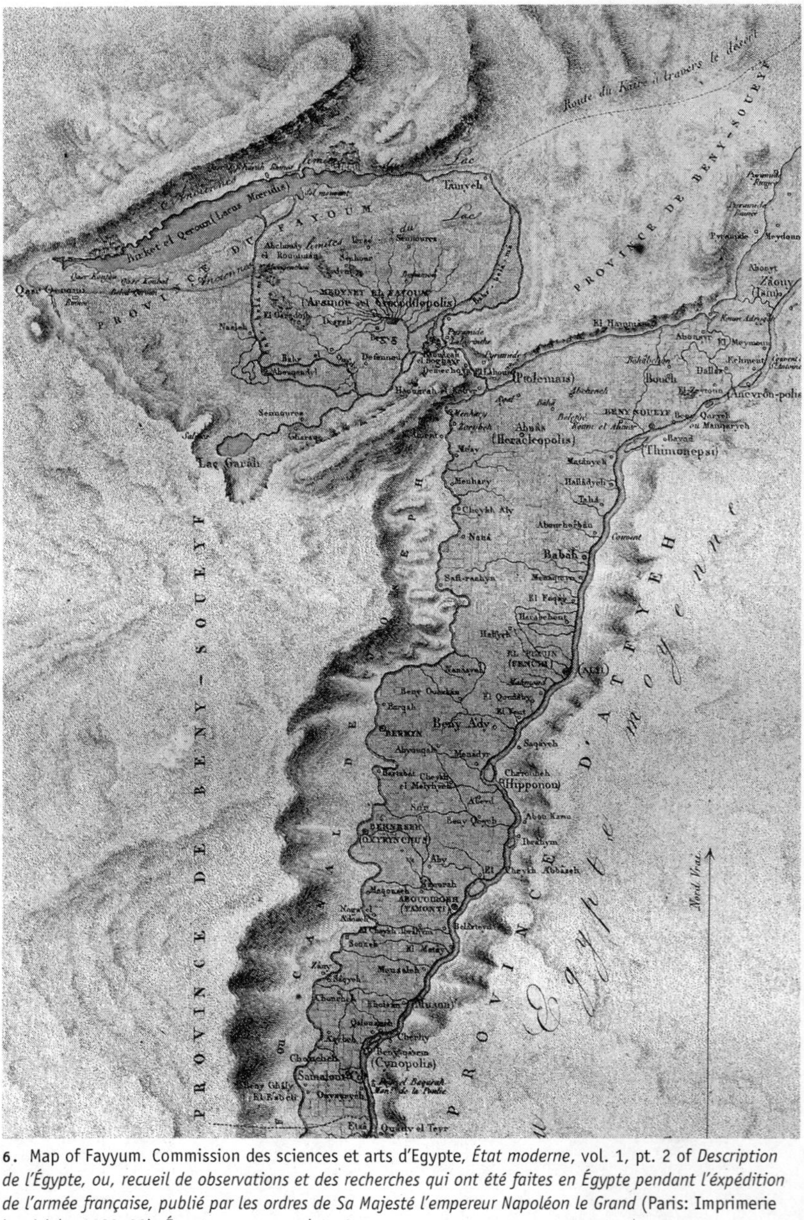

6. Map of Fayyum. Commission des sciences et arts d'Egypte, *État moderne*, vol. 1, pt. 2 of *Description de l'Égypte, ou, recueil de observations et des recherches qui ont été faites en Égypte pendant l'expédition de l'armée française, publié par les ordres de Sa Majesté l'empereur Napoléon le Grand* (Paris: Imprimerie impériale, 1809–28), Égypte moyenne, plate 6.

resource management and how they were able to use their privileged positions as possessors of knowledge of the local environment to affect the actions of the imperial government in Egypt, often against the interests of the state's own imperial officials. A primary goal throughout is to show how the relationship between Fayyum's rural residents and the Ottoman state functioned through a shared interest in irrigation and the cultivation of food destined for various parts of the empire.

Fayyum and Its Environment

Unlike most population centers in Egypt, Fayyum is in neither the Nile valley nor the Nile delta.[41] It lies in a large natural depression southwest of Cairo in Egypt's western desert, known as the Libyan Desert.[42] As described by one traveler to the region, "Fayoum appears as the bud of the great lotus plant of Egypt, growing out of one side of the Nile stem, just below the Delta blossom."[43] The area of Fayyum is roughly 1,733 square kilometers (669 square miles), nearly all of which is suitable for cultivation.[44] The northernmost section of the depression is entirely below sea level and bounded by a large brackish lake known as Birkat Qārūn, whose surface lies 45 meters (147.64 feet) below sea level.[45] The surface area of the lake is 214 square kilometers (82.63 square miles).[46] Although Fayyum is usually described as an oasis, it is technically not because it relies entirely on the Nile for its usable water.[47] Fayyum is separated from the Nile valley to the east by a high ridge of loose stone and dirt.[48] This ridge is pierced at only one point by a natural cut, and through this opening a canal brings Fayyum all of its water (save scant amounts from rain) for irrigation and other purposes. This extremely important waterway is known as Baḥr Yūsuf and branches off from the Nile at Beni Suef.[49] The canal leaves the Nile valley at an area known as al-Lāhūn and enters Fayyum 5 miles later at a point known as al-Hawara. It is literally the lifeline of Fayyum, because without this single waterway the region would be completely cut off from the Nile or any other freshwater source. The distribution and allotment of all of the region's water was thus controlled and manipulated at its entrance into Fayyum.[50]

The two most important irrigation features on Baḥr Yūsuf were the regulating dike of al-Lāhūn and the seawall-like dam of al-Gharaq.[51] The dike of al-Lāhūn was built at the narrow gap that allowed Baḥr Yūsuf into Fayyum and was thus vital to regulating the flow of the canal and the amount of water that entered the region.[52] The dam of al-Gharaq was farther along the canal and was much larger in surface area than the dike. Ottoman sources from the period describe it as a huge dam (*sedd-i azīm*) of impressive stature (*bina-i cesīm*) that had been in existence since times of old (*kadim ül-eyyamdan*).[53] It was built low in the Fayyum depression to trap the waters of the Nile behind its long wall in a huge lake (*buheyre-i azīme*) that was meant to supply the people of Fayyum with water for the entire year until the time of the next flood. It likewise prevented this water from spreading and branching out (*intişār ve inşiʿāb*) uncontrollably to flood villages and agricultural lands around Fayyum.

The people of Fayyum were thus completely dependent—for irrigation, drinking water, bathing water, and so forth—on the proper function of just a few irrigation works.[54] If for any reason one of them broke or was not functioning properly, great harm and hardship would be visited on the people of Fayyum, and, in turn, overall levels of food production in Egypt would drop.[55] Such a scenario was a constant fear in the eighteenth century.[56] Over and over again in petitions and imperial responses we find a standard rhetoric of complaint and of impending doom and danger related to the breaking of one of these irrigation structures and to the resulting decreases in food production. Because Fayyum was situated in a deep depression, the waters that reached it from the Nile gained great force as they moved down from higher elevations (al-Lāhūn was 23 meters, or 75.46 feet, above sea level). Irrigation features like the dam of al-Gharaq were thus constantly—"day after day [*yevmen feyevmen*]," in the words of one document of repair—exposed to the force of the Nile's crashing waves (*hurūş-i telātum-i emvāc*) and the rush of its deluge (*tezāhüm-i cereyan*).[57] It is therefore not surprising that the regulating dike of al-Lāhūn, the dam of al-Gharaq, and various other embankments, dikes, and dams on Baḥr Yūsuf were in constant need of repair. In the first half of the eighteenth century alone, nearly twenty major repairs were carried out on the dam of al-Gharaq.

Because Fayyum's topographical and geographical features meant that it relied on a single source of water to irrigate its entire area, the history of the region is a perfect case study of how changes to a single irrigation work carried significant and often dire consequences for wide swaths of agricultural land and for populations throughout Egypt and in other parts of the Ottoman Empire. If one dam broke on the canal of Baḥr Yūsuf, all of Fayyum's agricultural output for an entire year could be ruined.[58] The consequences of such an event for Egypt and for populations throughout the Ottoman Empire were great indeed. The manner in which Egyptian peasants and Ottoman bureaucrats dealt with the challenges posed by Fayyum's natural landscape shows just how invested peasant Fayyumis and the Ottoman state were in the collective maintenance of irrigation works in Egypt and how sensitive the environment of Fayyum was to changes in its irrigation network.

Calories of Empire, 1709–11

Fayyum did not immediately fall under direct Ottoman administration in 1517, when most of Egypt became a province of the empire; it was only in the early seventeenth century that the region came to be directly administered, most likely by the subprovincial governor of Jirja.[59] Once brought within the administrative purview of the empire, Fayyum became one of the most concentrated and productive areas of cultivation in the Ottoman Empire's agriculturally richest province, supplying calories to stomachs across the empire. Water diverted from the Nile to Fayyum through the canal of Baḥr Yūsuf was used to grow food for people in Egypt, Istanbul, Mecca, and Medina. Quite logically, then, the maintenance of irrigation features on the canal of Baḥr Yūsuf was a matter of imperial concern. Between 1709 and 1711, the Ottoman state bureaucracy made the importance of the maintenance of the canal's irrigation features to imperial fortunes abundantly clear with its vigorous efforts to repair the dam of al-Gharaq and the dike of al-Lāhūn, both of which had suffered from a recent lack of maintenance and high flood levels.[60] At the heart of a series of orders for repairs from 1709 to 1711 was the imperial bureaucracy's understanding of the significance and self-sufficiency of the "rim" relationship between Fayyum and the Hijaz. The movement of food from Egypt to the Hijaz was operationally and financially in the hands of the Egyptian peasants, merchants, and sailors who produced the food and managed the trade. Maintaining the water that grew this food, however, was the empire's responsibility. To emphasize this point, Ottoman orders repeatedly observe that grains grown in the soils of Fayyum were to be moved to the Hijaz to feed the thousands of yearly pilgrims.[61] Here, then, is an example of the central Ottoman administration ensuring a natural resource base needed to maintain a relationship between two of its provinces—and, of course, also needed to maintain a flow of taxes to the empire's coffers—that, after this initial imperial act, functioned largely outside of the state's hands.

In 1709, the divan of the Ottoman sultan Ahmet III sent a firman to the vali in Cairo noting that much of Fayyum's land had been found to be *sharāqī* (parched).[62] The region's fields were not receiving adequate amounts of water because the regulation mechanisms on the canal of Baḥr Yūsuf had fallen into a state of disrepair. From its imperial perspective, the sultan's council wrote that it was not only Fayyum that was directly affected by this deterioration in the irrigation network. Repeating a well-worn idea in repair cases of this sort, the bureaucracy made clear that the great destruction to agricultural lands that might result should this situation remain unaddressed would hurt Egypt's overall agricultural yield and thus greatly reduce the amount of tax revenue available to the state. The sultan therefore ordered his vali to send to Fayyum engineers of sound judgment (*mühendisin-i sahih ül-tahmin*) to oversee the reconstruction of these dams as quickly as possible (*alavechitta'cil*).[63] To fund this work, eleven Egyptian purses (*kise-i Mısrī*) were made available from the *irsaliye* (annual tribute) of 1708–9.[64]

ʿAbd al-Raḥman al-Jabartī's chronicle of Ottoman Egypt, one of the most important and widely cited Arabic narrative accounts of this period, makes only a fleeting reference to this firman.[65] It says simply that an imperial decree arrived in Cairo ordering the construction of a weir (*qanṭara*) at al-Lāhūn in Fayyum.[66] In the absence of Islamic court records from Fayyum, this passing phrase by al-Jabartī is one of the few references we have in any surviving Arabic source, published or unpublished, to this irrigation repair work in Fayyum in the summer of 1709.[67] Its history exists almost solely in the Ottoman archival record.

These 1709 repairs were apparently inadequate, because a little over one year later sections of the very wide brick and stone (*acır ve ahcar*) wall of the dam of al-Gharaq were still weak and leaning toward the water of Baḥr Yūsuf, on the brink of total collapse (*münhedim ve harab*).[68] As in the previous year, a new firman insisted that if this situation were allowed to continue, the people of Fayyum would soon find themselves deprived of even a single drop of water (*bir katre āb gelmeyūb*), and their lands would become *sharāqī* once again. In addition, as before, this lack of water and, in turn, food would not only mean great turmoil and distress (*perişanlık*) for the people of Fayyum but—and more important from the perspective of the Ottoman state—would also lead to the disruption and loss (*kasr ve noksan*) of state monies.[69] Before the promulgation of this sultanic order, the *kāşif* of Fayyum and the *qāḍī* (judge) of Beni Suef, along with other officials in the area, had been dispatched to inspect the dam.[70] On the basis of a report (*keşf*) summarizing their findings, the imperial administration released an additional eleven Egyptian purses and 11,600 *para* from the annual tribute of 1709–10 for the dam's repairs.[71] Like the previous firman, this decree ends with a warning against spending more than the allotted sum and against any laziness or inefficiency in the repair work.

In these two firmans from 1709 and 1710, the Ottoman sultan and his government made clear that their concerns with irrigation in Fayyum stemmed first from the state's need for cash and second from their interest in providing grains to feed imperial subjects, the bureaucratic elite, and pilgrims traveling to Mecca and Medina each year. The dam of al-Gharaq and the dike of al-Lāhūn remained in a decrepit state of disrepair into 1711, threatening the people of the Hijaz with acute food shortages.[72] The Nile flood that year was exceptionally large and powerful, such that it partially destroyed many sections of the two irrigation structures, both those parts repaired in the previous two years and those that had not been repaired at all. As a result, the works were near total collapse, threatening to deprive Fayyum of any form of irrigation or watering (*reyy ve irtivadan ātıl olmak*). Of the dangers posed to irrigation and cultivation in and around Fayyum by damage to the dam and dike over this three-year period, those in 1711 were by far the worst.

The report commissioned to inspect the canal measured 27,234 square *zirāʿ* (10,809.2 square meters) of damage to al-Gharaq's walls.[73] The damaged areas of the foundation (*paye*), walls (*divarlar*), and support girdle (*kemer*) of the Lāhūn dike measured a total of 9,980 square *zirāʿ* (3,961.1 square meters). Between the two structures, the total area in need of repair was thus 37,214 square *zirāʿ* (14,770.2 square meters). It was estimated (*alavechittahmin*) that the cost of repairing each square *zirāʿ* (0.3969 square meters) would be 15 *para*, bringing the total cost to twenty-two Egyptian purses and 8,510 *para*. The bulk of these funds would go toward purchasing the necessary building materials—mainly lime (*kireç*), wooden supports (*şecār*), and stone (*taş*)—and would be taken out of the Egyptian annual tribute of 1710–11.

Despite its difficulties and costs, the work required to return these irrigation features to what the inspection report calls their previously strong and rigid (*kadimden metin ve müstahkem*) condition was unavoidable. The Ottoman imperial system of natural resource management was predicated on all parts of this highly coordinated system functioning efficiently. The empire thus found its hand largely forced by the dictates of the river's flow and the man-made structures it oversaw to harness that flow. Once the physical reality of irrigation in the countryside was cast in the form of certain structural features, changing this rural world's infrastructural and imagined shape would have required a major investment of effort, money, and time. More important, the empire's hand was also forced by the historical precedent of the independently functioning relationship between Fayyum and the Hijaz. After all, Fayyum had for centuries sent its grain directly to the Hijaz for the sustenance of pilgrims and the populace of the holy cities. The large amount of money and effort put toward the repair of irrigation features on the canal and the fact that these repairs were carried out over three consecutive years show just how difficult and complex maintaining a steady supply of water was in Fayyum and how important these projects were both to the Hijaz and to the Ottoman state. Thus, the need to convert the sun's energy into calories in Egypt's irrigated soils drove much of the Ottoman Empire's rule in the province.

A Local Empire, 1727

The connections enabled by Egypt's agricultural productivity and maintained by the Ottoman Empire's imperially coordinated system of natural resource management could only work by relying on the knowledge and practices of those who knew Egypt, its water, and its soils best. This connection between the local and the imperial was one that moved in both directions as Egyptian peasants and village leaders alternately tugged on and were pulled by the imperial capital. With their accumulated knowledge and experience, Egyptian peasants in Fayyum and elsewhere provided the Ottoman administration with the local expertise it needed to function productively on the ground. In considering the role of peasants in the maintenance and repair of irrigation works in Fayyum, I argue that it was they—rather than the empire's bureaucrats—who pulled the strings of power in controlling the operation and use of dams, sluices, and weirs. The Ottoman Empire in many ways devolved authority in the realm of irrigation to Egyptian peasants and other villagers because they knew the local environments of Egypt—vital knowledge needed for repair projects involving water—better than anyone else.

Egyptian peasants, village elders, and representatives of the Ottoman Empire thus participated in and constituted a unified system of natural resource use and management. Local peasant understandings of rural environments did not reflect some idyllic state of nature. In Ottoman Egypt, as in all imperial contexts, they were infused materially and ethically with the economic and political demands, needs, and desires of the polity and society around them. Likewise, the history of the Ottoman Empire in Egypt and throughout its domains cannot be understood without seriously accounting for the influence of local practice and knowledge in the function of the state. We must, in short, keep both the imperial and the local in play. The story of the Ottoman Empire cannot be told without Egyptian peasants, and the story of Fayyum and of Egypt as a whole cannot be told without the empire.

At the end of June 1727, the imperial palace in Istanbul sent a firman to the Ottoman vali in Cairo, Mehmed Paşa, about the current state of disrepair of the dike of al-Lāhūn and the dam of al-Gharaq.[74] This decree came in response to an earlier missive from the Egyptian vali that was, in turn, prompted by numerous petitions from Egyptian subjects complaining that if the two irrigation works in Fayyum were not fixed immediately, they would collapse, destroying much of the agriculturally rich region and leading to massive and expensive reconstruction efforts.[75] According to the two Arabic documents (*ḥicce*) drafted for the vali by local peasant Fayyumis and village elders with experience and knowledge (*ehl-i vukūf*) in Fayyum, the dams had to be repaired in the coming few months, before the onset of that year's flood. In response to these locals' warnings of a massive flood (*garīk*) and widespread destruction, the vali sent some of his men to Fayyum, along with an architect and engineer, to inspect the damage. According to their report (*keşf*), the dams would cost a total of 347,195 *para* (or thirteen Egyptian purses and 22,195 *para*) to repair.

All of this information, together with the two original Arabic documents drafted for the vali, was forwarded to the palace in Istanbul. It was on the basis of these documents that the sultan and his retinue composed the firman sent to Cairo at the end of June 1727. Arriving in Egypt only a few weeks before the beginning of the flood season in late summer, this firman conveys the great foreboding and fear associated with the impending danger posed by the annual flood. As was customary, the funds needed for these repairs were to be taken from the province's annual tribute (in this case, of the year 1726–27). The sultan's order, however, stipulated that only 300,000 *para* (or twelve Egyptian purses) and not the full amount requested by the vali (347,195 *para*) were to be made available for this construction work. A dependable and trustworthy man (*muʿtemed ve sadākatkār*) was to be appointed to oversee these repairs and to inspect them once completed, and an engineer (*mühendis*), an architect (*miʿmar*), and a group of experts (*ehl-i vukūf*) were also to be consulted. As in similar cases, these men, especially the *ehl-i vukūf* and possibly the engineer as well, were in all likelihood locals from Fayyum who had risen in the imperial bureaucracy to become functionaries of the state.

This case shows that it was the *ahālī* of Fayyum who were most crucial to the timely and proper execution of these repairs. As the eyes and ears of the Ottoman Empire on the ground, peasant actors, working in concert with the *ehl-i vukūf* of that area, informed imperial authorities of the condition of the dams and of how best to proceed with their reconstruction. These men were locals who had probably spent most of their lives in Fayyum; they knew the dams and the normal course of the flood in their area, and they understood the soils and topographies of their villages' fields. Because the environment of the countryside and irrigation within it could be understood only through years of intimate local interactions and experience, the Ottoman bureaucracy of Egypt had to rely on Egyptian peasants like the ones in this case to maintain the province's rural irrigation network. The local knowledge, experience, and expertise of these men influenced not only how the dams in Fayyum were rebuilt but also ultimately how the Ottoman Empire managed irrigation works throughout Egypt. With the two Arabic reports composed in Fayyum and sent to Istanbul, the local knowledge of these Egyptians literally entered the bureaucracy of the Ottoman state to influence and sustain its rule. By insisting that fixing irrigation works would serve not only their own interests but also those of the state, these men were able to make the state work for them or, at the very least, to make the state pay for this work with money diverted from Egypt's annual tribute.

Ottoman administration in the Egyptian countryside was thus not some distant abstraction commanding what it may from another continent. On the contrary, it operated at a very local level through its own bureaucrats and, even more important, through the labor and knowledge of local Egyptians. Peasants, village leaders, bureaucrats, and the Ottoman capital city were accordingly connected through an imperial network in which each played an important role in the repair of irrigation works in the Egyptian countryside and in which the local knowledge of peasants in one part of the ecosystem impacted and sometimes even dictated the actions of others elsewhere. In short, Fayyum's dams affected economies, stomachs, ecosystems, and administrative structures in Istanbul, Cairo, the Hijaz, and elsewhere in the Ottoman Empire, and, conversely, Cairo and Istanbul affected these dams.

Bureaucratic Natural Disasters, 1741–46

Within the eco-community created by irrigation in rural Egypt, peasants very often interacted directly with the Ottoman imperial state, even against the personal interests of the empire's own bureaucrats. Peasants in a place like Fayyum were frequently closely aligned with the Ottoman state, which trusted them to protect it from the subterfuge of its own functionaries dispatched to the region. This dynamic of imperial connection and interaction again shows the importance of local peasant knowledge and practice to the workings of the entire empire, as well as the manner in which the use and maintenance of a natural resource like water tied rural communities in Egypt and beyond to centers of Ottoman power in Istanbul and Cairo. Telling examples of this interaction are instances in which Egyptian peasant leaders' complaints about the machinations of thieving or otherwise conniving Ottoman bureaucrats led to the removal of these imperial functionaries. In such cases, the information provided by peasants was very often adopted and executed as imperial policy.

In 1744, the *ahālī* of Fayyum sent a number of petitions to the governor's divan in Cairo complaining that the dam of al-Gharaq's walls were in a near constant state of disrepair.[76] There had been various attempts to repair the dam in 1741 and 1742, but these had proved wholly inadequate in 1743, when that year's powerful flood destroyed large sections of the dam.[77] These locals' letters rehashed much of the same standard rhetoric of complaint and imminent disaster that we saw previously. In response to these entreaties by a group of rural imperial subjects, the imperial government instructed Mehmed Paşa, the newly appointed vali of Egypt, to move swiftly to repair the structure.[78] The dam soon proved to be one of the most important issues of the new vali's two-year appointment. The sultan directed him to personally oversee its repair, bypassing the customary appointment of an overseer of construction (*bina emîni*). The sultan's firman also concedes that two *hatt-ı hümayun*s had been issued previously with respect to these repairs but without effect. In 1744, Mehmed Paşa was thus instructed to appoint a group of state officials, engineers, architects, and local men of knowledge (*erbab-ı vukūf*) to inspect the dam's damage and to establish the costs of the needed repairs. They determined the size of the damaged area to be 130,300 square *zirā*ʿ (51,716.1 square meters) and the cost of the repairs to total eighty-seven Egyptian purses and nine thousand *para*. As in previous repair cases, the necessary funds were to come from the Egyptian annual tribute (here from the year 1742–43).

Mehmed Paşa soon reported to the sultan that he had carried out the repairs as required.[79] His story, however, was contradicted by the correspondence of a group of Fayyumis who wrote to the imperial palace complaining that Mehmed Paşa had repaired only a small portion of the dam and that even this abbreviated repair work had not been done adequately. His failure to fix the dam properly thus left the structure weak and ill prepared to control the rushing waters of the flood. In 1745, the date of these letters, the Nile's floodwaters indeed destroyed both the damaged sections of the dam that Mehmed Paşa had failed to repair and, even more worrisome, other parts of the dam that had previously been fully functional. The state of the dam was therefore now worse than it had been before Mehmed Paşa's repairs, and, according to the letters from Fayyum, if the entire dam were not repaired properly, then all of it would soon be completely washed away.

In addition to the looming danger of the structure's total collapse, it was soon discovered that Mehmed Paşa had stolen some of the money allocated to him for the repair work. A review of the Egyptian divan's accounts revealed that the dam's repair cost only eighty Egyptian purses. There was thus a discrepancy of seven Egyptian purses and nine thousand *para* between what had been given to Mehmed Paşa and what he spent on the repairs. This combination of ineptitude and the theft of state funds led to his replacement as vali of Egypt by Mehmed Raghib Paşa in the middle of June 1745.[80] When Raghib Paşa's predecessor was approached about the missing funds, he conceded that he owed the difference to the state as a debt (*benim zimmetimdir*) and swore (*edā ederim*) that he would repay it in a timely fashion.

To gain a better understanding of the condition of the dam of al-Gharaq, Raghib Paşa sent a group of state officials, engineers, architects, and local Fayyumis to inspect the damaged structure. These men concentrated on three sections of the dam's long wall: formerly damaged areas that Mehmed Paşa had inadequately repaired, damaged areas he did not repair, and parts of the dam destroyed after Mehmed Paşa's repair work. Sections of the dam that had been improperly repaired measured 89,610 square *zirā*ʿ (35,566.2 square meters). Damaged areas that were supposed to have been repaired by Mehmed Paşa but that were left in ruin measured 46,362.5 square *zirā*ʿ (18,401.3 square meters). And formerly sturdy portions of the dam that had been destroyed by the most recent flood measured 20,025 square *zirā*ʿ (7,947.9 square meters). The total area in need of repair was thus 155,997.5 square *zirā*ʿ (61,915.4 square meters). As before, 15 *para* were required to repair each square *zirā*ʿ (0.3969 square meters). The total cost of these repairs—which included 35,000 *para* needed for other unspecified matters related to the construction (*mühimmat-ı lāzime bina*)—was therefore forty-one Egyptian purses and 5,812 *para*. This sum was in addition to the eighty-seven Egyptian purses and 9,000 *para* previously allocated for the repair work.[81]

Mehmed Paşa's attempted thievery hurt both Egyptian peasants by saddling them with the ecological consequences of the dam's disrepair and the Ottoman state by robbing it of essential food and revenue. The letters from the *ahālī* of Fayyum to the imperial divan document how waterworks in Egypt created an irrigation network across the province that bound Ottoman bureaucrats, imperial elites in Istanbul and Cairo, Egyptian peasant cultivators, their village leaders, and rural groups such as the *ehl-i vukūf* into one cooperative eco-community in which even the smallest changes in any one part of the network affected all those in the system. Because so many people and so much revenue in various locations throughout the Ottoman Empire were tied up in the proper function of a single irrigation work like the dam of al-Gharaq in Fayyum, the Ottoman bureaucracy of Egypt was forced to deal with the natural forces of the flood and the bureaucratic disasters caused by provincial governors who stole from the state. Moreover, because they were the ones most directly affected by and aware of Mehmed Paşa's illegal actions, peasants and village leaders in Fayyum were the first to bring his theft to the attention of the Ottoman state. It was thus these very direct links between peasants and other locals in the Egyptian countryside and Ottoman bureaucrats in Cairo and Istanbul that enabled the people of Fayyum to report Mehmed Paşa's dishonesty to the central state and that eventually led to the dam's proper repair.[82]

Conclusion

Through a consideration of the Ottoman archival record of repairs to irrigation works in Ottoman Fayyum in the first half of the eighteenth century, in conjunction with other Arabic archival and chronicler sources of the period, this chapter has suggested a way of conceptualizing the Ottoman Empire that avoids—or at least forces us to reevaluate—the model of center and periphery that remains the dominant mode of explaining relationships between Istanbul and the empire's provinces. Instead of a picture of the empire as a set of radiating spokes that diminish in strength as one moves farther away from the hub of Istanbul, the Ottoman Empire as seen through the history of irrigation in Ottoman Fayyum proves to be a much more complicated political and ecological combination of centripetal and centrifugal forces. Centripetally, the empire certainly pulled on its provinces for resources and taxes, but the pull was in both directions. Fayyum's peasants regularly informed the Ottoman state about irrigation works in rural Egypt and affected how the state dealt with these works. They were the ones who most often initiated repairs and who directed the state as to how best to undertake them. Moreover, the word of local Fayyumis was considered paramount in the removal of cheating state bureaucrats.

At the same time that Fayyum was in constant back-and-forth engagement with the Ottoman capital, its administrative, economic, and caloric relationships also moved centrifugally outward in many directions away from the center. The most important of Fayyum's centrifugal relationships in the Ottoman period was its role as a food supplier to the Hijaz. It is important to remember that this relationship, like that between Egypt and other regions, also functioned—as evidenced by the records of Islamic courts throughout Egypt—largely outside of the direct administrative purview of the Ottoman state. More fundamentally, it operated in the context of the global economic, ecological, and epidemiological relationships Egypt maintained in the first half of the eighteenth century. The overall point here, then, is that to understand the history of Ottoman Fayyum—and by analogy other rural regions of the empire and, in turn, the history of the empire as a whole—we must take into account its multiple relationships to various areas often at some distance from Egypt. The existence and exploitation of multiple kinds of sources for the narration of the history of Ottoman Egypt thus can only take us so far. We must also be open to allowing these sources to lead us toward conceptualizations of the Ottoman Empire that do not sit comfortably with our current thinking. The resulting payoff is potentially great. As I have shown in this chapter, we begin to be able to imagine Ottoman and Egyptian history in new ways that question prevailing assumptions about the nature and geography of imperial rule, the role of peasants and other locals in that rule, and the importance of nonhuman agents such as water and silt in shaping the empire.

Part Two: Work

4 Rural Muscle

The last half millennium witnessed a massive sea change in the relationship between labor and environment in Egypt. From the time of the Ottoman Empire's conquest of Egypt in 1517 until the end of the eighteenth century, Egypt's rural cultivators worked to maintain irrigation and other environmental manipulation technologies through small-scale repair and infrastructure projects. Most often, several or a dozen workers devoted a few days to repairing canals, waterwheels, sluice gates, and embankments that helped to provide water to their local communities. They lived at home during these repairs, were often paid, and generally managed this work on their own.

At the turn of the nineteenth century, this system of coordinated localized work began to change.[1] Soon rural labor exploded beyond the local. Massive numbers of peasant workers were forcibly moved great distances to repair irrigation and other infrastructural works that did not serve their families or villages. The close connection between local peoples and their environments was no longer the driving force behind the rural manipulation projects that had for millennia made Egypt one of the most lucrative and productive agricultural zones in all of Eurasia and Africa. Abstract notions of resource extraction, commercialization, and state-building came to control the relationship between rural labor and the environment. This chapter explains this change in the relationship between labor and the environment over the last half millennium. It shows how forced labor, deleterious environmental exploitation, extractive economics, and population movements emerged at the end of the eighteenth century and how they have come to characterize the relationship between work and the environment in rural Egypt from that period until today.

The reasons behind this fundamental transformation in Egypt's labor regime were political. Beginning around 1750 and extending until today, Egyptian politics has been dominated by a concentrated group of rural (and, more recently, urban) elites with deep connections to Egypt's rulers.[2] A small group of people control economic resources in Egypt and have used this control to "develop" the country along lines that benefit themselves and their crony partners. This system first began in the second half of the eighteenth century and was the primary factor that changed how farmers, the overwhelming majority of Egyptians, interacted with environmental resources.

Situating these shifting human relationships with the environment in the middle of the eighteenth century flies in the face of most stories about Egypt's transition to modernity—a transition effected, we are told, either by the French invasion of Egypt, the genius of Mehmet ʿAli (again, Egypt's Ottoman governor-general from 1805 to 1847), or the sprouting roots of European-defined enlightenment long buried in Egyptian earth.[3] The general consensus that emerges from all sides within this contested historiographic field is that the political and economic changes that impacted labor, and most other forms of politics, economics, and social formations in Egypt, originated around 1800. In truth, they began earlier. This chapter's mid-eighteenth-century account of the emergence and effects of large-scale public works projects indeed provides evidence that the kinds of centralized administrative structures that are usually only associated with the early nineteenth-century state of Mehmet ʿAli actually began decades before. This serves as an important corrective to the periodization of the last five hundred years of Egyptian history.

The Russo-Ottoman wars of the last half of the eighteenth century, the expanding corporatization of the Ottoman military, and inflationary pressures on the Ottoman economy all contributed to the realignment of power in various Ottoman provinces at the end of the eighteenth century.[4] In Egypt, the manifestation of these political and economic developments was the expanding power of provincial governors. Unlike the classical system of Ottoman governance whereby imperial officials were rotated among various positions, often in different provinces, to prevent them from gaining any sort of localized power, at the end of the eighteenth century a series of Ottoman governors in Egypt monopolized power for decades and succeeded in carving out for themselves and their associates a space of autonomy from the empire.[5] They withheld tax revenue, seized resources, and eventually created their own military forces. The most prominent of these governors were ʿAlī Bey al-Kabīr

7. ʿAlī Bey al-Kabīr. J.C. Bentinck, *Aly Bey Roy d'Egypte 1773*. Yale University Art Gallery, 1980.14.5. http://artgallery.yale.edu/collections/objects/aly-bey-roy-degypte-1773.

and Muḥammad Bey Abū al-Dhahab.[6] In the 1760s and 1770s, they seized wide tracts of land and used these properties to raise revenues to pay soldiers and to patronize political and economic partners in the province and elsewhere in the empire. These provincial elites reached the height of their power in the 1770s when they led several invasions of Ottoman territory in Greater Syria.[7] They held territory for a few years before being pushed back to Egypt. As part of this growing autonomy of Egypt from the rest of the empire, several important Mamluk families in Egypt assembled large estates that included farms, pious endowments, and other fixed properties.[8] The Ottomans had treated Egypt as one big tax farm for about two and a half centuries. At the end of the eighteenth century, various interests wanted slices of this very big pie.

One of the primary results of these political and economic plays for more power in Egypt was the concentration of labor and environmental resources in just a few wealthy hands. This process of resource concentration was helped along by demographic shifts related to climate and disease that reduced human and animal populations in the countryside.[9] Emigration from the countryside to cities, the abandonment of farms, and the outright seizure of property allowed for the creation of large hierarchical estates that began producing agricultural commodities in commercial quantities for increasingly globalizing markets. As these estates grew in size, extraction efficiency, and productivity, their demands for increased and different kinds of labor grew in lockstep. Devoting more land and more resources to the production of more food (and increasingly more nonconsumable goods) for a widening market meant more labor. Thus, as part of this process of emergent commercial agriculture, estate holders undertook increasingly larger and more labor-intensive infrastructural manipulation projects on their land.[10] This shift from partially marketized subsistence to intensive commercial agriculture was therefore the key factor that transformed Egypt's labor regime in the decades before 1800. And because labor was a dynamic and capacious social phenomenon linked to politics and economics, tracking changes in the history of labor in Egypt in the second half of the eighteenth century helps to shift our focus away from the early nineteenth century as the temporal locus of epochal change.

To fully understand the complex connections between political and economic changes on the one hand and the relationships between rural labor and the environment on the other, this chapter will first sketch a picture of rural environmental manipulation projects in the early modern period to show their small-scale nature, local control over them, and their relatively light touch on the environment. It will then show

how such projects and the character of work on them fundamentally changed *before* and continued to change through the turn of the nineteenth century. Forced labor, large-scale and deleterious environmental exploitation, extractive economics, and population movements came to characterize the relationship between work and the environment in rural Egypt in the nineteenth and twentieth centuries. Furthermore, the consequences of these changes from the second half of the eighteenth century are still with us. Thus, to understand the history of labor, politics, economics, and the environment in Egypt over the last two centuries, we must understand the fundamentally transformative shifts that occurred at the turn of the nineteenth century.[11]

Let's Work!

Rural labor was usually a small-scale affair in early modern Ottoman Egypt. Groups of up to a few dozen peasants worked to repair irrigation and other infrastructural features in and around their villages. Their labor benefited themselves and their families directly. This work was very often initiated by rural cultivators who came to one of the Ottoman Empire's courts in Egypt—the most immediate presence of the empire in the province—to complain about a canal or embankment.[12] Imperial functionaries would then undertake an inspection of this repair request to determine whether or not it warranted the empire's financial resources. If it did, local peasants would then invest their sweat equity to implement the repairs. This coordinated localism ensured that the empire's rural infrastructure was efficiently and quickly maintained. Local communities provided the labor and expertise the imperial bureaucracy relied upon to maintain water flowing in Egypt; the empire provided the cash.

Although the vast majority of these repair jobs were of a small-scale and very local nature, their consequences were clearly imperial. The proper function of irrigation features, even in the remotest areas of rural Egypt, was of deep imperial concern since, as we saw with the example of Fayyum, the food these waterworks helped to grow fed people all over the Ottoman Empire. The labor of a very small number of Egyptian peasants (and nonhuman animals) on minor dikes and canals in the Egyptian countryside thus connected them to peoples and locales at great distances from themselves.[13] Their labor was, in other words, part of an Ottoman imperial economy spread across the Middle East, southeastern Europe, and North Africa. Tied together through the work of environmental resource management, the empire needed peasants, and peasants needed the empire.

Laborers performed many different tasks in rural Ottoman Egypt. In the early modern period, they were most often identified as "people of the village" (*ahālī al-nāḥiya*) or "those who benefit from its [for example, a particular irrigation feature's or road's] use" (*man yastaʿīnunuhu bihi*). As these phrases suggest, the most salient feature of human labor between roughly 1500 and 1750 was its extremely localized and small-scale character. "People of the village" worked to repair irrigation features and other structures that directly benefited themselves, their families, and their agricultural lands.[14] Their incentive and desire to work stemmed from their vested interests in the proper function of the irrigation features on which they so profoundly relied.

Perhaps the most common category of laborer on irrigation projects in Ottoman Egypt was ditchdiggers.[15] These men cleared excess mud, dirt, branches, and leaves that often collected in canals and along embankments. They were also charged with clearing some of the mud and silt that collected along the shallow bottoms of canals; this rich dirt was then often put toward reinforcing embankments or fertilizing fields. Other collective categories of workers were also responsible for dredging canals and making sure dirt was properly used in the reinforcement of embankments.[16] Repair and construction projects also needed able bodies to act as lifters and carriers to move all sorts of items: building materials to worksites and dirt and garbage away from them.[17]

A group known literally as "possessors of discernment" or simply "those of knowledge, expertise, and experience" were also regularly involved in irrigation projects.[18] These were older men who did not do a great deal of manual labor per se but who served a rather more advisory and directive role in these projects. Given their years in the countryside, they understood the specific topography and geography of worksites and knew what areas needed the most attention. Their experience endowed them with sophisticated local knowledge about the irrigation works of a particular area, thus making them important sources of local expertise for the Ottoman state. As was often repeated in repair cases, all of these laborers, whatever their title or role in maintenance and repair work, were to labor "day and night" until "maximum irrigation was fully achieved."[19]

In 1652, these and other classes of laborers were employed in the work of opening canals that fed off of the "sea" formed by the waters of the major canal of al-Baḥr al-Ṣaghīr in the subprovince of al-Manṣūra in the northeast delta.[20] As was customary, all of the men used in the project were expressly ordered to be acquired from villages directly impacted by this set of canals. Likewise, in another case from 1707 about the reinforcement of al-Baḥr al-Ṣaghīr's embankments, Ottoman officials ordered the extraction of beasts of burden, peasants, and those with experience from villages near the canal to clean and repair the waterway.[21] In contrast to what would come later, local labor in the early modern period was frequently compensated, sometimes in cash, but more often in kind. To repair a dike in the village of Mīt ʿĀfiyya in the subprovince of al-Gharbiyya, for example, the head of this village was instructed to collect nine men for the job and to provide each with provisions and shelter for the one night the work required.[22] Even though this was a very small amount of work—only one night—the nine men were compensated for their labor, which, again, went toward the betterment of their own community. In another case from the subprovince of al-Gharbiyya, twelve men charged with dredging 1,150 *qaṣaba* (4.6 kilometers) of a canal were provided with food and shelter for the one evening they worked.[23] Of the necessity and morality of these early modern wages, the respected eighteenth-century Egyptian imam Muḥammad ibn Sālim al-Ḥifnāwī wrote, "The worker who brings his task to completion deserves the fee for his labor."[24]

As the foregoing examples and thousands of others show, a primary component of the established system of early modern rural infrastructural labor was the use of local workers of all kinds. These workers, moreover, were very often afforded some sort of compensation. Diggers, reinforcers, heavy-lifters, and local experts were collectively employed as the most efficacious and efficient means of preserving and protecting Egypt's irrigation works, the vital organs of the imperial body's ecological health. Thus, a primary principle undergirding the logic of rural labor in the early modern period was that those who lived near irrigation works, used them, relied upon them, and benefited most from them were also responsible for their upkeep and repair. For both the Ottoman state and its rural subjects, this devolution of authority to local communities was the most effective, expedient, and indeed easiest way of maintaining the complex infrastructure that made Egypt the wealthiest province of the empire.

Of course, not all work in the early modern period was voluntary and paid. Corvée (forced labor) certainly existed.[25] Between the sixteenth and the middle of the eighteenth centuries, corvées were only employed on a small minority of lands—those known as *ūsya*. This was land that had once been part of a tax farm (*iltizām*) but that had been given over to the personal management and control of a tax farmer (*multaztim*), usually because the land had fallen into disuse and was therefore deemed of negligible tax potential.[26] *Ūsya* land was hence tax-exempt and functioned more or less as the private property of tax farmers. At the end of the eighteenth century, the high-water mark for this kind of land, *ūsya* made up only 10 percent of all of Egypt's arable land and was confined mostly to the delta and Middle Egypt.[27] Thus, only a tenth of early modern Egypt's fields were open to corvée. Moreover, when compared to what would come later, these corvées usually consisted of relatively simple tasks—clearing mud from canals, plowing, or bringing in crops.

One of the best descriptions of early modern corvée comes from the pen of Yūsuf al-Shirbīnī. Composed in the seventeenth century, al-Shirbīnī's text is a satirical and often bawdy portrait of the countryside consisting of forty-seven lines of poetry supposedly written by an Egyptian peasant, followed by al-Shirbīnī's extensive commentary on the verse. Of corvée, he writes:

> This occurs when activities that require it, such as the excavations of [pits for] waterwheels, the gathering of crops, and the digging of canals, take place. The corvée is found only in those *multazims*' villages that include *ūsya* land. . . . The *multazim* sends oxen, timber, plows, and whatever else is needed and appoints an agent to take charge of it and prepares a place for the timber and animals belonging to it. . . . He also delegates someone to spend money on the upkeep of the animals, etc., and to keep careful accounts. . . . In some villages the corvée applies to a number of men, fixed by household, for example. Thus they say, "From such and such a household one man is to go, and from such and such two" according to the quota set for them in the distant or more recent past.[28]

Al-Shirbīnī continues on to explain that laborers were, as we saw above, almost exclusively "people in the village."[29] Moreover, he devotes great energy to explaining the etymology of the word used for corvée in the early modern period—*al-ʿauna*. This word is derived from an Arabic root related to help or assistance and is important to note since it suggests something of the cooperative and even nonviolent nature of this brand of early modern labor.[30] In the nineteenth century, both the nature of corvée and the words used to signify it would change drastically. In explaining the use of the word *al-ʿauna* for the early modern practice, al-Shirbīnī writes:

> The ʿ*auna* (sic) . . . is so called because it derives from *muʿāwana* ("helping one another") because it consists of a group who go out to help one another in the *multazim*'s work and so on. Or it may be that it is the name for a group that cooperates in doing something, which is why they say "They fucked So-and-so last night *as a group* [ʿ*auna* (*sic*)]", i.e., they all cooperated in fucking him at one go in the byre or the granary. . . . Or it is from *māʿūn*, a name for the big storage jar.[31]

His raunchy mockery of classic Arab grammarians aside, al-Shirbīnī's point is clear—corvée was a cooperative activity of shared labor and interest. Local laborers helped one another to improve their collective local situation of environmental resource management.

So as not to romanticize this kind of labor—there was, of course, nothing romantic about it—al-Shirbīnī notes that many peasants attempted to escape from corvées. He offers the following verse: "And on the day when the corvée descends on the people in the village / Umm Waṭīf hides me in the oven."[32] Corvée on *ūsya* land was moreover hereditary. "A person subject to the corvée cannot be released from it, and if he dies they impose it on his son and so on. It is a great tribulation for the peasants."[33]

For millennia before the Ottomans ever stepped foot in Egypt, Egyptians had, of course, been forced to work. Although there were some commonalities over time, the nature of this labor was specific to each period of Egyptian history. In the early modern period, the Ottoman imperial system created a network of coordinated localism in which tax farmers, village elders, and very often peasants themselves participated in managing the labor needed to keep canals flowing and food growing. This system was the result of an understanding that the best way to oversee the ecological complexities and specificities of thousands of villages throughout Egypt was to allow locals to control canals, sluice gates, and fields. As long as they paid taxes, acquiesced to imperial demands, and continued to export food out of Egypt, local peasants were afforded an amount of autonomy over their land, water, and labor.

You Work!

All of this would begin to change in the second half of the eighteenth century. As part of their schemes to increase and then maintain their autonomy from the Ottoman Empire, Egyptian elites began undertaking larger and ever more complex repair and infrastructure manipulation projects. To show how these infrastructural efforts reflected fundamental changes in the use and conceptualization of labor and the environment in Egypt, I will concentrate on three large-scale building projects in the period from the middle of the eighteenth century to the beginning of the nineteenth. These three construction ventures track the exponential growth in the magnitude and intensity of labor utilization in Egypt. Indeed, in contrast to the earlier concept of *al-ʿauna*, forced labor at the turn of the nineteenth century came to be identified almost exclusively as *al-sukhra* (or *al-taskhīr*), deriving from an Arabic root for words related to derision, subservience, servitude, and exploitation.[34] Projects grew bigger and bigger, used more and more workers from farther and farther away, and became increasingly violent and exacting in their use of human laborers. Labor became alienated from person and place. These changes to Egypt's labor regime in the decades before 1800 show that forms of administrative centralization and invasive state authority began before the reign of Mehmet ʿAli, and they help to set the stage for the subsequent history of labor's relationship to the Egyptian environment in the nineteenth and twentieth centuries.

The repair of a grain storage facility in Rosetta in 1747 offers a particularly good example of the beginnings of this changing labor regime, pointing to how repair work in the middle of the eighteenth century was becoming larger and more complex.[35] Grain storage facilities were crucial institutions for the management of Egypt's ecological wealth. Egypt was the most agriculturally productive and lucrative province of the Ottoman Empire.[36] Indeed, throughout its history, Egypt's ecology was perhaps the primary reason for its geopolitical importance. To get food out of Egypt, the Ottomans, like other political powers, relied on a series of storage facilities in Egypt's Mediterranean and Red Sea port cities to collect and house grain to prepare it for shipping.[37] If human and animal labor helped harness the Nile's water to irrigate land to grow food, grain storage facilities allowed this caloric sustenance to find stomachs in North Africa, the Middle East, Anatolia, and southeastern Europe.[38] The supply chain from Nile to stomach passed through these institutions. Economic and political change in Egypt almost always involved transformations in both food supplies and the irrigation system that sustained them. To this point, the enormous amounts of labor, money, bureaucratic energy, and managerial acumen put toward organizing work on grain storage facilities at the end of the eighteenth century was part and parcel of the political and economic machinations of elites for power in this period.

The grain storage depot repaired in Rosetta was part of the pious endowment (*waqf*) complex of a man named Ibrāhīm Aghā. It was in very bad shape in 1747. The vaulted arches of some of the storage rooms on the southern side of the facility were broken; parts of its southern wall had collapsed; the walls, roofs, and long wooden roof supports of some of the upper stories needed reinforcement; and associated storage areas, a coffeehouse, and a group of taverns—all of which served the facility's patrons—also had to be fixed.

These many complex and labor-intensive repairs were carried out by a few hundred workers using a wide variety of materials. Both these men and the materials they used were carefully counted, organized, and assigned to various sections of the project. After an introductory paragraph explaining the necessary repairs, setting the length of the period of work, and naming those responsible for overseeing it, this very long repair case goes on to enumerate in list form (as in an expense sheet) all the materials and kinds of labor needed, along with their attendant costs. The building supplies included plaster, tiles, lime, nails, red brick, stones, and fourteen different varieties of wood.[39]

The hundreds of laborers needed for this repair work were to devote eighty-four days to the project. The written form of this case—itself an important innovation in this period—underlies the bureaucratic conceptualization of the workforce needed for the repair job. The case consists of a daily entry for each of the eighty-four days of labor allotted for the project. Each entry is first labeled with the date and day of the week. Listed underneath this top row of information are the kinds of laborers employed for that day (stone lifters, builders, wood carriers, masons, and so on) along with the number of workers from each category.[40] Below that is the total cost for that type of laborer on that day. On Saturday, 31 December 1746, for example, one engineer, seven masons, sixteen unspecified kinds of workers, one carpenter, three water carriers, and three other individuals were used.[41] In addition, each entry records the amount of coffee allotted for that day's work. Coffee and water were the only forms of sustenance provided these men.[42] The last bit of information in each of these daily entries is the total cost of men and materials for that day's work. So on the same day in December 1746, the one engineer cost 17 *niṣf faḍḍa* (*para*), the seven masons 98 *niṣf faḍḍa*, the sixteen workers 113 *niṣf faḍḍa*, the carpenter 14 *niṣf faḍḍa*, the three water carriers 17 *niṣf faḍḍa*, and the three other individuals 10 *niṣf faḍḍa*. Including another 4 *niṣf faḍḍa* paid out for coffee, the total cost of labor for that one day was thus 273 *niṣf faḍḍa*. The price for each of these classes of laborer remained fairly consistent throughout the eighty-four-day period of work.

Most of the workers employed in this grain storage depot reconstruction were brought to the worksite from elsewhere.[43] There were those who transported materials from quite far away—workers from Cairo who sailed to Rosetta with building supplies and lumberjacks from outside of Egypt who brought wood to the city, for example. The majority of the workers, though, were forcibly moved from villages near Rosetta. No explicit mention is made of their lodgings, so presumably none were afforded them. Perhaps some workers constructed temporary accommodations on their own; perhaps some others were able to sleep in their homes provided they reported to work at a certain hour each morning. Whatever the case may have been, the total cost of labor on this project was 16,278 *niṣf faḍḍa*, nearly 29 percent of the total of 56,290 *niṣf faḍḍa* spent on the repairs.

The scale of the repairs evidenced in this case starkly differentiates it from the kinds of early modern construction jobs discussed above. Instead of a few workers, hundreds were used. Instead of a day or two of work, this job took nearly three months. Workers were moved significant distances away from their families to work on a project that held little direct benefit for themselves. They were not paid for their work and were given only coffee to sustain their bodies. What is perhaps most striking about this case of grain supply management is its casting of peasant labor on the pages of the court's register. Workers were very carefully counted, organized, and assigned to work tasks on each of this job's eighty-four days. Their provisions were allotted in precise quantities, and the costs of all these expenses were dutifully calculated. In scale, level of detail, and wrenching human effects, this case and others like it represented the beginnings of new conceptualizations of labor and bureaucracy in the middle of the eighteenth century. Laborers were plugged into an emerging bureaucratic structure in which they existed as parts of categories of abstract workers, groups whose actions were now scheduled, arranged, and micromanaged by an administrative mentality of charts and metrics. Workers literally became numbers on a page.

At the end of the eighteenth century, then, labor was no longer characterized by the extreme locality of the early modern period, workers were generally not paid, and projects were demanding much more labor. With this amplification in scale and complexity came both a proportional increase in the hardship and suffering experienced by laborers and ever greater deleterious effects on those environments being manipulated to produce more. These consequences on humans and environments are seen most clearly in the period's many large canal projects.

One of the most important of these was the repair of the canal of Banī Kalb in the southern subprovince of Manfalūṭ in 1808.[44] Work on this canal lasted seventy-nine days, from Thursday, 5 May 1808, to Friday, 21 July 1808, and required 33,412 worker-days (the overall number of individual work slots required during the seventy-nine days; there was, of course, some worker turnover from day to day, but how much is impossible to calculate). On any given day, 260 to 560 men were put to work; the average daily number of workers over the seventy-nine-day period was 424. The sheer enormity of these numbers already suggests that this repair work was clearly of a very different sort than anything that had come before it—certainly orders of magnitude larger than the smaller projects of the sixteenth and seventeenth centuries and even more gargantuan than the kind of repair work undertaken on the grain depot in Rosetta in 1747. Like this case, however, the (presumably) thousands of workers used in Manfalūṭ in 1808 were very precisely organized, measured, and enumerated. Every individual peasant laborer was counted (but not named) on each day of work. Laborers were fed but were not paid beyond that.

There were no breaks during the seventy-nine days of work on the canal.[45] Although there is no clear indication of the waterway's dimensions, the length of the work period and the number of workers enumerated suggest an undeniably large project. The scale of this environmental manipulation was thus much more interventionist and ecologically transformative than that of earlier canal repair cases. The location and timing of this labor is also noteworthy. The two and a half months required for the project were spread over May, June, and July in Manfalūṭ, a southern Egyptian town where summer temperatures regularly surpassed 100°F.[46] Taking place in one of the hottest locations in Egypt during the hottest season of the year, work on this project was no doubt grueling; and although there is no specific mention of any worker deaths or injuries, based on similar repair jobs, we can safely assume that some of the thousands of men in this case sustained injuries or died.[47] Likewise, there is no information about how all these workers were brought to the canal. Perhaps some came without resistance, but certainly the majority had to be brought by force to the worksite. These laborers, in other words, clearly experienced some amount of hardship during their work.

As in the previous case from Rosetta, here too the bureaucratic positioning and casting of workers in the written record took the form of a chart enumerating the daily amount of labor used during the repair. For each of the seventy-nine entries, first the day of the week and then the date are listed. Underneath this information, the total number of workers for that day is specified. There is no other information given. The 33,412 worker-days were not further delineated into classes of laborers as they had been in the Rosetta repairs. All the workers were simply referred to as *nafar* (an individual man). The mass of humanity that toiled, sweat, and experienced pain during this canal project had again become an empty set of numbers on a page—large numbers, to say the least, but numbers still. This process of turning a group of individual men (*anfār*, pl. of *nafar*) into numbers, of course, involved a great deal of force, social rupture, and violence.[48] These thousands of workers in early nineteenth-century Manfalūṭ thus functioned as "fictitious facts-on-paper" with no identity other than that of being from among *al-anfār*.[49] No other information is given about their origins or lives. Where were their home villages? How were they brought to work on the project? What did they do after finishing a day's work? How and where did they live while working? As numbers on a page, these workers were first stripped of their subjective identity and then recast as parts of an abstract and faceless category—that of *anfār*-laborer.

In the period of this enormous project in Manfalūṭ, it was clear enough to Egypt's majority of rural cultivators that the world around them, and the place of their labor in that world, was radically changing. After a few decades of increasingly harsh and labor-intensive projects like the one in Manfalūṭ in 1808, peasants looked to the growing state bureaucracy of Mehmet ʿAli—Egypt's Ottoman governor-general from 1805 to 1847 and the most successful of the period's provincial leaders seeking autonomy from the empire—to be their savior from the brutality and raw greed of rural elites. As we saw in Manfalūṭ, by the early nineteenth century it had become

> customary for the tax farmer or his deputy to have the peasants summoned by the watchman on the evening before the morning they were needed. Those peasants who failed to appear on some pretext or another, the watchman or the supervisor dragged forth with abuse, curses and a sound beating. This was called forced labor or corvée. The peasants had grown accustomed to this; indeed they considered it to be a legitimate duty.[50]

In 1812, Mehmet ʿAli issued a decree legalizing his personal seizure of all of Egypt's tax farms, essentially making him Egypt's only sanctioned tax farmer.[51] Peasants took this as a promising sign that the power of conniving and cruel tax farmers would now finally be checked. This expectation is what allowed a group of peasants in May 1814 to respond to a local tax farmer's demands of labor from them with the following statement: "Find someone else! I am busy with my own work! There is nothing left for you in this country! Your days are over, and we have become the pasha's [Mehmet ʿAli's] peasants."[52] Only a few years earlier, it would have been unthinkable for the peasants of Manfalūṭ to express such a sentiment to the rural elites who forced them to work. Still, any expectation that Mehmet ʿAli would be a less greedy and brutal tax farmer than his predecessors was soon buried in the dirt of the largest environmental manipulation project Egypt had ever seen, the rebuilding of the Maḥmūdiyya Canal.

This canal reconstruction project embodied the multitude of transformations that had been wrought in Egypt's rural labor regime between the middle of the eighteenth century and the beginning of the nineteenth. In this period, Alexandria, Egypt's second city, was not connected to the Nile valley via waterway.[53] To reach the port city from Cairo, ships had to sail down the Nile to the Mediterranean and then hug the coast west, an arduous and often dangerous undertaking.[54] The advantages to be derived from connecting this hub of Mediterranean trade and commerce to Egypt's waterway network were obvious, and there had indeed been various attempts since antiquity to construct such a canal.[55] Moreover, a freshwater conduit to Alexandria from the Nile was also desirable as a source of what proved to be all-too-often-scarce drinking water for the city's residents.[56] For all these reasons, as well as some of his own, at the end of the 1810s, Mehmet ʿAli took to reviv-

8. Maḥmūdiyya Canal. Commission des sciences et arts d'Egypte, *État moderne*, vol. 2, pt. 2 of *Description de l'Égypte*, Alexandrie, plate 99.

ing earlier attempts to connect the Nile's Rosetta branch to Alexandria. His version of the canal would be named the Maḥmūdiyya Canal, after Ottoman sultan Mahmud II (r. 1808–39).[57]

Not only was the Maḥmūdiyya the largest canal-building project ever undertaken in Egypt to that point, but it was also one of the most deadly. Estimates for the canal's corvée labor range between 315,000 and 360,000 individuals.[58] To put this number in perspective, consider that the population of Cairo in 1821 was 218,560 and that that of Egypt as a whole was around 4.5 million in 1800 and 5 million in 1830.[59] Thus, more men than the total population of Cairo were forcibly moved to work on the Maḥmūdiyya Canal. Perhaps more shockingly, nearly a third of these workers—100,000 people—died during the canal's construction.[60] This was 2 percent of Egypt's overall population. The equivalent number of relative fatalities in the United States today would be 6 million people.

Accounts of the canal construction work make obvious why so many people died. The Egyptian chronicler al-Jabartī, a contemporary of the project, offers the following description of labor on the canal in August 1819:

> He [Mehmet ʿAli] ordered the governors of the rural districts to assemble the peasants for work, and this command was executed. They were roped together and delivered by boats, thus missing the cultivation of sorghum, which is their sustenance. This time they suffered hardship over and above what they had originally suffered. Many died from cold and fatigue. Dirt from the excavation was dumped on every peasant who fell, even if he were still alive. When the peasants had been sent back to their villages for the harvest, money was demanded from them plus a camel-load of straw for every *faddān*, and a *kayl* each of wheat and beans. They had to sell their grain at a low price but at a full measure. No sooner had they done this than they were called back to work on the canal in order to drain the extremely saline water which continued to spring from the ground. The first time they had suffered from extreme cold; now, from extreme heat and scarcity of potable water.[61]

This was a new world of labor utilization.[62]

In addition to its tens of thousands of deadly individual outcomes, the massive environmental engineering project that was the Maḥmūdiyya Canal had many other consequences as well. In a kind of infrastructural ripple effect, a whole series of other irrigation features had to be constructed or repaired in the vicinity of the canal to support its eighty kilometers. Many basins were dug at various points along its length; there were numerous areas that needed to be reinforced with strong embankments; the lake at the canal's end was to have three new embankments built around it to go along with repairs to the six already there; numerous dams and bridges were also needed; and a series of levees and dams had to be constructed along the Mediterranean coast to the north of the canal to protect it and its precious sweet water from sea-surge flooding.[63] Proof of just how massive an undertaking all this was, the construction of the canal and its supporting irrigation works was the primary reason behind the establishment of the Egyptian School of Engineering in the fall of 1816.[64] Almost all of the school's early graduates worked on the Maḥmūdiyya project.[65] Further, the financial costs of the canal were obviously enormous, as both cash and building supplies were diverted to the construction work.[66] And the wrenching demographic human costs of the project went beyond even the corvée's gargantuan number of more than three hundred thousand workers, since many of these men came to the canal site with their families.

The environmental impacts of the construction work were also numerous. One of the main goals of the canal was to provide sweet drinking water for the people of Alexandria. The canal's repair, however, actually increased salinization levels in both the canal's water and the soils around it.[67] Much of the canal ran through land that had once been under the Mediterranean, and there were therefore high levels of salt already in the soil. Digging a few meters down into the earth helped to release these salts, which were then carried and distributed throughout the canal's waters. Moreover, because of the inadequate packing of various embankment walls along the canal's length, salty water regularly spilled out of the waterway onto otherwise fertile soil. Water was also lost to the canal's subsurface because of seepage through its bottom and walls. Losing salty water through the walls of the Maḥmūdiyya not only made the canal shallower, but it also contributed to increasing salinity levels in nearby soil. A related problem was siltation. Loosely packed walls, a weak current, and windy conditions caused the canal to silt up very quickly, making the passage of ships through its shallow waters exceptionally tricky.[68] Indeed, throughout the early 1820s ships frequently wrecked and ran aground in the canal.[69]

The most significant aspect of the Maḥmūdiyya project for the subsequent history of public works and the environment in Egypt was the state imagination it embodied and reflected. In the face of such enormous costs in resources, human lives, and environmental destruction, it was only the imagination of what a canal or dam could and would do for Egypt that sustained and justified such work. This imagination permeates the correspondence between Mehmet ʿAli and Mahmud II about the canal.[70] The Maḥmūdiyya would protect Alexandria from food and water shortages, helping to make the city a vital center of commerce around the Mediterranean.[71] This new Alexandria—along, of course, with its pleasant climate and general agreeability—would attract merchants, intellectuals, and others to settle in the city. Indeed, Ottoman governors would soon make it—not Cairo—their seat of power in Egypt. Thus, the Maḥmūdiyya, the sultan went on to say, would change the entire meaning of Egypt itself. Instead of the word *Mısır* (the proper name for Egypt in Turkish) signifying the city of Cairo, a common convention, it would come to mean Alexandria.[72] The Maḥmūdiyya would thus make Alexandria Egypt's first city and capital. As it comes through in this correspondence, then, public works had the power literally to redefine both the imagination and material reality of Egypt.

Conclusion

The enormous social, economic, political, and environmental effects of the shifts in Egypt's labor regime that culminated with the Maḥmūdiyya project shaped Egypt for the next two centuries. Roughly every fifty years since the middle of the eighteenth century and projects like the massive 1747 repair of the grain storage facility in Rosetta, various Egyptian governments—regardless of their purported political bent, whether khedival, colonial, socialist, nationalist, neoliberal, or otherwise—have undertaken an enormous public works scheme that, like the Maḥmūdiyya project in the early nineteenth century, was meant to change both the imagination and the reality of Egypt. The Suez Canal (1850s and 1860s),[73] the first Aswan Dam (1890s),[74] the Aswan High Dam (1950s and 1960s),[75] and the recent Toshka scheme[76] are only the most prominent of these, but there are dozens of others that have garnered less fame but have had no less of an impact on Egypt's workers and environment.

The commonalities of these projects are striking. Each embodied the imagined potential of public works—and usually of water—to change both the environmental and political landscape of Egypt. In each case, Egypt was to become something radically different after the completion of a canal or dam. Another common feature of these projects was their failure. While they did in some ways meet their original intentions, their huge human, economic, and environmental costs usually far outweighed their returns. We already discussed the Maḥmūdiyya's human, ecological, and other costs. Suez exploded Egypt's foreign debt and was a contributing factor to its colonization at the end of the nineteenth century.[77] The political, economic, human, and environmental price tag of the Aswan High Dam was likewise enormous: the destruction of much of historic Nubia, increased reliance on chemical fertilizers to overcome salinization, shrinkage of the Nile delta, overmining of the adjacent granite hills, and massive water losses due to evaporation from Lake Nasser (to name only the most obvious problems).[78] The failure of the recent Toshka scheme demonstrates that Egyptian governments have yet to grasp—or, perhaps more accurately and troublesome, have yet to be concerned about—the colossal ecological and human costs of such massive environmental engineering ventures.[79] What is more—and this is perhaps the most nefarious result of these projects—all of these canals and dams further alienated Egyptian peasants from their lands, their labor, and the environments their families had cultivated for generations. As these projects grew exponentially in size and complexity from the Maḥmūdiyya to Toshka, bulldozers came to replace pickaxes and shovels; multinational companies replaced local village leaders; foreign laborers replaced Egyptian ones; and peasants' knowledge and experience of canals, embankments, and soils were deemed increasingly less important for the management and harnessing of natural resources.

Thus, as political and economic power concentrated in fewer and fewer hands beginning in the second half of the eighteenth century, Egypt's peasant laborers became increasingly separated from their environments, accustomed means of work, and social forms and norms.[80] The creation and maintenance of an autonomous, centralized, and powerful state has been the primary objective of political power in Egypt, and elsewhere of course, since the late eighteenth century.[81] One of the consequences of this model of governance is the increasing distance between those making decisions and the impacts of those decisions. As this chapter shows, the accelerating divergence between government and governed, embodied here, perhaps unexpectedly, in the transition from Ottoman imperial rule to a centralized Egyptian state apparatus, increased the deleterious, unforeseen, and costly effects of faraway political decisions on local peoples and environments. Older forms of labor practice and the intimate localized knowledge of thousands of village environments were gathered up, monopolized, transformed, and largely misconstrued by a modernizing, centralizing state to be put toward new massive environmental manipulation schemes. Thus, the history of Egypt over the last half millennium shows how environmental labor became a primary site of contestation over, and indeed a functional key to, political and economic power.

While certainly an enormous bureaucratic structure in its own right, a model of political belonging like that of the Ottoman imperial system of environmental resource management afforded Egyptian peasants the autonomy and space to control irrigation and ecology in their own local environments. This system was not a function of the largesse of the imperial state, but rather developed as a kind of negotiated settlement between the Ottomans and their peasant subjects reflecting their collective understanding that local control was the most efficient and effective means to govern natural resources and landscapes in Egypt. Such a form of government is now largely lost to us. It is, however, perhaps making a comeback. Environmentalists, government agencies, various nongovernmental organizations, and many other entities and individuals now stress the importance of local control and autonomy over natural resource management and environmental stewardship. Without romanticizing the local above all else, most today seem to agree that one size clearly cannot fit all. In contrast to what has dominated in Egypt since the turn of the nineteenth century, the Ottoman model of labor and natural resource management, much like some of those ideas advanced today, recognized the importance and autonomy of local environmental knowledge, expertise, and work.[82] While, of course, we cannot somehow return to an Ottoman schema of labor and environmental resource management, we should recognize that other models beyond our world exist and that they might even be better.

5 Expert Measures

The rural world was made. There is nothing natural about a field, a river, or a forest. Each has a history of negotiation, choice, manipulation, accident, conflict, and compromise. One of the great insights of environmental history is to demonstrate the complex ways in which humans, nonhumans, and geophysical and climatic processes have participated in forging the pasts of all environments and thus the histories of the societies these environments supported. These observations are particularly salient for Ottoman and Middle East historians. The lands that would become the Ottoman Empire were subject to human manipulation for millennia prior to the Ottoman conquests, and, unlike many regions of the globe, the documentary evidence of these changes exists for historians to be able to narrate these changes. When the Ottomans entered the Middle East, the Balkans, and North Africa, they did not find an untouched pristine wilderness. On the contrary, they encountered an intricately arranged and cultivated world, one with a particular historical and ecological order that had been forged through millennia of environmental management techniques. Over the centuries of Ottoman rule, these regions would continue to be manipulated and remade through imperial, environmental, economic, and social processes.

Focusing on just one aspect of this vast and complex

history, this chapter explores the role of rural engineers in Ottoman Egypt between the seventeenth and nineteenth centuries. Identified as *mühendis* (*muhandis* in Arabic) in the archival record, these individuals were integral to the manipulation of rural environments and helped to maintain and develop constructive relationships between local communities and the Ottoman imperial administration. Until the early nineteenth century, a shared interest in the maintenance and proper function of infrastructure including canals, embankments, roads, bridges, and quays kept rural peoples and the Ottoman state in regular communication over complicated matters related to the repair, manipulation, and expansion of these public works.[1] Engineers were particularly important because they functioned as a kind of middleman between peasant interests and imperial concerns. Judges and other imperial officials relied on them to preserve a balanced relationship between local needs and imperial desires in rural Ottoman Egypt. The history of these engineers thus illuminates something of the logic and function of Ottoman governance in rural areas of the empire in the early modern period. This backdrop proves essential in considering what came of these engineers and of engineering expertise in the early nineteenth century. As the last chapter explained, the nature of labor in rural Ottoman Egypt changed drastically at the turn of the nineteenth century. This chapter ends with a consideration of what these transformations meant for engineers in Egypt.

Downstream, 1664

In the summer of 1664, a problem emerged with a canal embankment in the village of Shārimsāḥ in the subprovince of al-Daqahliyya in the northeast of the delta.[2] Hoping to find a solution, the head of the village brought the problem to imperial administrators at the subprovincial court of al-Manṣūra, the seat of al-Daqahliyya. Three sections of the face of the embankment had deteriorated and become disjointed (*takhalkhala*), falling into the canal's water. All three sections had been eroded to the point that the dried mud and clay of the internal portions of the embankment had become as soft as manure (*ṣāra sibākhan*). Water was thus allowed to spill wastefully out of the canal instead of being properly channeled to fields, and the land behind the embankment had become a soppy (*ghamīqa*), muddy mess. There was, moreover, a real possibility that these damaged sections would be completely destroyed by the rushing waters of the next season's flood, obviously a

9. Village in al-Daqahliyya on the Damietta Branch of the Nile. Commission des sciences et arts d'Egypte, *État moderne*, vol. 1, pt. 2 of *Description de l'Égypte*, Basse Égypte, plate 76.

cause of enormous concern for both those living near the embankment and those farther downstream.

The judge sitting in his court in al-Manṣūra was not an expert in irrigation works or infrastructural repair. In order for him to make a decision about what to do in this case, he needed expert testimony, particularly to ensure that any monies disbursed for repairs were properly and effectively spent. The imperative for expertise was not simply a bureaucratic and financial requirement to fulfill the court's legal function. The correct and timely repair of the embankment had real-world consequences. Peasants from various downstream villages on the canal came to the court to testify that failure to repair the embankment would cause their fields, and hence their lives and livelihoods, great harm. They added that neither they nor their village leaders had the financial or technical wherewithal to help repair the upstream embankment in Shārimsāḥ that controlled so much of their agricultural fate. Thus, for different reasons, the court, local leaders, and peasant communities were all in need of someone to provide the expert information required to fix the broken embankment.

Enter the engineer. Al-Muʿallim ʿAṭāʾ Allah was identified in this case, and in others, as the engineer of al-Manṣūra (*al-muhandis bil-Manṣūra*) and was, as his title suggests, likely a local Egyptian who resided somewhere in the subprovince of al-Daqahliyya. He was summoned to the court, made aware of the situation, and then dispatched to the field to gather the information needed to repair the embankment and restore the canal's proper flow. He went right to work.

ʿAṭāʾ Allah's first task was to consult with rural cultivators in Shārimsāḥ about the state of the embankment and then to talk to people in the downstream villages about the possible effects of its disrepair on their communities.[3] Furthering their earlier testimony, all explained to him the potentially immense deleterious consequences of failing to fix the embankment. The yearly flood was only weeks away. Should it arrive with its full force before the embankment was repaired, the resulting damage to both fields and the canal would be enormous. The uncontrolled water would surely sweep away much of the embankment, along with other irrigation structures, and the waterway would be left in an unfixable state (*lā qudra li-aḥad ʿalā saddihi*). Everyone on the canal thus implored ʿAṭāʾ Allah to do all he could to fix the embankment quickly and sturdily.

Armed with his initial charge from the court and now with this corroborating information regarding the existing situation, ʿAṭāʾ Allah and his assistants set out to survey the damage to the embankment. Three parts of the structure had been destroyed. The first broken section measured 13 by 1.5 *qaṣaba*s, and was opposite a group of three waterwheels. The second damaged portion was near a basin known as al-Waḥdāniyya and measured 8 by 1.5 *qaṣaba*s. The third damaged area was 9 by 1 *qaṣaba*s. Based on these measurements, ʿAṭāʾ Allah estimated the total cost of the repairs to be 200,000 *niṣf fiḍḍa*. This sum included the needed repair materials, their transport to the construction site, the specialized tools required to move them, and the necessary labor.

He then returned to the court to report his findings to the judge and the head of Shārimsāḥ. He relayed what locals around and below the embankment had told him about its disrepair and summarized his measurements of the damaged structure and his estimates for the cost of its repair. The total of 200,000 *niṣf fiḍḍa* needed to fix the embankment was a huge sum. Other repair jobs in the subprovince in this period usually cost orders of magnitude less. In 1646, for example, 18,120 *niṣf fiḍḍa* was spent on a series of repairs carried out on a canal and its embankments in the city of al-Manṣūra.[4] In Shārimsāḥ in 1664, costs were high, but there was simply no way to avoid the enormous sum. The court's expert witness had reported his findings, and the consequences of not implementing his recommendations would be much greater than

200,000 *niṣf fiḍḍa*. The court recorded what ʿAṭāʾ Allah had to say and approved the repairs.

Engineer ʿAṭāʾ Allah was one of the most powerful parties in this case. The whole project to repair the canal's embankments rested on his expert recommendations. His consultation with locals, his measurements, and his cost estimates moved the repair process forward. Without him, the embankment would have languished in disrepair, fields would have flooded, and tax revenues would have fallen. His authority over most of the case's other parties is evidenced by the court's decision to ignore the village head's initial recommendation to reinforce the embankment with another surface structure (*raṣīf*). He presented this idea to the court before ʿAṭāʾ Allah was summoned for his expertise, but ultimately the engineer's ideas trumped the village head's. The testimony of local peasant cultivators was also an important component of this case. Their recommendations, though, were ultimately filtered through ʿAṭāʾ Allah and so were presented to the court as part of his package of ideas. The judge in this case deferred to him as well. The engineer ʿAṭāʾ Allah's privileged knowledge and expertise was thus the most significant factor in the repair of the embankment and consequent massive expenditure of imperial funds.

His expertise was the crucial link between the institution of the court and the many peasant communities directly affected by imperial actions on the canal. The logic governing the Ottoman management of irrigation in rural Egypt recognized that water usage linked peasants in the Egyptian countryside not only to other, often distant, villages, but also to the palace and to areas of the empire far beyond Egypt. Grains grown by the Nile's irrigated waters and the tax revenues they helped to raise connected Egyptian peasants to Istanbul and elsewhere. ʿAṭāʾ Allah was an intermediary in the relationship between al-Manṣūra and Istanbul, between a small-scale canal's particular ecology and the empire it supported. He went back and forth coordinating between the court and the countryside and between Shārimsāḥ and its downstream villages. Just as these Egyptian villages were connected to Shārimsāḥ through their shared use of the canal, so Istanbul, the Hijaz, and other areas of the empire that consumed Egyptian grains were also the village's downstream communities. That is, the status of a canal embankment in a village like Shārimsāḥ deeply impacted many other places outside of Egypt. Engineers like ʿAṭāʾ Allah were therefore not simply fixing irrigation works but also working to keep the connective tissue between empire and community healthy and functional.

Scale, 1680

Engineers were such central actors in rural Ottoman Egypt in part because of the sheer scale of the projects to which they contributed. Regularly assisting the imperial administration in infrastructural work of such immense size and importance secured their place in the function of the empire. Indeed, the information and expertise provided by engineers in rural Ottoman Egypt controlled massive amounts of resources, cash, labor, and effort.

In 1680, the tax farmer (*multazim*) of the village of Ṭunāmil, Muḥammad Aghā, came to the court of al-Manṣūra to register a problem with a canal known as Baḥr al-Fuḍālī.[5] This canal, which served as the main source of water for his village and ten others, was barely flowing because of the enormous amounts of silt and plant matter that had accumulated in it that year.[6] Muḥammad thus asked the court to dredge and clean the canal. To corroborate Muḥammad's claims and to offer further details, the court asked the engineer of al-Manṣūra (again *al-muhandis bil-Manṣūra*), al-Muʿallim Ḥasan, to inspect the canal. After measuring it—it was eighty *zirāʿ* in length and an average of twenty *zirāʿ* in width—Ḥasan reported back to the court that the canal was indeed in dire need of dredging.[7]

While at the canal site, Ḥasan also noticed a problem on one of the banks of the waterway. Measuring the distance between the canal and a large orchard in Ṭunāmil, he and his assistants found that there was not enough room for people to pass on the canal bank. One could, however, avoid this narrow section of canal bank by passing to the other side of the canal and then back again on two bridges that spanned the canal on either side of the orchard. Each bridge was five *zirāʿ* wide, twenty *zirāʿ* long, and ten *zirāʿ* high, and both were in need of repair. According to Ḥasan's report, their reconstruction would require one hundred thousand mud bricks (*tūba*) and an unspecified enormous amount of stone. The work needed to clean the canal and fix the two bridges would last thirty days, and on each day forty workers would be required. These men were to be paid the going rate for this kind of work and be given provisions of food. On the basis of Ḥasan's report, the court authorized both the dredging of the canal and the added work of repairing the two bridges.

Ḥasan's authority to direct the repairs in this case is obvious. On his word alone, the court not only approved the work it had originally been asked to do on the canal but also undertook repairs that went much further than those brought to its attention by the tax farmer of Ṭunāmil.

More significant, this case highlights the enormous scale of the sort of irrigation work that was regularly pursued in rural Ottoman Egypt. The numbers are telling—eleven villages, one hundred thousand mud bricks, and wages and food for forty workers for thirty days. The capital, resources, and organization demanded by such numbers underscore both the level of environmental, social, and economic manipulation such projects entailed and the enormous trust and power invested in the person of the engineer. As ʿAṭāʾ Allah had done in Shārimsāḥ a few years earlier, Ḥasan advanced the project on Baḥr al-Fuḍālī. He measured the canal, informed the court about the two broken bridges, and offered estimates of the work's cost and labor demands. Projects of this scale could only be entrusted to and managed by people with knowledge, expertise, and proven reliability. These people were engineers like Ḥasan.

Knowledge, 1705

A case from roughly twenty-five years later offers further information about how the Ottoman administration organized and utilized the expertise of engineers.[8] Also from the court of al-Manṣūra, this case concerns the repair of a deteriorating embankment at the mouth of the large canal of al-Baḥr al-Ṣaghīr that flowed east to the Lake of al-Manzala in the subprovince of al-Daqahliyya. In 1705, after years of neglect, the

10. Fishermen on Lake Manzala. Commission des sciences et arts d'Egypte, *État moderne*, vol. 1, pt. 2 of *Description de l'Égypte*, Basse Égypte, plate 76.

embankment was in desperate need of maintenance. Parts of it had broken off and fallen into the canal, while other sections were badly silted up. Because state funds (*al-māl al-mīrī*) were to be used to fix the embankment, the heads of the seven military blocs in al-Daqahliyya, judges from the subprovince, and other local elites came to the court to discuss these repairs.[9] As in the previous cases, they consulted with the engineer of al-Manṣūra about the repairs, ultimately deferring to his expertise.

In 1705, the engineer was a man named al-Ḥājj Shāhīn, and his title was head of the corporation of engineers of al-Manṣūra (*shaykh ṭā'ifat al-muhandisīn bil-Manṣūra*).[10] This title tellingly points to the presence of an organized institution of engineers in rural Ottoman Egypt. Like the members of similar guild formations, engineers understood the power they could derive from their knowledge and expertise and adopted corporate organization in order to advocate and protect their interests.[11] This was an early modern knowledge economy.

The commodification of their collective engineering expertise is abundantly clear in this case. Shāhīn and his associates were brought from Damietta—over forty miles away from al-Manṣūra—to inspect the damaged embankment and offer advice about how to repair it. The case unequivocally states that these men were valued for their expertise (*li-kaun anna lihum khibra wa ma'rifa*).[12] Their "importation" from Damietta is mentioned repeatedly throughout the case. Moreover, although never stated explicitly in the text, there was likely a cost associated with bringing these men from Damietta. Perhaps they were even paid for their services. These men are therefore most usefully thought of as traveling consultants who were employed by the Ottoman state for their technical knowledge and expertise.

Conceptualizing these engineers' knowledge as a commodity also helps explain their movement across the subprovince of al-Daqahliyya. The Ottomans were masters of comparative advantage.[13] They regularly organized the transport of resources between different parts of the empire to achieve optimal configurations of effort and capital. In the case of Egypt, this meant moving the province's excess grain to other parts of the empire and moving wood from southern Anatolia and elsewhere to the timber-bereft Nile valley.[14] Like grain and wood, albeit on a more local scale, the engineering know-how of Shāhīn and his fellow engineers was moved to areas where it was needed most.

And what did these men find once they got to the canal of al-Baḥr al-Ṣaghīr? Their first recommendation to the imperial bureaucracy was to remove the massive quantity of silt that had collected in the canal.

The total volume that had to be dredged was 906.5 cubic *zirāʿ*. The cost of dredging a single cubic *zirāʿ* was 80 *niṣf fiḍḍa*, so the total cost of this part of the repairs was 72,520 *niṣf fiḍḍa*.[15] In the embankment itself, a section measuring 700 square *zirāʿ* had to be replaced. The cost of repairing a single square *zirāʿ* of the embankment was again 80 *niṣf fiḍḍa*, bringing the cost of this portion of the work to 56,000 *niṣf fiḍḍa*. Another 10,000 *niṣf fiḍḍa* was needed to purchase dirt, stones, and other building materials. Thus, the grand total estimated for this work by the engineer Shāhīn and his associates was 138,520 *niṣf fiḍḍa*. Throughout the text of this case, all of these measurements and repair costs were reported as being on the authority of the engineers (*al-takhmīn bi-maʿrifat al-muhandisīn*). Subsequent cases recorded after the repairs had been completed confirmed the accuracy of the engineers' estimates.[16]

Rural engineers' authority and the value of their expertise was thus in large part a function of the accuracy of their estimates. Running way over cost or overestimating the amount of time needed for a repair job obviously did not help engineers to sell their knowledge to the Ottoman state. Perhaps the Ottoman administration of Egypt undertook the added complexity of bringing Shāhīn and his associates from Damietta (rather than using local engineers) precisely because they were known to give accurate estimates in their repair work or, more generally, because of their proven integrity and honesty. Reputations clearly mattered in Ottoman Egypt's knowledge economy, and engineers like Shāhīn used this fact to their advantage.

Unlike some other cases we have seen, peasant communities did not have much input in the repair of al-Baḥr al-Ṣaghīr's embankment in 1705. This was primarily because the canal was part of the state's imperial irrigation infrastructure. There were two classes of canal in rural Egypt: *sulṭānī* (imperial) and *baladī* (local).[17] If a canal served a large group of peasants, contributed to the common good, or promoted equality among peasants, it was considered a *sulṭānī* canal, and the responsibilities for such canals fell to the Ottoman state in Egypt. *Baladī* canals, by contrast, were those that served the irrigation needs of a particular community. These were to be maintained by local leaders. Although both *sulṭānī* and *baladī* canals ultimately remained the property of the state, their upkeep was very often entrusted to locals—engineers, peasants, and rural elites—since they were the ones who used and most directly relied on these waterways. Since al-Baḥr al-Ṣaghīr was a *sulṭānī* canal, the responsibility to repair it fell squarely on the shoulders of the state, hence the use of imperial funds (*al-māl al-mīrī*) to repair it.

In the end, it did not much matter if a canal was classed as *sulṭānī* or *baladī*. Engineers were key figures in the maintenance and repair of all of Egypt's canals.

Soundness, 1709

As in the subprovince of al-Daqahliyya, state funds were likewise regularly used to repair *sulṭānī* irrigation works in Fayyum throughout the first half of the eighteenth century, and engineers played similarly prominent roles in these cases. In 1709, the divan of the Ottoman sultan Ahmet III sent a firman to the vali in Cairo about ongoing damage to the regulation mechanisms on the important canal of Baḥr Yūsuf in Fayyum.[18] This single canal was the lifeline of Fayyum. As we saw in chapter 3, the region, which lies in a depression southwest of Cairo, is unique in Egypt since it is the only major agricultural zone located in neither the Nile valley nor the delta. In the eighteenth century, Baḥr Yūsuf provided all of Fayyum's water. Problems of the sort that arose in 1709 thus led to widespread environmental stress throughout the region.

According to the firman in this case, much of Fayyum was currently unwatered (*sharāqī*) because of the failure of Baḥr Yūsuf's irrigation mechanisms. From its downstream perspective, the sultan's council strongly asserted that it was not only Fayyum that was directly affected by this deterioration in the irrigation network. Repeating an idea we have seen already, this firman emphasized the great destruction to agricultural lands that might result should this situation remain unaddressed, which would hurt Egypt's overall agricultural output and thus greatly reduce the amount of tax revenue available to the state. To stave off these deleterious consequences, the imperial bureaucracy once again turned to engineers. The sultan ordered the vali of Egypt to send to Fayyum a group of engineers of sound judgment (*mühendisin-i sahih ül-tahmin*) to oversee the reconstruction of the canal's dams and embankments as quickly as possible (*alavechitta'cil*). To fund this work, eleven Egyptian purses (*kise-i Mısrī*) were made available from the annual tribute of 1708–9.

Inspections and repair work on the canal continued for several years under the collective direction and management of engineers in Fayyum. In 1711, they registered a report with the imperial administration about the deteriorated state of many of the canal's irrigation works.[19] In the major dam of al-Gharaq, for instance, there was a damaged area measuring 27,234 square *zirā'*. Broken sections in the foundation (*paye*), walls (*divarlar*), and support girdle (*kemer*) of the dike of al-Lāhūn totaled 9,980 square *zirāʿ*. Between just these two structures (there were other damaged ones as well), the total area in need of repair was thus 37,214 square *zirāʿ*. It was estimated (*alavechittahmin*) that the cost of repairing one square *zirāʿ* would be 15 *para*, bringing the total cost of fixing these two irrigation works to twenty-two Egyptian purses and 8,510 *para*. The bulk of these funds were needed to purchase building materials—mainly lime (*kireç*), wooden supports (*şecār*), and stone (*taş*). Istanbul directed that these monies be taken from the Egyptian annual tribute of 1710–11.

The engineers of Fayyum thus worked in conjunction with the imperial administration to fix problems on Baḥr Yūsuf. Engineers were a clearly identified and recognized group in Fayyum whose trustworthiness and soundness of mind were both confirmed by the Ottoman state and relied upon to carry out vital infrastructural work.

Intermediary, 1713

The role of the engineer as intermediary between the imperial and the local is further demonstrated by cases of irrigation repair that, like the previous example from Fayyum, climbed up the bureaucratic ladder to reach the palace in Istanbul. Such cases usually involved very large irrigation structures whose damage or destruction would have had serious consequences for imperial governance in Egypt and indeed throughout the empire. In 1713, a petition from Manfalūṭ in southern Egypt reached the sultan's court.[20] The supports of a weir on a canal in the village of Waḥīshāt near Manfalūṭ had been broken by the force of the water's incessant pounding and fallen into a dangerous state of disrepair. If the supports gave way entirely, water would rush uncontrollably through the canal, and the many villages that relied on it would lose their primary source of water. Agricultural lands would remain parched and dry and food supplies and tax income would suffer.

This threatening situation was a cause of great concern for the Ottoman sultan and his retinue. They understood the downstream implications of a massive loss of agricultural resources and revenues from an area of high cultivation like Manfalūṭ. Indeed the seriousness of the situation is evidenced by the fact that this petition was not handled by the court in Manfalūṭ, but instead bubbled all the way up to the palace itself. In response, the sultan issued a firman instructing his vali to hire, yet again, an engineer and other local experts (*ehl-i hibre ve erbab-i vukuf*) to inspect the situation.[21] The engineer's name was el-Hâc Mehmet, and, quite tellingly, of all the men of technical knowledge identified in this case, he was the only one specifically named. Thus, again, we find a collective body of technical expertise in the Egyptian countryside that took the lead in inspecting and repairing irrigation works.

After Mehmet and his associates visited the deteriorating weir and completed their measurements, they reported back to the palace, through the vali, that the area in need of repair measured 9,110 square *zirāʿ* and would cost 18,130 *para* to fix. After some deliberation, and with the stated goal of properly reinforcing the weir to prevent future damage, funds were made available to undertake the repairs. The firman issued to release these funds again repeatedly invoked the authority of the engineer as proof of the urgency of the repair work and justification for the costs.

Like previous cases, this one turned on the expertise of the engineer. Mehmet was the one who directed how much the state was to spend on repairing the damaged weir. Indeed, the line of communication in this case ran very clearly from a particular petitioning community in Egypt through the engineer to the sultan. It was the engineer's technical knowledge that integrated him into the imperial administration, that made him essential to the empire and its governance of Egypt. The sultan in many ways devolved authority over technical matters such as the repair of a weir to local experts like Mehmet who through years of working in the countryside had proved both their acumen and their trustworthiness. Although Mehmet could not fix irrigation works without the money provided by the state, the state could not fix them without the knowledge provided by experts like Mehmet. This was why engineers were so indispensable to Ottoman governance and why Ottoman governance was so indispensable to engineers.

Persistence, 1816

The expertise of local Egyptian engineers like ʿAṭāʾ Allah, Ḥasan, Shāhīn, and Mehmet would remain important well into the period of Mehmet ʿAli's reforms in the first half of the nineteenth century. Mehmet ʿAli's major innovation in the realm of engineering was the establishment of a School of Engineering in the fall of 1816. Despite importing European teachers, textbooks, and training for the school, Mehmet ʿAli continued to privilege the knowledge and technical skill of Egyptian engineers. Local engineers still drove how Ottoman rulers used engineering expertise to fix Egypt's infrastructure. Even the story of the founding logic of the engineering school speaks to the role of local technical expertise. As the Egyptian chronicler al-Jabartī relates:

> A Cairene named Ḥusayn Çelebi ʿAjūwa had the idea of a wheel to use in stripping rice. He made a model of it in tinplate which revolved with great ease, so that whereas the conventional apparatus required four oxen to drive it, his needed only two. The pasha admired this model when it was presented to him, so he gave Ḥusayn some money and ordered him to build a wheel in Damietta crafted according to his knowledge of engineering. With a decree authorizing the wood, iron, and whatever money he needed, Ḥusayn made the machine, thus verifying what he had claimed. After constructing another at Rosetta, he gained renown.
>
> The pasha became convinced, based on Ḥusayn Çelebi's feat, that Egyptians have a superior aptitude for the sciences. Accordingly he ordered that a school be built in the courtyard of his palace in which a group of natives and the pasha's mamluks were enrolled under the teacher Ḥasan Efendi, known as al-Darwīsh al-Mawṣilī. With the collaboration of a Turk named Rūḥ al-Dīn Efendi and several Europeans the principles of accounting and engineering were taught, as well as arithmetic, geometry, and trigonometry, and algebra.[22] Various technical instruments of English manufacture were provided, with which the students could measure distance, elevation, and area. Provided monthly stipends and yearly clothing allowances, they met regularly in this school, which was called the School of Engineering, every morning of the week until shortly past noon, when they returned to their homes. Some days they made field trips to the open country to study surveying. In fact, knowledge of surveying was the pasha's main goal.[23]

The institutionalization of engineering knowledge in a formal school was something novel in Ottoman Egypt.[24] What is even more striking, however, is the extent to which early modern engineers and their modes of knowledge persisted and remained central into the nineteenth century. Engineers continued to communicate directly with Ottoman officials about the possibilities of improving rural technologies, and the imperial state continued to give them money to facilitate their work—as is made clear in the story of Ḥusayn Çelebi. Reputation, moreover, remained fundamental to the economy of engineering in the early nineteenth century. It was Ḥusayn Çelebi's "renown" that proved his worth. The Ottoman administration of Egypt thus still clearly relied on local Egyptian engineers.

Mehmet ʿAli's belief "that Egyptians have a superior aptitude for the sciences" came from the sustained role engineers played in Egyptian society. Mehmet ʿAli's school was a formal recognition of this fact and an attempt to institutionalize these engineers' knowledge. Many of the students in the school, those referred to as "natives" by al-Jabartī, were engineers like ʿAṭāʾ Allah, Ḥasan, Shāhīn, and Mehmet—locals from rural communities throughout Egypt who were brought to Cairo in an effort to centralize their expertise for the benefit of Mehmet ʿAli's government. And even though others now participated much more directly in the development and utilization of engineering expertise in Egypt—Europeans, Ottoman officials, Mehmet ʿAli's own underlings, and government technocrats—local Egyptian engineers, the *ehl-i hibre ve erbab-i vukuf* of the countryside, still offered the best advice and most expert knowledge.

The persistence of their central role as skilled engineering professionals came through clearly when Mehmet ʿAli's son, Ibrahim Paşa, undertook a cadastral survey in 1821. He assembled several groups of surveyors and engineers: officials from the School of Engineering, sixty engineers from Upper Egyptian towns and villages, a group of European engineers, and a number of Coptic surveyors led by the Coptic notable al-Muʿallim Ghālī. Each of these groups of engineers represented a different tradition and mode of understanding technical knowledge, and each claimed the right to administer the cadastre themselves. In order to choose which group of experts to use for the survey, Ibrahim Paşa arranged a kind of engineering contest. "He announced that he wanted precision combined with speed and set a test on a plot of land which would demonstrate precision and variations."[25] The winners of this competition were the Coptic engineers. If we accept al-Jabartī's claim that the School of Engineering was founded primarily for purposes of surveying, then, at least in this instance, the school seems to have failed in its mission. Its graduates were defeated in a surveying contest by Egyptian engineers with no connection to the school. Thus, despite the emergence of European and other forms and tools of scientific knowledge, local Egyptian engineering know-how was still deemed the most useful. A local tradition of engineering expertise persisted.[26]

Conclusion

We are accustomed to seeing the decades around 1800 as a period of complete rupture—in Egypt, in the Ottoman Empire, around the world. The early modern centuries were, we are told, vastly different from the nineteenth century, and in multiple realms they indeed were, as we saw in the last chapter. In the face of this temporal divide, however, the persistence of engineering knowledge in Ottoman Egypt serves as an important reminder of some of the many continuities between these supposedly incommensurable periods. While the founding of the School of Engineering, often taken as a marker of innovation and rupture, was of course a significant event in and of itself, it did not radically alter the role or status of local engineering knowledge in Egypt. Engineers drawn from local communities throughout Egypt still directed the state in its management and manipulation of the countryside.

This kind of before-and-after-1800 historiography of the Ottoman Empire has a corollary in the field of environmental history that is worth considering in this context. Environmental historians have identified a notion of "pristine nature" as both a fiction and a trap.[27] Pristine nature is the idea that somehow nature existed in a perfect state of harmony, balance, and sustainability before humans came to destroy it. Embedded in this idea is the specter of ecological "decline"—another conceptual fiction that Ottoman historians are all too aware of. The declensionist narrative of environmental history posits that the overwhelming majority of human interactions with nature have been detrimental.[28] Humans have depleted, mangled, and scarred environments in various irreversible ways, forever ruining what could only have been a pristine, because now lost, version of nature. Critics of these ideas have emphasized that the relationships between humans and the rest of nature are much more complex than this simple story of one-way decline and ruin. Environments shape humans, humans then reshape environments, these new environments offer a new set of limits for humans, and so on and so forth.[29] It is this dialectical relationship that we must understand to fully grasp environmental history. The idea of pristine nature thus in many ways takes ecology out of history. Environmental history works to put ecology back into history and history back into ecology.

These ideas and their critiques are instructive for Ottoman historians. The empire's early modern history was not a pristine moment of unvariegated state and society relations waiting patiently—outside of history, as it were—for the forceful ideas, actors, and wars of the nineteenth century. As I have tried to show in this chapter, and throughout this book, the early modern centuries were a dynamic and conflicted period in their own right, not a mere empty stage awaiting the main event of the nineteenth century. The notion of a pristine early modernity thus sets up a false dichotomy between what preceded 1800 and what followed it. The divide is not so unassailable.

Rural engineers were an important component of this story of continuity. As their example shows, certain kinds of actors and forms of expertise persisted across the purportedly absolute temporal divide of 1800. Indeed, the ways in which engineering knowledge came to shape nineteenth-century rural Egypt followed patterns set much earlier. As we saw with the examples of ʿAṭāʾ Allah, Ḥasan, Shāhīn, and Mehmet, engineers were the experts on whom the Ottoman administration relied to defuse the tension and conflict inherent in the management of irrigation works and, ultimately, to properly build and repair early modern Egypt's rural infrastructure. Their authority derived from their expertise, direct personal experience of the countryside, reputation, and ability to provide accurate estimates of repair dimensions and costs. They served as crucial intermediaries between imperial desires and ambitions and local ecological realties and economic interests. Engineers helped to make the rural world.

Part Three: Animal

6

Animal Capital

On 16 May 1792, al-Ḥājj Muṣṭafā ibn al-Marḥūm al-Ḥājj Darwīsh ʿIzzat ibn ʿUbayd from the city of Damanhūr in the northwestern Egyptian subprovince of al-Baḥayra came to that city's court to file a claim against al-Shābb ʿAlī ibn Sīdī Aḥmad al-Muzayyin from the nearby village of Surunbāy.[1] Five days earlier, al-Ḥājj Muṣṭafā had purchased an ox (*thaur*) from al-Shābb ʿAlī for the price of thirty-five *riyāl*s. Upon taking this ox home with him later that day, al-Ḥājj Muṣṭafā immediately put the powerful animal to work on his waterwheel. He secured the animal into the waterwheel's harness and watched it move the enormous weight of the wheel a few turns. Without such an animal, moving water from a well or basin into irrigation ditches was nearly impossible. Confident that the ox understood its new laboring role, al-Ḥājj Muṣṭafā retired for the evening with the hope that his fields would be adequately watered by the time he awoke the next morning.

Much to his dismay, however, al-Ḥājj Muṣṭafā arose from his slumber to find his newly purchased ox dead. According to the plaintiff al-Ḥājj Muṣṭafā's testimony in the court's records, during the night the animal had fallen into the small pond around the waterwheels that brought water to his rice fields (*hawr min sawāqī al-aruzz*). Either because of the impact of the fall itself or as a result of drowning once in the water or because of a combina-

11. Ox on waterwheel. Commission des sciences et arts d'Egypte, *État moderne*, vol. 2, pt. 2 of *Description de l'Égypte*, Arts et métiers, plate 5.

tion of the two, the ox had died. Seeking redress, al-Ḥājj Muṣṭafā appeared in his local Islamic law court that spring to request repayment of the thirty-five *riyāl*s (a significant amount of money for a peasant in this period) with which he had originally purchased the animal from the defendant al-Shābb ʿAlī. The text of this case does not cite any specific reason as to why al-Ḥājj Muṣṭafā made this rather weak request in the court. He perhaps meant to imply that there was some undisclosed defect with the animal at the time of purchase. Or perhaps he felt that since he had only purchased the animal a few hours before it died, he could somehow link the animal's death to its seller.[2] Whatever the case may have been, the defendant al-Shābb ʿAlī, not surprisingly, denied any responsibility for the fate of the animal after its sale and demanded that the plaintiff bring forth proof (*thubūt*) that he, al-Shābb ʿAlī, was somehow implicated in the animal's death. When al-Ḥājj Muṣṭafā was unable to make any logical connection between the seller of the animal and its death, the judge ruled that the plaintiff al-Ḥājj Muṣṭafā had no right to the thirty-five *riyāl*s with which he had purchased the ox and that he was not to make any further claims on al-Shābb ʿAlī.

Though this case is legally quite straightforward—a plaintiff made a claim, the defendant denied the claim and asked the plaintiff to prove it, the plaintiff could not, and the judge sided with the defendant—it encapsulates many of the most significant aspects of the history of animals in the early modern Muslim world. In a pre-fossil-fuel society without steam engines, trucks, or other mechanized forms of technology, animals served as the heavy-lifters, stores of energy, and long-distance transporters needed to power the economy of the Ottoman Empire and other early modern polities. These agricultural actors addressed some of the most fundamental problems in the Egyptian countryside: how to move water from canals and streams to fields, how to plow, and how to transport grains, foodstuffs, information, and people from fields to towns and markets. As such, roads, doorways, gates, and various other structures were constructed to accommodate the passage of animals; waterwheels were designed and built to fit around the size of these large mammals' bodies; and volumes of food shipments were limited by the amounts animals could pull or carry on their backs.

Through a consideration of the vast amount of information about animals found in the thousands of registers produced by rural courts in seventeenth- and eighteenth-century Ottoman Egypt, this chapter and the next argue that historians of the Ottoman Empire and of the early modern Muslim world more generally must account for the economic centrality of animals as forms of property. Though social and economic historians of the Ottoman Empire have long worked to elucidate the immense wealth accrued by imperial subjects and merchants through the ownership of and trade in various commodities, we have yet to take seriously the study of livestock in the economy of Ottoman Egypt or anywhere else in the empire. Paper, silk, coffee, and myriad other material goods have all rightly and usefully been considered as key components in the economic history of the empire.[3] Animals deserve no less. This chapter thus seeks to add to this literature on commodities and property by describing and analyzing the enormous economic significance of animals in Ottoman Egypt. While livestock (a term that in and of itself points to the economic importance of animals) were similar to these other commodities, they were also very different in several respects. As living, moving, sentient, and procreating forms of property, animals indeed challenged—as we shall see below—numerous aspects of legal practice concerning property and ownership rights in Ottoman Egypt.

Food production lay at the heart of why animals were so important in Ottoman Egypt and, moreover, why Ottoman Egypt—the largest grain-producing region of the empire—was vital to the entire imperial system of the Ottoman state. The utilization of nonhuman forms of labor in rural Egypt was an integral part of the effort to sustain human life throughout the Ottoman Empire since the need to grow and transport food demanded the efforts and caloric output of *both* animals and humans.[4] Getting food into the stomachs of people across the expanse of the Ottoman Empire was a driving force behind imperial rule in Egypt and behind this early modern empire's employment of animal power and technology (in the forms of plows, waterwheels, harnesses, and the like) to maximize its overall levels of agricultural production.

At the same time, animals were also key symbols of status in rural Ottoman Egypt and represented significant investments of capital. Peasants in the Ottoman Empire did not own land as property; the state owned all agricultural land in the empire while peasants held the usufruct rights to cultivate parcels of that land. Because peasants could not purchase land and seldom owned slaves or other large forms of property, animals were some of the largest and most productive forms of agricultural capital in the Ottoman Empire and represented much of the economic base of Egyptian rural society. Animals were markers of status and wealth since they did work that could not otherwise be done in the rural worlds of the early modern Ottoman Empire.

As it is impossible to cover the entire range of how animals were used in early modern Ottoman Egypt, I focus my analysis in this chapter on domesticated animals as property in the agricultural realm.[5] My interest in this subject comes from the ubiquity of references to animals in the archival record of Islamic courts throughout Egypt in the early modern period. I consider animal use in two regions of high agricultural output in Ottoman Egypt—the Nile delta and Upper Egypt (al-Ṣaʿīd). The majority of the examples in this chapter come from the court records of cities in the former region (Damanhūr, al-Manṣūra, and Rosetta) rather than from those in the latter (I use the records of Isnā and Manfalūṭ), reflecting the delta's higher levels of agricultural productivity and output during the Ottoman period. More peasants, more fields, and more canals meant there were more animals in the delta than in Upper Egypt. As I will outline below, animals were central to legal disputes over financial transactions (as in the case that opened this chapter), estate inventories, thefts, marriages, and public works projects. It is not enough, however, to simply register the widespread presence of animals in the agricultural economy of early modern Egypt. Thus, in the second half of this chapter, I will first explain how a consideration of animal work furthers our understanding both of labor in rural Egypt and of how the Ottoman state conceived of that labor. I will then show how animals functioned and were conceptualized as distinct kinds of property. Because they could move, die, lactate, and procreate, disputes involving domesticated ungulates—the majority of animals employed in rural Ottoman Egypt—represent some of the most challenging and unique cases concerning property in the archival record. My examination of animals as property in Ottoman Egypt is thus also meant to contribute more generally to our understanding of notions of property and of nonhuman nature in the early modern Muslim world.

The Economic Hierarchy of Animals in Ottoman Egypt

The vast archival record of eighteenth-century rural Ottoman Egypt testifies to the fact that domesticated ungulates were extremely valuable economic and social commodities in the Egyptian countryside and often constituted the single most expensive item peasants owned.[6] As already noted, the high cost of these animals came first and foremost from the valuable services they performed in the countryside—they plowed fields, produced milk, turned waterwheels, hauled cargo, and served as forms of transportation. At the top of the economic hierarchy of these quadrupeds was the enormous Egyptian *jāmūsa*, or buffalo cow.[7] The *jāmūsa* was by far the most expensive and valuable animal of frequent use in the Egyptian countryside.[8] In a case of inheritance, for example, a buffalo cow was valued at over three times the price of a donkey.[9] In another case involving the estate of the deceased Muṣṭafā ʿAbd Rabbu, a *jāmūsa* was again priced at over three and a half times the value of a donkey—the *jāmūsa* cost sixty-five *niṣf faḍḍa* while the donkey only cost eighteen *niṣf faḍḍa*.[10] In an estate inventory from the court of al-Baḥayra from the year 1795, we find that both a camel and a *jāmūsa* were each about six times more valuable than an ox.[11] Indeed, the camel and the *jāmūsa* were among the most expensive single items in this entire estate belonging to the deceased al-Shaykh Ibrāhīm ibn al-Shaykh Ramaḍān al-Shayūnī al-Khaḍrāwī. In another case, a *jāmūsa* was valued at two and a half times the price of a cow and at over three times the price of an ox.[12]

When al-Ḥājj Ḥasan ibn al-Marḥūm al-Sayyid ʿAlī al-Sharīf from the village of Bīwīṭ in al-Baḥayra died in the year 1794, nearly his entire inheritable estate consisted of animals.[13] Of the 298 *niṣf faḍḍa* that made up the total value of this estate, animals accounted for 294 *niṣf faḍḍa*; the other 4 *niṣf faḍḍa* consisted of a quantity of rice.[14] As seen in the accompanying chart, oxen and *jāmūsa*s were the most valued animals in this particular estate.[15] The value of the other animals was markedly below that of these two.

On the rare occasion that horses appeared in the estates of deceased peasants or residents of subprovincial cities, they were sometimes judged to be more valuable than *jāmūsa*s.[16] Often, however, a *jāmūsa*

Table 1 Animals in the estate of al-Ḥājj Ḥasan ibn al-Marḥūm al-Sayyid ʿAlī al-Sharīf

Animal	Quantity	Total value (*niṣf faḍḍa*)	Unit value (*niṣf faḍḍa*)
Ox	2	131	65.5
Jāmūsa	2	100	50
Cow	1	16	16
Camel	1	15	15
Ewe	8	32	4

was still considered more expensive than a horse.[17] In any case, horses were quite rare in the Egyptian countryside and were most usually possessed by the rich.[18] They were, moreover, also considered to be lavish gifts.[19] It was customary, for example, that new governors to Egypt were given horses as part of the gift package they received upon their arrival in the province.[20] This reward of animals partly served to compensate these imperial governors for the animals they were not allowed to bring by ship to their new post. Other animals were also commonly given as gifts.[21] For example, the very wealthy Bedouin Shaykh al-Humām from the village of Akhmīm in Upper Egypt gave a gift of three hundred camels to the pilgrimage procession every year.[22] Horses were also the preferred animal for riding between various regions of rural Egypt.

In the case of the estate inventory of the deceased Muḥammad ibn ʿAbd al-Raḥman ibn Muḥammad Sulīkar, the combined value of his *jāmūsa* and its calf was greater than all his other possessions, save his stocks of rice, wood, and salt.[23] Only these items possessed in large bulk quantities proved to be more expensive than this man's one buffalo cow and its calf. The *jāmūsa* was so expensive an animal that peasants often had to pool their resources to be able to purchase one. In these cases, peasants bought shares in a *jāmūsa* allowing them partial rights to the animal's labor, milk, and other productive capacities. In a case of inheritance, for example, a particularly wealthy man bequeathed to each of three different heirs a third of one *jāmūsa* and to each of two other heirs half of another *jāmūsa*. Tellingly, each of these shares was more expensive than the full price of the man's red cow (*baqara ḥamrāʾ*) and its calf.[24] In another case, two men came to court each to buy half of a *jāmūsa* from another man.[25] Splitting shares in an animal was not limited to *jāmūsa*s alone and commonly occurred with other animals as well.[26]

In this economic hierarchy of animals in the Egyptian countryside, donkeys, cows, and camels came beneath she-camels (s. *nāqa*) and *jā-*

mūsas.[27] Whereas both of the latter produced milk, among the former it was only the cow that could produce milk. Unlike she-camels and the *jāmūsa*, though, cows were not extremely useful as work animals. They were big and slow, not easily ridden, and required a great deal of food and space to maintain.[28] It was primarily because they could produce milk that Egyptian peasants considered them of value and generally more expensive than male camels and donkeys, both of which were available in great numbers throughout Egypt and were, thus, very easy to acquire and cheaper than cows.[29] To sum up, then, based on the relative prices of these animals in Ottoman Egyptian estate inventories, the hierarchy of animals in the Egyptian countryside was crowned by the *jāmūsa*, the ox, and the she-camel (and occasionally the horse). Next came the cow and then the donkey, the male camel, and calves. For example, a particular probate case lists a cow at eight times the price of a donkey, two times the price of a calf, and sixteen times the price of a sheep.[30]

The accompanying four tables outline the rather large estate of ʿAlī Muṣṭafā ʿĪsāwī of Isnā, which contained many shares in animals.[31] The large animal holdings in his estate give us further details about the relative values of animals of moderate expense in the middle of the eighteenth century. As seen here, cows were both the most common and the most expensive animal in ʿAlī Muṣṭafā ʿĪsāwī's estate and were subject to similar sorts of sharing agreements as other animals.[32]

Table 2 Cow ownership of ʿAlī Muṣṭafā ʿĪsāwī

Percentage of cow owned	Price (*niṣf faḍḍa*)
Half	175
Half (plus the right to the cow's offspring)	310
Fourth	82
Half	225
Half	300
Third	150
Fourth	82
Half	310
Fourth	125
Fourth	50
Half	320
Half	230
Half	200
Third	150
5 and 2/3 cows	2,709

Table 3 Donkey ownership of ʿAlī Muṣṭafā ʿĪsāwī

Percentage of donkey owned	Price (*niṣf faḍḍa*)
Whole	60
Third	60
Half	75
1 and 5/6 donkeys	195

Table 4 Calf ownership of ʿAlī Muṣṭafā ʿĪsāwī

Percentage of calf owned	Price (*niṣf faḍḍa*)
Fourth	50
Half	30
Fourth	50
1 calf	130

Table 5 Average value per animal

Animal	Average price (*niṣf faḍḍa*)
Cow	478
Donkey	106
Calf	130

The Comparative Value of Animals

Looking at other probate cases gives us important information about the costs of animals relative to other nonliving things and even to humans, and it thus allows us to develop a fuller sense of the economic worth of animals in Ottoman Egypt. In the case of cows, for instance, we find that they—like the *jāmūsa*—were much more valuable than most household items. In a case from 1813, for example, the combined value of a cow and a young calf owned by one Muṣṭafā al-Faqīr ibn Muḥammad ʿAqaṣ was 4,430 *niṣf faḍḍa*.[33] This price is striking in comparison to some of the other items in his estate. A copper kettle (*dist nuḥās*), for instance, was valued at only 90 *niṣf faḍḍa*, an antique *ṭarbūsh* was priced at 120 *niṣf faḍḍa*, a red shawl (*shāl aḥmar*) cost 20 *niṣf faḍḍa*, and a *zaʿabūṭ* (a common woolen peasant garment) was priced at 60 *niṣf faḍḍa*. Another estate consisted of a 20 percent share in a donkey, various kinds of eggs, and a set of wooden spoons.[34] Of these items, the price of the man's donkey shares was much higher than the price of the spoons or of the large number of eggs in his estate.

On the rare occasion that probate cases in Egypt's rural courts dealt with estates that included humans, we get an important glimpse as to how animals, humans, and inanimate objects were situated relative to one another in an early modern economic and social hierarchy of value. One such case comes from 1741 from the court of the coastal Mediterranean city of Rosetta. The wealthy deceased Aḥmad ibn Aḥmad, who was also known as al-Sayyid Yūnī, owned a great deal of rice, two cows, a black concubine, some dates and figs, bricks, and lime.[35] Of these, the most expensive item by far was the large amount of rice he owned at the time of his death. The second most expensive listing in the estate was his black concubine, who cost 2,000 *niṣf faḍḍa*. The price of this concubine, though, was dwarfed not by another inheritable item, but by an expense paid out by the estate: the preparation of the deceased's body and its burial (*tajhīz wa takfīn*). This postmortem expense, which occasionally appears as part of probate cases (presumably when the deceased made provisions for it), totaled 5,000 *niṣf faḍḍa* in this instance. The combined cost of his two cows was 1,000 *niṣf faḍḍa*, and together the figs and dates cost 500 *niṣf faḍḍa*. The brick and lime were also listed together as costing 500 *niṣf faḍḍa*.

In a statement about the relative value of humans, animals, and things, this case lists the human concubine alongside all the other items in the estate without any distinguishing marker meant to set her apart from the nonhuman entries.[36] The concubine, not surprisingly given her legal status, was considered to be property like the rest, four times as expensive as a cow.[37] Similar to a shawl or a donkey, even this concubine's color was identified. Thus, within an economic arrangement in which human owners controlled both human and nonhuman property, this example clearly indicates that any sort of divide between the human and the nonhuman did not inform the legal treatment of owned slaves. Concubines, slaves, and animals alike were all entered into the realm of the commodity to be evaluated, priced, bought, and sold. In other words, not only were slaves and concubines simply not free, but they were conceived of in literally the same terms as animals and inanimate objects.

Consuming Animals

As seen above, animals that produced milk and could be utilized as efficient laborers in agricultural cultivation—animals like the *jāmūsa* and the she-camel—were the most valuable beasts of burden in rural life.[38] Thus, animals that produced milk but were less useful as agricultural workers, animals like cows, followed she-camels and *jāmūsa*s in their economic value in the Egyptian countryside.[39] As with human peasants, animals were considered to be highly valuable because their labor contributed to the production of food. Unlike Egyptian peasants, however, animals also possessed the added economic potential of producing milk, which could be made into various consumable goods. Indeed, she-camels were some of the most expensive animals to be owned in Ottoman Egypt because they served as sources of milk for peasant communities.[40] For example, the most expensive item in the estate of the deceased ʿAlī ʿAbd al-Qādir al-Ḥamdanī was a she-camel worth 1,133 *niṣf faḍḍa*.[41] Containing large amounts of potassium, iron, and other vitamins and generally considered to be more nutritious than cow's milk, camel milk was an important part of the Egyptian peasant diet.[42] Camels were, moreover, historically some of the most prized living possessions in all of Egypt, especially among the Bedouin.[43]

Animals were also, of course, killed for their meat, though, this seems to have been a rather rare occurrence in the Egyptian countryside, and peasants seldom, if ever, killed their own animals for their meat.[44] Eating meat was a luxury few could afford, and thus eating animal flesh came to be associated with the wealthy.[45] Animals were, for example, always slaughtered as part of the celebration in honor of a new Ottoman vali arriving in Cairo.[46] And we are told as well that the vali received 180 *okke* (509.1 pounds) of veal every day from the slaughterhouse (*silihāne*) of Cairo.[47]

Elsewhere, we read that a rich man in Isnā endowed a *waqf* before his death on the island of Aṣfūn.[48] As part of the *waqf*'s deed, he instructed that horses, camels, cows, *jāmūsa*s, and sheep be kept on hand to be killed and served as food to those guests who came to the *waqf* on special occasions.[49] The endowment of this rather grand complex and the provisions the endower made for the feeding of guests were a sure means of preserving prestige and status in his community after his death.[50]

It was not only land animals that were consumed as food. Fish and birds were also eaten throughout the Egyptian countryside. There was, for instance, a successful fish market in Rosetta[51] and another in Old

Cairo.[52] There was also a very lucrative *muqāṭaʿa* known as Baḥayrat al-Samak that was charged with the regulation of fishing and hunting rights on a large lake near Damietta.[53] A similar *muqāṭaʿa* was established by Ḥasan Paşa in 1786–87 to regulate the Lake of Maṭariyya near Cairo.[54] In return for this tax farm, the area's multazim had to pay the state two hundred thousand *para* annually.[55] The lake's circumference was a distance of several days' journey, and the people who lived around it earned their income from the sale of different kinds of fish from the lake and of the many sorts of seabirds that frequented the area.[56] Fishing was also regulated on the shores of Bulaq and Old Cairo, in canals in and around Cairo, at the mouth of the Nile's branches in Rosetta and Damietta, in lakes formed by the Nile's overflow in Cairo, and in Lake Maʿdiyya in Alexandria.[57] The multazims of these areas usually collected two fish from each fisherman as a payment and were entitled to 15 to 20 percent of the revenues raised from the sale of fish in their area. There was also a corporation of fishmongers in the markets of Cairo.[58] Despite this evidence, the centrality of the Nile to peasant life, and the high amounts of protein and other nutrients in fish, it remains curious that these animals are not mentioned as a source of food more often in the early modern period. Pigeons were also kept by numerous peasants for food and for the market.[59] Nevertheless, in a rural society where most food was grown in the soil, the luxury of eating meat of any kind was usually one enjoyed by the rich or reserved for special occasions such as feasts or weddings.

Animals at Work

Like their abilities to produce food, animals' useful labor made them coveted and valuable possessions for both peasants and the Ottoman administration of Egypt. For example, the exceptionally rich Bedouin *shaykh* and *amir* (notable) Humām ibn Yūsuf ibn Aḥmad ibn Muḥammad ibn Humām ibn Ṣubayḥ ibn Sībīh al-Hawwārī of Upper Egypt put over twelve thousand head of cattle to work in the cultivation of sugarcane on his estate.[60] In addition to this impressive number, he also employed cattle for the threshing of grain, for the function of mills, and for the turning of waterwheels, and his estate also included many water buffalo and dairy cows. As in the case that opened this chapter, here again animals were used in the maintenance and function of Egypt's irrigation infrastructure.

Moreover, animals were very often used in the digging and dredging of canals and in the reinforcement of canal embankments. In August

1774, for instance, camels, donkeys, and horses were employed in the redigging and redredging of a series of canals in the province of al-Daqahliyya that ran to the villages of Minyyat Ṭalkhā and Minyyat Ḥaḍr from the large canal known as al-Baḥr al-Ṣaghīr.[61] Camels were also often used to clear debris and mud that hindered the proper function of wells throughout the countryside.[62] In another case, *jāmūsa*s were utilized in the maintenance of a canal's embankments in a village near Sandūb in the subprovince of al-Daqahliyya.[63] The expenses incurred in this repair were to be paid out by the villagers overseeing this work, who were then to be reimbursed by the imperial bureaucracy. The largest expense these villagers reported was that of renting the *jāmūsa*s for the heavy labor needed during the repair work.

In another case concerning the repair of two waterwheels in Damanhūr in the subprovince of al-Baḥayra, we see how *both* animals and humans came to be conceptualized by the Ottoman administration of Egypt as the same unit of expense—laborer.[64] This case lists expenses incurred during the four years required for the repair of these two waterwheels, from September 1790 to the summer of 1794. Much of the over thirty-seven *niṣf faḍḍa* spent on this work went to cover the costs of bricks, wood, and lime needed in these repairs. Human and animal workers, though, also each received an entry in this expense list: workers and builders were needed to move and arrange materials, and donkeys were used to haul mud to and from the worksite. Similar to the conceptual propinquity of concubines and animals in the probate record discussed above, in this case human and animal laborers were listed in *almost* the same breath as part of the same analytical category of laborer.

A similar dynamic was at play in the repair of a dam (*ḥibs*) in the village of Shūbar in the subprovince of al-Gharbiyya.[65] This village's entire water supply came from the amount of the flood's deluge this dam could capture and store during the annual inundation. The repairs outlined in this case called for both human and animal labor: one hundred men (*rajul*) and sixty-two oxen (*thaur*). These laborers were to devote themselves to the hauling of straw and dirt and to properly using these materials to reinforce the dam. This work required bags (s. *kīs*) to carry the straw and dirt and rakes (*jarārīf*) to move and collect the dirt. Provisions of beans (*fūl*) were also made for these human and animal laborers to sustain them as they worked. The casting of these laborers on the pages of this register, the fact that food was not differentiated as going only to peasants or to oxen, and the fact that these two kinds of workers were treated as a single unit of labor again suggest that in the realm of work on irrigation features in the Egyptian countryside, the Ottoman bureaucracy did not distinguish between human and animal workers. These human and nonhuman laborers were abstracted and recast by the Ottoman bureaucracy of Egypt as possessions of the state to be enumerated, moved around, and configured as demanded by the needs of Egypt's irrigation network. From the imperial perspective of these repair cases, the basic fact was that both humans and animals were considered to be workers. Indeed, whether they had hands or hooves, laborers were laborers.

Not only did animals perform the arduous tasks of transporting building materials and hauling unwanted dirt from worksites, but they were also essential for the actual function of many irrigation works, most especially the waterwheel. As we have seen already, the muscles of large domesticated agricultural animals were used to turn waterwheels that brought water to the canals and fields of rural Egypt. Without the vital energy provided by these animals, most of Egypt's waterwheels could not have functioned. Thus, as evidenced by numerous cases from rural Egyptian courts, part of the reason why animals like oxen and *jāmūsa*s were very expensive and coveted possessions was their ability to move waterwheels.[66] Camels were rarely used to turn waterwheels and seem to have been reserved for hauling and transporting items. They were the animals most commonly used, for example, to haul lumber through the desert between Cairo and Suez for the construction of boats in the Red Sea port city.[67] These animals were also often employed on irrigation worksites and in the course of other construction projects to remove excess dirt and materials.[68]

Animals in Question: Fraud and Theft

Thus far in this chapter, I have shown how domesticated animals were in widespread use in a variety of arenas of rural life in early modern Ottoman Egypt. Not only were animals used in this agricultural economy to produce food and to carry out useful work, but they also represented some of the most expensive items owned by Egyptian peasants. As coveted items of economic and social status, animals were thus subject to property disputes, fraud, theft, and all sorts of other peasant attempts to gain a comparative advantage through the possession of animals. We now turn to what some of these struggles tell us about notions of property, law, and ownership in Ottoman Egypt.

Some of the most common property disputes involving animals and other possessions emerged during the execution of wills.[69] In a case from the southern city of Manfalūṭ, for example, two brothers wrangled over a cow, a donkey, and two sheep that belonged to their deceased father.[70] Inheritors also often sold animals that were part of a deceased's estate. For instance, in the summer of 1797 after the death of his wife, the amir Sulaymān Jurbajī Murād sold a quarter of a calf and half a *jāmūsa* that had belonged to her.[71] Marriages and divorces were other legal occasions that involved the transfer of large amounts of animal capital.[72]

In addition to these very common kinds of disputes in the courts of Ottoman Egypt, other more complex and rarer instances of crimes involving animals provide us with a more nuanced understanding of notions of property in this period. One such case was the purchase of a camel by an unnamed peasant (*rajul fallāḥ*) from an unnamed seller (*bāʾiʿ*) in al-Baḥayra in the year 1752.[73] The supposed fraud involved in this transaction came to light through the testimony of a third man, who asserted that he had entrusted the above seller to watch over his horses and camels while he was occupied with another matter. This man thus claimed in court that the sale of the camel—*his* camel—to the peasant was illegal since none of these animals were ever in the legal possession of the seller. In the court's ruling, however, the judge instructed that the peasant could keep the camel he purchased from the seller because the peasant had solid legal proof (*bayyina sharʿiyya*)—usually either the testimony of two qualified male witnesses or some sort of written legal instrument—stating that the camel was indeed the property of the seller. Thus, it was the sale and not the purchase of the animal that was deemed illegal.

Cases of animal theft offer us further insight into how the ownership of animals—and hence of other kinds of property too—was actually established and protected by rural Egyptian courts. The most common kind of animal theft in rural Egypt was that carried out by Bedouins (*ʿarab* or *ʿurbān*)[74] or by roaming groups of bandits (*ashqiyāʾ* in Arabic, *eşkıya* in Ottoman Turkish).[75] These outsiders to village communities most often stole animals for their economic value and for the food and other essentials their carcasses provided.[76] Villagers stealing animals from their fellow villagers was not as common an occurrence in the Egyptian countryside for several reasons. First, the sheer size and clumsiness of large domesticated animals made it difficult to steal them in the first place and then to conceal them once stolen. Moreover, since the majority of peasants lived in villages where almost everyone knew everyone else and had some sense of their property holdings and animals, a theft was a very obvious act—as we shall soon see—since it would be clear when a peasant no longer had a particular animal and even clearer when someone else did. Nevertheless, when thefts did occur and, on the even rarer occasion, when these cases were adjudicated in a local Islamic law court and not through some other out-of-court settlement, they show the extent to which the possession of animals was a highly coveted sign of social status and how far people were willing to go to regain their animals.[77]

In al-Baḥayra in 1787, ʿAlī ibn Aḥmad Muḥammad—who is described as being one of the elders (*al-mashāʾikh*) of the village of al-Raḥmāniyya—accused ʿAlī ibn ʿAlī al-Ḥamd from the same village of stealing from him two steers (*thaurayn*).[78] According to the plaintiff's testimony, nineteen days earlier the defendant had entered his home at night and stolen the two steers: one red (*aḥmar*) and the other piebald (*ablaq*).[79] To rectify the situation, the plaintiff requested from the court that the defendant be made to immediately return the stolen animals to him. When prompted for a response to these accusations, the defendant denied them. To counter this denial and to prove his own case further, the plaintiff brought two male witnesses—again, sufficient legal proof—from al-Raḥmāniyya to the court: Yūsuf ibn Ḥasan ʿAṭīya and ʿAlī ibn Dāwūd al-Khawlī. Each gave his testimony independently of the other (*addā kullun minhumā shahādatahu ʿalā infirādihā*) and in the presence of the defendant (*bi-wajh al-muddaʿā ʿalayhi*). Each of the two witnesses claimed to have seen the defendant with the two steers seventeen days earlier (in other words, two days after the alleged theft) walking in a nighttime wedding procession (*zifāf*) in al-Raḥmāniyya. According to the text of this case, the witnesses claimed to have seen the defendant at the wedding pulling the two steers from their bridles (*bi-zimāmihimā*). The witnesses' descriptions of the animals' colors and coat patterns matched those of the plaintiff's stolen animals. Weighing the defendant's denial against the testimony of these two witnesses, the judge ruled in favor of the plaintiff and ordered the defendant to return the two steers to their rightful owner. Upon hearing this verdict, the defendant broke down in court and confessed to stealing the two animals. He added, however, that he had later given them to a man named Ḥasan Shukr, who, we are told, was also present at court that day.

The crucial event in the proper adjudication of this case was the fact that these steers—once stolen—were paraded in public through the village. As a very public event, a wedding procession was an opportunity for people in a village to assert their social rank and to display their possessions. As important forms of capital, large animals like these two steers were key symbols of wealth and local prestige.[80] Indeed, the testimony in this case was clear: the thief held these animals by the reins to make sure everyone present at the wedding would know they were his (even though legally they were not). The case makes the point that he was the literal leader (*huwa qāʾid*) of these two animals during the wedding procession. Furthermore, the description of the two steers in the full detail of their color and coat patterns was an important factor in identifying these animals led by the defendant in the wedding procession as the plaintiff's. Indeed, this was the key piece of information that allowed the witnesses to make the connection that led to the judge's verdict. They described the steers they saw with the defendant at the wedding as the two aforementioned described steers (*al-thaurayn al-madhkūrayn al-mawṣūfayn*).

Did the defendant in this case really believe he could get away with parading around his village for all to see during a wedding procession pulling the very same two steers that were stolen from a notable member of the village two days earlier? Did the thief perhaps know all along that the steers would be returned to their rightful owner? Did he, in other words, ignore the legal repercussions he surely knew would befall him in favor of the temporary social status that would be accorded to him as the possessor of these animals? Perhaps he was put up to the task by the mysterious Ḥasan Shukr, who would later take the steers himself and who sought to pay the thief for his illegal act by allowing him to use the steers in the wedding procession. Or was the defendant's claim that he had later given the steers to Ḥasan Shukr a vain attempt to put the blame on someone else to rescue his name from shame in the village?

Whatever the case may have been, the defendant's motives are not explicitly clear. What is clear, though, is that the thief and the victim were from very different social classes in the village. The plaintiff was a notable while the thief was not; he is specifically mentioned as being from among *al-mashāʾikh* while the defendant was not; he owned two steers and probably many more animals while, though it remains unstated, the thief likely did not. Perhaps the plaintiff even knew the judge presiding in the case and the others assembled in the courtroom. Surely most of those in the wedding procession of which his steers were a part knew the plaintiff. The different social and economic backgrounds of the two parties in this case—measured in part by their possession or lack of animals—was a key determining factor in its legal outcome.[81]

Animals on the Move

As seen in these cases of fraud and theft, establishing the proper ownership of animals was an issue of grave importance and intense contestation precisely because they were so valuable and useful as forms of property in rural Ottoman Egypt. It was with respect to this very question of establishing the legal ownership of animals that these living creatures' abilities to procreate and move set them apart from other forms of property. The function of Islamic law courts and wider legal, social, and economic notions of property in Egypt thus had to adjust to the unique attributes of these living forms of property, especially given the lack of fenced space in the Egyptian countryside.

In an example of this from Manfalūṭ, a man named Rushdān ʿAbd al-Rasūl from the village of Jalda came to the court of Manfalūṭ in an attempt to regain his donkey.[82] He alleged that the green she-donkey (*al-ḥimāra al-khaḍrāʾ*), now in the possession of a woman named ʿAzīza bint Ḥasan Abū Aḥmad, was actually the female offspring of one of his donkeys that had wandered off about six months ago and was thus his legal property since, according to Islamic law, any offspring of an animal was the legal possession of the original birthing animal's owner.[83] For her part, ʿAzīza denied that the animal was Rushdān's, asserting that she had legally bought the young calf from the sale of her brother's estate a few months earlier. To counter ʿAzīza's claim and to strengthen his own, Rushdān asked to have the disputed newborn donkey brought to court. Along the lines of an early modern kind of genetic testing, he produced solid legal proof in the form of two witnesses who testified that this young donkey was indeed the offspring of Rushdān's donkey and therefore his legal property.[84] Finding this evidence sufficient, the court ordered that the young donkey be returned to Rushdān.[85]

Because the offspring of one's animals was considered one's own property, one could sell an animal and still retain rights to the offspring of that animal. It was therefore often specified when an animal was sold whether or not the buyer was also purchasing the rights to the animal's offspring. For example, a case was brought to the court of al-Baḥayra in 1786 concerning the shares of half a *jāmūsa* that had recently given birth to two calves.[86] The man who had previously sold the *jāmūsa* came to court to assert his right to its two calves since he had only sold his shares in the calves' mother but none of his shares in the offspring. This arrangement was not only in place to ensure legal and financial control over animals, but also served to regulate breeding. By specifying the ownership of offspring and, hence, the breeding patterns of their animals, Egyptian peasants were able to control and select the breeding of their domesticated ungulates so as to privilege desirable traits and diminish undesirable ones.

Another case about a wandering animal helps to elucidate the nature of the ownership of moving forms of property and how this ownership was established. In January 1794, ʿAlī ibn al-Marḥūm ʿUthmān Abū Bakr from the village of Sunṭīs came to the court of al-Baḥayra seeking to regain possession of his wandering cow.[87] As the plaintiff in this case, ʿAlī claimed that he was the rightful owner of a brown cow (*baqara samrāʾ al-lawn*) that he had legally bought from one ʿAlī ʿUmar eight years ago. About a year after this purchase, the plaintiff began to let his cow wander freely with other animals around the fields of their village community in Sunṭīs (*sāriḥa maʿa al-bahāʾim bi-nāḥiyat Sunṭīs*). Recently, however—and this was the immediate reason for the case being brought to court—it had come to the attention of the plaintiff ʿAlī that his cow was now in the possession (*taḥt yad*) of the defendant in this case, a man named al-Sayyid ʿAbd Allāh ibn Sālim Qarqūr from the village of Zāwiyat Naʿīm, and, significantly, that it had given birth to a new calf.[88] ʿAlī thus came to court to regain from ʿAbd Allāh possession of both the cow and its new offspring.

In response to ʿAlī's claims, ʿAbd Allāh retorted that in the previous year he had rightfully purchased this cow for the price of fourteen *riyāl*s and one *jāmūsa* from Ḥasan Abū ʿĀliyya of the village of Ruzzāfa, who was the representative (*al-wakīl*) of a man named Ḥasan Abu Ḥamīda from the same village.[89] He furthermore denied that this cow had recently given birth. To strengthen his case, he brought the aforementioned Ḥasan Abū ʿĀliyya to the court to testify that he had indeed sold the cow in question to ʿAbd Allāh a year ago. For his part, this witness Ḥasan also told the court that he himself had bought the cow five years earlier from a now-deceased peasant from the village of Qabr al-Umarāʾ. In response to these witnesses' testimony, the plaintiff ʿAlī produced two witnesses of his own from Sunṭīs who swore to the court that ʿAlī was the rightful owner of the cow and that beginning seven years ago he had let this animal roam freely around the village. After a period of deliberation, the judge ruled that the animal was to be returned to the plaintiff ʿAlī. He added, though, that ʿAbd Allāh was entitled to receive from Ḥasan repayment of the amount with which he had purchased the animal—fourteen *riyāl*s and a *jāmūsa*. Thus, as in a previous ruling we saw, it was not the purchase of the animal that was deemed illegal but rather its sale, since the cow was never in the legal possession of the person who had sold it to ʿAbd Allāh.

The cow in this case was allowed to roam freely in and around its village for a period of seven years. From the perspective of an animal's owner, there was good reason to let one's bovine possession wander. First, by allowing his animal to scavenge around for its own food, an owner could avoid the costs of planting and storing or otherwise acquiring feed.[90] Second, there were few feral or wild animals in the Egyptian countryside. Thus, the threats of degenerate breeding or of falling prey to another animal were nearly nonexistent in rural Ottoman Egypt.[91] Moreover, ʿAlī was presumably able to keep tabs on his domestic animal as long as it did not stray too far out of the village. As evidenced by this case, the major threat to this wandering animal's status was not another animal but rather its seizure by an individual other than its owner. Indeed, it was only when ʿAbd Allāh attempted to claim ownership of this animal in violation of ʿAlī's legal rights that this case came to court.

Moreover, not only did ʿAbd Allāh's seizure of this animal challenge ʿAlī's legal ownership of the cow, but it also struck at the collective rights of the community in which the animal roamed. Because the cow was allowed to wander freely in its village, peasants in the area saw it on a regular basis and recognized it as ʿAlī's legal property. Moreover, they most probably also used the cow for various tasks, consumed its milk, and collectively took care of its upkeep and safety.[92] This cow, in other words, came to function over the course of seven years as a kind of collective possession of the entire village community. This in no way, of course, ever changed the legal reality of ʿAlī's ownership of the animal, but it did mean that those in ʿAlī's village benefited from the presence of his cow and at the same time were implicated in its protection and upkeep.[93] I stress this here to make the point that the history of this cow's presence and function within a wider village community existed largely outside of its strict juridical status as the property of ʿAlī. Thus, ʿAlī's assertion of his individual legal right of ownership of this animal was important not only to him but also to the collective (albeit legally irrelevant) rights of his wider village community.

Conclusion

By centering the role of animals as property, this chapter asserts that no history of Ottoman Egypt—or for that matter anywhere else in the Ottoman Empire or the wider early modern world—can afford to ignore the role of animals in that history.[94] As an early modern agrarian empire whose calories and revenues were almost entirely based on the cultivation of foodstuffs, the Ottoman Empire was a polity that relied on the labor of thousands of humans and animals in its midst to cultivate crops. Exported from Egypt to areas all over the Ottoman Empire, this food served as the connective tissue between animal labor in rural Egypt and peoples across the Mediterranean and Middle East. As this chapter shows, animals were thus deeply enmeshed in the empire's social, economic, cultural, and architectural structures through their high value as laborers, as sources of food and drink, and as stores of energy for transport and even heat in homes during the winter. This fundamental reliance on the power, productivity, and work of animals—an economic dependence on their very bodies—produced a rural world of a particular order and physical geography—one that would no doubt have been of a radically different shape without animals or with other kinds of animals in it.

The utilization of animals in multiple spheres of rural life by nearly all members of rural society made them one of the most important economic commodities in the Egyptian countryside; indeed, rural estate inventories make clear that animals often constituted the largest single source of wealth for Egyptian peasants and rural elites alike.[95] The early modern Egyptian countryside was thus a world characterized by cooperative, constructive, and deeply intertwined relationships between humans and animals mediated mostly through their shared labor in the agricultural realm and through the human ownership of animals. If any changes were to occur in the numbers, the quality, or the kinds of animals that made up this agricultural economic order, the consequences would obviously be colossal. Simply put, animals were valuable because they provided the energy that humans needed to undertake their daily labor and production needs.

7

Brute Force

Until the middle of the eighteenth century, the human-animal relationship was fundamental to economic and social life in Ottoman Egypt and throughout the world. As we saw in the previous chapter, rural communities, in particular, were completely dependent on animal labor. Between roughly 1750 and 1850, however, that relationship began to change around the globe, as nearly all early modern agrarian societies experienced some form of intense transformation involving the commercialization and modernization of their economies. In the vast historical literature on these transitions from subsistence agriculture to market-driven commercial agriculture, land, human labor, and capital accumulation are the usual protagonists. Few would disagree with John F. Richards's early twenty-first-century summary: "Human management of land everywhere has become more centralized, more intrusive, and more instrumentally effective. Large-scale capitalist forms of agriculture and resource extraction are prominent throughout the world."[1]

Two of the most recent and productive attempts to tackle the problem of this particular economic transformation are the group of publications that emerged from what came to be known as the Brenner debate and newer scholarship on the question of divergence between China and Europe.[2] Much of the work that has come out of these two historiographical literatures revolves around questions of natural resources and human labor. Brenner understood the early modern agrarian revolution in Western Europe to be a result of shifts in labor and class relations, while his critics stressed the role of population increases and improved agricultural technology and cultivation techniques. Similarly, natural resources and labor also figure prominently in Kenneth Pomeranz's study of divergence. Very simply summarized, Pomeranz argues that Europe was able to surpass China economically because its exploitation of resources in the Americas allowed it to overcome existing ecological constraints and because of the fortunate geographic accident of its proximity to coal deposits. Those two factors, in turn, allowed Europe to create much more efficient means of maximizing labor than China.

Humans were not the only living creatures driving the transformations that are tracked in the Brenner debate and the divergence literature. Because of their ubiquity and centrality in the early modern world, both as natural commodities and as laborers, domesticated animals also played a pivotal role. As part of the transition from subsistence agriculture to commodity-based commercial economies, animal labor was replaced by human labor and subsequently machine labor. To fully understand the transformations of agrarian societies, we must therefore understand the historically significant causes and consequences of the shift away from a world characterized primarily by widespread dependence on animal labor and intense human-animal interactions.

The history of human and animal labor in rural Ottoman Egypt makes the importance of all this abundantly clear. As we have seen, historians of Egypt and the Ottoman Empire have been drawn to the decades between 1780 and 1810 in seeking to understand and explain some of the many significant social, economic, and political changes that characterized the Middle East at the turn of the nineteenth century. For example, it was at the end of the eighteenth century that the massive accumulation of land by a small group of rural leaders began to transform Egypt and other parts of the empire.[3] In the years that followed, the most centralized government in centuries emerged in Cairo. It undertook various wars of expansion; eventually founded schools, ministries, hospitals, military colleges, and other institutions of state; and built roads, bridges, canals, and other infrastructural works.[4] Between 1780 and 1810 as well, the populations of Egypt's two major cities—Cairo and Alexandria—increased rapidly.[5] Why all or any of this happened has proved to be one of the most vexing and pressing questions facing historians of Ottoman Egypt. As Brenner, Pomeranz,

Richards, and many others make clear, similar processes of political and social transformation and of the commercialization of early modern agrarian economies occurred across the globe in this period. The Egyptian example is therefore instructive in our efforts to understand these worldwide phenomena because it illuminates the specific mechanisms through which rural economies transitioned away from animal labor.

Against the backdrop of environmental histories of the human-animal relationship, the literature on transitions from subsistence to commercial agriculture in the early modern period, and Ottoman economic and social history, a reading of archival sources from the Ottoman central bureaucracy in Istanbul, various rural Egyptian courts, and contemporary British imperial institutions shows that the changing economic status of domesticated animals during the decades between 1780 and 1810 played a central role in the transformation of Egypt from a predominantly agrarian province of the Ottoman Empire to a centralized bureaucratic protostate with a looser affiliation to the empire.

As we saw in the previous chapter, in the early modern period, one of the primary sources of wealth in rural Ottoman Egypt was the ownership of large domesticated animals—water buffalo, oxen, cows, donkeys, and camels.[6] Beginning in 1780, Egypt was hit by a series of climatic and disease events that decimated both human and nonhuman animal populations. The deaths of so many animals in such a short period of time forced a realignment of wealth in the countryside. With a massive portion of the rural economy wiped out, rural Egyptian elites turned to other forms of wealth. The most important of these was land. The final decades of the eighteenth century therefore witnessed the massive accumulation of land by a small group of rural leaders, who then began to use that wealth as a means of political power in Egypt and elsewhere in the Ottoman Empire.[7] To make this land productive, the new landholders needed effective labor to replace the many animals that had been killed by disease and drought in this period. They met this laboring deficit through the corvée of the Egyptian peasantry.

The shift away from animal power to human and eventually machine power represented a fundamental transformation in the energy regime of Ottoman Egypt.[8] Before coal and steam, animal power—including, of course, that of human muscle—was the primary purveyor of usable energy in Egypt.[9] Analyzing where and how societies get their sources of energy is basic to understanding the history of any economy, culture, or ecosystem.[10] It would be nearly impossible, for example, to understand the history of the twentieth century without at least some consideration of fossil fuels. We clearly could not tell the history of the American city without accounting for the role of the automobile, for instance.[11] Likewise, to analyze the economics of early modern rural Egypt, or almost anywhere else in the preindustrial world, without considering the role of animal power would be to miss a major part of the story.[12] The question of why so much changed in Egypt at the end of the eighteenth century, therefore, can be answered through an analysis of the dialectical relationship that developed in this period between social changes that led to the formation of large rural estates and changes in the demands of Egyptians for more and different kinds of energy sources.

Furthermore, the Egyptian case offers a useful elaboration on much of the literature in both environmental history and economic history that identifies the year 1800 as the crucial moment of transition between the only two energy regimes in human history: from solar energy to fossil fuels.[13] The case of Ottoman Egypt roughly follows this chronology, but is important in showing that a critical intermediate stage in that transition was a move from animal to human labor. As animal populations dwindled, an energy deficit developed in the availability of labor power, forcing new rural elites to seek out other workers to meet their demands for ever-greater amounts of reproducible labor, discipline, efficiency, and work capacity to expand their estates. The labor and energy output of animals, which in the early modern period was wholly sufficient to achieve most tasks in the countryside, was now found to be inadequate in this new landscape of centralized and expanding property. To overcome this energy gap, large landholders, and eventually the Egyptian state itself, turned to human power. Thus, much as coal and steam machines can be seen as representing a comparative step forward over the previous caloric motors of biological power, humans in early nineteenth-century Egypt came to be regarded as better working machines than animals.[14] In showing how the vital economic roles of animal and human labor changed a rural economy, the Egyptian example therefore sheds light on the fundamental part that animals played in the economic and energetic transformations that made possible the transition from a primarily agrarian subsistence economy to a market-driven commercial economy—a transition that was crucial to the history of all rural societies.

Survival of the Fittest

As we saw in the previous chapter, early modern Ottoman Egypt ran on animals. This situation of intense human-animal interaction and mutual reliance, however, largely began to change at the end of the eighteenth century. Much animal wealth would soon be wiped out, and a new energy regime would eventually be sought to replace the resulting deficit of animal power. The main cause of this shift away from animal wealth and energy was a vast reduction in domesticated animal populations at the end of the eighteenth century that coalesced with growing political efforts by local Egyptian elites to pull away from the central authority of the Ottoman bureaucracy in Istanbul. From a crucible of death, disease, drought, and destruction at century's end, a new economic landscape emerged in the countryside, one in which animals played a diminished role. Furthermore, this new economic and social topography, achieved by decades of rural distress and elite economic and political maneuvering, set Egypt on an altered, unprecedented, and irreversible course, in which human labor and land—not animals—became the keys to rural wealth and political influence.

Beginning in the 1780s, several major outbreaks of disease, relentless food shortages, and severe weather had a devastating impact on cattle and other domestic animal populations in rural Egypt.[15] In the summer of 1784, a widespread plague outbreak decimated livestock in southern Egypt.[16] This plague continued into the fall of 1784 and then reappeared in the fall of 1785.[17] Reduced animal populations in these years, coupled with lower than expected Nile floods that left soil dry and unfertilized, led to food and fodder shortages throughout the countryside as fields could no longer be plowed effectively. After a low flood in late 1784 led to intense food scarcity throughout the countryside and Cairo, both humans and animals suffered greatly for more than a year. Many of Cairo's residents were forced to comb through the city's garbage to find food—mostly melon rinds and other discarded items.[18] The situation was so dire that the starvation and death of an animal was considered an extraordinary treat for hungry humans. When the ruling elite or members of the military threw away the carcass of a diseased donkey, camel, or horse, city residents thronged to it, grabbing whatever they could from the body. According to the Egyptian chronicler al-Jabartī, people's hunger was so intense that, rather than starve to death, many of them ate these infected animals' meat raw.[19]

A few years later, in 1787, another strong epizootic spread among livestock populations throughout Cairo and the delta. Animal carcasses lay all over Cairo's streets, piled up where they had fallen. "Cattle collapsed in the streets and in the pastures," wrote al-Jabartī, filling the countryside with "the stench of decaying carcasses." One wealthy cattle owner lost 160 of his cattle to the disease.[20] Evidence of the fact that livestock were "the basis of the economy" for both rich and poor alike, during the epizootic of 1787, "peasants bewailed the loss of their cattle, realizing what a blessing its possession had been for them."[21] That summer was especially difficult. Not only had animal populations dropped precipitously, but many humans were also dying from a new plague epidemic that had come to Egypt that previous spring. In addition, the province was in the grip of a severe drought. Fields were parched, and the crops that did grow were quickly consumed by rats, which—like their human and other animal counterparts—were searching for food wherever they could find it. Because so many donkeys, mules, camels, and horses had died from disease that year, prices for the remaining animals "rose very high." This epizootic persisted throughout 1787, its virulence spiking again in October.[22]

Coincident human and animal plagues continued their unrelenting rampage during the final years of the eighteenth century.[23] The troubles of 1787 only grew worse in 1788.[24] And in the spring of 1791, an extremely powerful plague outbreak followed directly on the heels of an epizootic that had wiped out large numbers of livestock the previous year.[25] The 1791 plague was said to have killed more than three hundred thousand men, women, and children, and its destruction of rural human and animal capital plunged Egypt into chaos (*perişanlık*).[26] By all accounts, that year's plague was perhaps the worst of the last two decades of the century.[27] It affected not just Egypt but also many other parts of the eastern Mediterranean, including Istanbul, the Morea, Venice, Libyan Tripoli, Izmir, and other cities on the Anatolian coast.[28] Deadly diseases of both humans and animals continued in Egypt into 1792 along with severe drought, creating a situation in which humans, cattle, rats, and other animals were once again competing for dwindling food supplies.[29] With the flood many months away, and with no significant rainfall, fields were dry and very difficult to farm.[30] When rural cultivators did try to plow their fields, they found only worms and rats. These circumstances benefited at least one group of animals: rat populations actually seem to have increased during that year, thanks to the prevalence of worms and small weeds. There were so many rats, in fact, that they moved from eating worms to consuming some of the few fruits that were growing in the branches of trees, thereby robbing humans of even more potential food. Indeed, "Whatever crops were saved from the worms the rats ate."[31] Humans could not even find straw and began eating weeds. All of this obviously affected cattle as well, since they did not have sufficient foodstuffs or herbage for themselves.

These instances of drought, disease, and food shortages for humans, animals, insects, and rodents created a particularly dire situation of competition for scarce resources. With another low Nile flood in the fall of 1792, scant amounts of grain were produced, reserves were held back from the market, shortages ensued, and prices skyrocketed.[32] The paltry quantities of cereals and clover that were grown were quickly devoured by worms. Many of the poor scavenged for dry grass and other herbage to pass off as straw for animal fodder. They either sold these bundles in Cairo for great sums or were robbed of them by soldiers and other troublemakers. In a sign of just how critical crop shortages had become, some vendors felt it necessary to sell their fodder from behind locked doors to protect themselves against attack and theft. Both humans and animals had nothing to eat and were literally starving to death. Animal carcasses littered the streets of Cairo, to the point that "one could hardly put one's foot down without stepping on creatures lying dead in the alleyways." Starved and hungry humans would immediately pounce on a fallen animal carcass, "even if it stank," wrote al-Jabartī, attempting to salvage some amount of consumable meat from it. Things were so critical that people "would have eaten [human] babies."[33]

Fodder and food shortages would persist throughout the last decades of the eighteenth century and the early years of the nineteenth. Adding to the problem, between April 1799 and February 1800 there was another severe plague epidemic.[34] Things got so bad in March 1800 that cattle were dying of starvation because no straw, beans, barley, or dried clover could be found.[35] Some people tried to sell their livestock off before they completely wasted away, but no one was willing to buy such emaciated animals, even at rock-bottom prices.[36] On the rare occasions when fodder was available, its scarcity and the high demand for it made it prohibitively expensive for most.[37] Warfare in this period was also often a cause of disruptions to food supplies for both humans and animals. In late March 1800, during the French occupation, residents of Cairo faced especially trying circumstances as fighting, plundering, looting, and all forms of mayhem gripped the city.[38] Low supplies of food were already being monopolized by Ottoman and French soldiers, contributing to the mass starvation of the human poor and animals alike.[39] And in 1801 and 1802, the central Egyptian city of Jirja was again hit by both a human plague epidemic and a cattle epizootic that claimed many human and animal lives and led to much rural distress.[40]

In addition to these human plagues, epizootics, and the many droughts and famines that hit Egypt between 1780 and 1810, the first few years of the nineteenth century were punctuated by numerous episodes of historically significant weather and wide fluctuations in temperature, all of which accentuated the already stark reality of massive numbers of animal and human deaths. In November 1804, for example, ominous storm clouds gathered over the city of Bilbays in the subprovince of al-Sharqiyya just to the northeast of Cairo.[41] An enormous thunderstorm ensued, and it was reported that lightning strikes killed about twenty people and numerous cows and sheep.[42] Four years later, in November 1808, hailstones "the size of hen eggs" killed numerous animals and caused widespread property damage in al-Maḥalla al-Kubrā in the subprovince of al-Gharbiyya. More productively from the perspective of the area's farmers, the hail also killed the worms that were feeding on the early crops in their fields. Similar hailstones, again described as being the size of hens' eggs, killed livestock in a storm in February 1809. The winter of 1813 saw especially cold temperatures in the delta, with freezing rain and even some snowfall.[43] Humans, cattle, crops, and fish all died because of these extreme conditions. Of course, these more spectacular weather events were interspersed with the regular destruction caused by the yearly flood, which consistently carried away both animals and crops, and often humans as well.[44]

Thus, while the three decades between 1780 and 1810 saw Napoleon invade Egypt and Mehmet ʿAli take power in the Nile valley, this period also witnessed a much more fundamental and historically consequential massive reduction in the number of domesticated animals in Egypt. Epizootics, extreme weather conditions, and fluctuations in flood and cultivation levels killed off Egyptian animals in huge numbers at the end of the eighteenth century, and it likely took several decades for these populations to recover to something even close to their precrisis numbers, if they ever recovered at all.[45] These population declines forced large landholders in the countryside to seek out other forms of caloric energy to replace animals. As a result, the composition and nature of the rural economy was forever changed.

Missing Animals

The social and economic consequences of a reduction in the number of laboring domestic animals in Egypt were immediately and acutely felt by Egypt's rural population. After the terrible epizootics of the spring of 1787, for example, peasants found themselves lacking the energy they needed to undertake their normal agricultural tasks.[46] Without cattle, farmers had to look for alternative means to thresh grain and turn waterwheels.[47] Many attempted to push waterwheels themselves, but given that these machines were built to be moved by the powerful muscles of water buffalo and other beasts of burden, humans simply did not

12. Infrastructural mechanics of animal labor. Commission des sciences et arts d'Egypte, *État moderne*, vol. 2, pt. 2 of *Description de l'Égypte*, Arts et métiers, plate 3.

possess the strength required to turn them effectively. Much of the infrastructure of the countryside was built for animals, not humans. Faced with this reality, some rural cultivators attempted to acquire donkeys, camels, and horses to replace their dead animals on waterwheels. But given the vast reduction in the number of animals in the countryside, the costs of purchasing and feeding these quadrupeds had increased dramatically, making it prohibitively expensive for most people to buy them.[48]

By the end of the eighteenth century, decades of diminishing animal populations had clearly created a very different rural landscape. It was now much more difficult and far more expensive for individual peasants to acquire beasts of burden to use in agriculture, on irrigation works, and for the various other tasks for which they had come to rely on animals. Wealthy rural elites who secured animals through purchase, force, or theft were thus at a marked advantage, with the ability to amass large landholdings and expand existing ones, thereby increasing the gap between themselves and Egypt's majority population of subsistence farmers. During the plague of 1785, for example, many amirs and other wealthy individuals took the opportunity of the mayhem caused by massive numbers of human dead to seize large numbers of cattle from both the deceased and the living.[49] Rich individuals who could acquire animals to work in fields, turn waterwheels, and transport goods to market were therefore able to take control of the most important energy source of the period. As a consequence, this period and the decades after it were characterized by the centralized accumulation of agricultural resources, foremost among them land, by rural elites who had been positioning themselves for just such an opportunity throughout the final decades of the eighteenth century.

Egypt's human and animal populations had of course been wiped out before by disease, environmental destruction, and war, but what made the late eighteenth-century demographic shifts both different from previous instances of population decline and fundamentally constitutive of later periods were the several preceding decades of attempts by provincial elites in Egypt and throughout the empire to pull away from the Ottoman imperial administration.[50] In the 1760s and 1770s, for example, Ottoman provincial governors, most notably ʿAlī Bey al-Kabīr and Muḥammad Bey Abū Dhahab, began forcibly seizing vast tracts of land in Egypt to distribute to their allies and to patronize potential economic and political partners.[51] In an even clearer affront to Ottoman imperial rule, and in what would prove to be a harbinger of later attempts by Mehmet ʿAli, these provincial elites also launched several, ultimately unsuccessful, invasions of Ottoman territory in Greater Syria in the 1770s.[52] Another example of an attempt to carve out some amount of autonomy from the empire was the creation of several large family estates by Mamluk grandees and other notables in Egypt at this time.[53] These economic arrangements built along familial lines included large landholdings, endowments of pious foundations in Egypt and elsewhere, and strategic ties with political powers throughout the province.[54]

Groups of local power brokers were thus already seeking to assemble or expand large landed estates by the time animal and human populations were reduced at the end of the eighteenth century. The population declines presented these power brokers with the opportunity they had been waiting for to take control of more land. It was this coupling of attempts to secure political and economic autonomy from the Ottoman Empire with demographic shifts in human and animal populations that made possible an irreversible phase change in the human-animal relationship in Egypt.[55] Once this transition had been achieved, Egypt was set on a new course, and there was no going back—even after animal populations had partially rebounded—to earlier social and economic formations in which intense human-animal interactions had dominated.[56]

The process of concentrating land in the hands of a wealthy few was helped along by the fact that reduced levels of animal ownership in the countryside created a class of animal-less, and hence eventually landless, peasants in need of work and income. No longer able to plow fields or move waterwheels, many of these peasants either moved to towns or cities (hence this period's urban population increases) or sought work on the large centralizing estates developing around them.[57] They thus represented a newly created labor pool from which large landholders could draw cheap agricultural workers. Those few emerging elites who could take on the initial high cost of acquiring some of the smaller number of animals in the countryside therefore became the large landholders who would eventually come to employ many peasants on their estates—the very peasants who had lost their animals and could not replace them.[58] Animal deaths were thus a crucial factor in a shifting agricultural order at the end of the eighteenth century that resulted in the eventual primacy of land as the foundation of wealth and prestige in Ottoman Egypt. It was this realignment of rural capital to replace animals—"the basis of the economy," to repeat al-Jabartī's phrase—that was the most significant outcome of the changing energy regime and political economy of animals at the turn of the nineteenth century.[59]

This accumulation of land by rural Egyptian elites would prove central to the subsequent history of nineteenth-century Egypt, with the Egyptian state itself eventually emerging as the largest of these landed estates.[60] As these estates increased in size, productivity, and output efficiency, the need to irrigate and cultivate land grew accordingly. Market-driven demands fueling land accumulation therefore also resulted in the further centralization of both human and nonhuman labor and other vital resources.[61] In turn, this concentration and attempted monopolization of the caloric power vested in the bodies of humans and animals led to both the increasing use of corvée on large estates at the end of the eighteenth century and a decrease in the number of animals used outside of these estates. Because of corvée, a drop in the overall productivity of the countryside, plague, and emigration to cities and towns, there were fewer independent farmers on the land. Thus, in addition to the collection of animals on large estates—and hence their removal from the more general rural economy—reductions in the human population meant there were fewer individual peasants seeking animals of their own to work alongside them on their small-scale farms. At the end of the eighteenth century, there was therefore less of an overall economic imperative placed on domesticated animals, the historic motors of the Egyptian rural economy.

As landed elites continued to expand their holdings—working to produce more food and other goods for an increasingly global market—they sought out more land, more resources, and more labor power. A key aspect of this process of emergent commercial agriculture was that estate holders undertook ever larger and more labor-intensive projects on their land. At the end of the eighteenth century, therefore, we begin to see both more complicated agricultural schemes and more projects to build irrigation and other public works.[62] As these rural manipulation efforts grew in complexity and size, and given the increased availability of low-cost peasant workers, the labor pools for such projects shifted away from the mix of humans and animals seen in the early modern period and became almost completely dominated by humans. Thus, even as animal populations recovered, nonhuman laborers were increasingly passed over for work on progressively more intricate projects that involved a great deal of capital. The fundamental social, economic, and energetic phase changes that had been achieved in the previous decades through political moves made possible by the ecological forces of disease and climate brought animal labor closer to the brink of obsolescence than it had ever been before in Egypt's agrarian economy. Of course, animals were still used when humans could not match their strength or stamina—to pull heavy materials or waste, to transport items to market, and so forth. But even these remaining economic functions would soon come to an end with the introduction of rail and coal in Egypt and with other advances in mechanization and machine technology in the first half of the nineteenth century.[63] The result of all this was the decreasing economic importance of animals, and therefore an overall reduction in the centrality of domesticated animals in Egypt.

Thus, Egypt's economic transformation from subsistence to commercial agriculture was fundamentally based on the replacement of animals by land as the largest and most basic source of rural wealth.[64] The history of land use and landownership in Ottoman Egypt has received a great deal of scholarly attention. The most recent extensive study of the issue in this period is Kenneth M. Cuno's *The Pasha's Peasants*. In this detailed account, Cuno makes clear that the intense economic stratification of land in Egypt began not under Mehmet ʿAli but earlier in the last half of the eighteenth century, as rural elites jockeyed to increase their power through the control of land and other resources.[65] Cuno's concerns are mostly with proving that market relations and commercial agriculture began to emerge before the middle of the nineteenth century. His evidence clearly shows this to be the case. What remains unexplored in Cuno's work, and in almost all of the literature on late eighteenth-century economic stratification (in Egypt and elsewhere), is how the decline in the most important traditional form of rural property (animals) led to the further concentration of an alternative form of wealth (land) in the hands of a few elite families in the countryside.[66]

At the end of the eighteenth century, the demise of the two powerful centralizing regimes of ʿAlī Bey al-Kabīr and Muḥammad Bey Abū Dhahab left a power vacuum in Egypt.[67] One result of this lack of centralized administration was that those individuals who had amassed sizable amounts of land under the rule of the two beys began to challenge one another for the control of resources.[68] It was this class of men who most benefited from and took advantage of the shrinking animal populations of the 1780s and 1790s. They were the ones who used these decades of disease and death to achieve a transition from animals to land as the primary source of wealth in rural Egypt. The archival record of estate inventories from the first two decades of the nineteenth century clearly shows land playing a far greater role, and animals a far smaller role, than ever before in Egyptian property transactions and disputes.[69]

While land was a constant source of wealth in rural Ottoman Egypt in the early modern period, ownership of land was not the basis for that wealth. Rather, until the end of the eighteenth century, the predominant rights Egyptian farmers had vis-à-vis land were usufruct rights to its products. And while at times these rights functioned as de facto ownership—making land open to sale, inheritance, gift, and rent—they could very easily be taken away.[70] Indeed, the seventeenth and eighteenth centuries are replete with examples of the seizure of land by both the Ottoman state and its functionaries.[71] Thus, what Egyptian regional elites were able to accomplish by the beginning of the nineteenth century, in part because of decreases in animal populations, was a move from usufruct to something more closely resembling landownership.

As a political marker of this economic transformation, this new form of landholding would come to be ensconced in Egyptian administrative practice in 1812, when Mehmet ʿAli seized all of Egypt's tax farms (*iltizāms*), making the legal ownership of land the primary means of staking claims to land. This legal innovation represented the final bureaucratic instantiation of the late eighteenth-century process of the concentration of land that emerged because of elite ambitions in the countryside made realizable by the deaths of massive numbers of animals. The concentration of landed wealth by a few individuals therefore contributed to the demise of the tax farming system, which had been the dominant form of land tenure in Egypt throughout the Ottoman period.[72] What is more, the very same large landholders who had emerged in the 1780s and 1790s as the wealthy rural elite were the ones who received most of the land that was seized and then redistributed by Mehmet ʿAli; they clearly had the necessary political connections and were the only ones deemed capable of cultivating land successfully. Thus, large estates grew larger, and more and more peasants were made landless and hence found themselves without their traditional modes of livelihood and in need of a source of income.

Given the vastly reduced number of animals available in rural Egypt between 1780 and 1810, only a few large estate holders could afford the inflated sums needed to acquire the animals necessary to overcome the minimum energy requirement for agricultural work.[73] Meeting this initial high laboring cost, however, ensured that these men would become the elite in Ottoman Egypt's changed rural economic and social landscape. In this recently remade world, newly animal-less and landless peasants were in abundance, and hence they provided a ready labor pool from which large landholders could draw the energy supplies they needed. Over the course of this period, together with the growing economic primacy of land, we therefore see a steady shift away from animals to humans as the primary energy suppliers for labor in the countryside. Animals were in short supply and expensive; human laborers were abundant and cheap.

New Beasts of Burden

With human workers now costing less than animals, corvée emerged as the most important kind of work on the expanding estates of the early nineteenth century. Not only were animals becoming far too expensive, but they were increasingly seen as too unwieldy and too unpredictable for the regular and reproducible work demanded by the rapidly bureaucratizing state in this period.[74] Indeed, as we saw in chapter 4, by the first two decades of the nineteenth century, a fundamental transformation had occurred in the character of labor in rural Egypt. Whereas humans and animals had previously worked together on smaller repair projects that served their own communities, now massive numbers of forced human workers became the preferred means of infrastructural construction and manipulation. This shift meant first that animals were gradually being stripped of their central economic role in the countryside and second that humans were becoming the most dominant component of Egypt's new energy regime. Whereas animal prices had shot up drastically immediately after the population declines of the last decades of the eighteenth century, animals became less valued in the new economy as the infrastructure of rural labor became increasingly centered on human workers.

Corvée had certainly been utilized in Ottoman Egypt before the early nineteenth century, but not to the same degree or with the same ferocity. In the seventeenth and eighteenth centuries, corvée was most commonly organized at the very local level of the tax farm.[75] It was mostly small-scale and temporary, undertaken and overseen by those long resident and immediately recognized within village communities, in accordance with established precedents, and without the full weight of a state bureaucracy behind it.[76] And even then, as we have seen, corvée in the early modern period was usually undertaken only on the small minority of lands over which tax farmers exercised direct usufruct control (about 10 percent of Egypt's total arable land).[77] Peasants were forced to clear mud from canals or to bring in crops—relatively easy tasks when compared to the massive infrastructural projects that would come later. Many peasants, however, not surprisingly, found even this local form of forced labor to be objectionable and attempted to escape from it.[78]

As discussed in chapter 4, the term most often used for this sort of corvée in the early modern period was *al-ʿauna*, deriving from an Arabic root related to assistance or help. In contrast, the term most often given to corvée in the nineteenth century was *al-sukhra*, from the Arabic root for words related to derision, subservience, servitude, and exploitation.[79] This semantic shift in the description of human labor is a telling marker of the difference between these two forms of corvée. While coercion is clearly integral to the nature of the practice itself, it is important to understand that in the earlier period, human and animal laborers worked directly to improve their local communities; in the later period, by contrast, peasants were much more solidly controlled and made to serve under the forceful power of a strong and faceless state. Indeed, between the sixteenth and the nineteenth centuries, as a general phenomenon, small-scale peasant labor moved from being under the local supervision and utilization of village tax farmers to being controlled by the centralized, more distant, and more abstract bureaucracy of Mehmet ʿAli.[80] This shift away from the local to the centrally bureaucratic is embodied in the response of a group of peasants in May 1814—two years after the abolition of the tax farming system—to their local tax farmer's demands of labor from them: "Your days are over, and we have become the pasha's [Mehmet ʿAli's] peasants."[81]

Indeed they had. And the pasha put his peasants to work. Throughout the first few decades of the nineteenth century, we clearly see a newly emergent and decidedly imbalanced preference for human over animal labor. Some of the best examples of this shift are the numerous repair, expansion, and building projects undertaken on Egypt's irrigation network in this period. At the beginning of Mehmet ʿAli's rule in 1805, the total length of Egypt's canals was approximately 514 miles. New construction efforts initiated by Mehmet ʿAli's administration more than doubled this total length, to roughly 1,200 miles by the end of his reign in 1847.[82] The new waterways led to an increase of about 18 percent in the amount of cultivatable land in Egypt between 1813 and the 1840s.[83] Keeping these canals clean was an enormous task. In the delta alone, approximately 20,730,118 cubic meters of silt had to be dredged from canals every year.[84] These projects of canal construction, expansion, and maintenance began very early in Mehmet ʿAli's reign and were soon occurring all over Egypt.[85] And it was humans, not animals, who dug these canals, kept them clean, hauled away excess dirt, and lifted the heavy tools and baskets needed to carry out this work.

As the intense and widespread canal work of the first quarter of the nineteenth century shows, corvée was not simply a temporary measure to replace those animals that had been removed from the labor pool between 1780 and 1810. Indeed, even when animals became available again after these decades of population losses, humans remained the preferred form of rural labor. The economy had shifted fundamentally away from animal to human labor, and not even a recovery in the availability of domestic animals could push it back to its former character.

Moreover, because corvée was so different from the human-animal laboring nexus that had existed in Egypt for millennia before the nineteenth century, and because it was used so extensively throughout Egypt, it had massive effects on the social and economic structure of rural people's everyday lives.[86] As an example of these transformations, consider the movement of peasant laborers entailed by corvée. The physical relocation of human caloric energy to fuel projects that did not benefit these rural people's own communities or families meant that they were further alienated from their own lands and that it was ever more difficult to stop the march toward the concentration of land and energy resources in just a few hands.

The details of this new system of human labor are telling. Peasants in the early nineteenth century contributed about sixty days of corvée labor every year, excluding the amount of time it took to travel to and from the area where they were working.[87] Workers were collected and delivered to a worksite by the heads of their villages and were assigned tasks as a village unit so that peasants from the same community worked side by side on construction projects.[88] Over the course of one year, depending on that year's projects, an average of 400,000 men could be forcibly moved to work.[89] To put this number in perspective, consider again that the population of Cairo in 1821 was 218,560, and that that of Egypt as a whole was around 4.5 million in 1800 and 5 million in 1830.[90] The demographic effects of this enormous annual movement of people—about 8 or 9 percent of Egypt's total population—were in actuality somewhat greater, since workers occasionally brought their families with them to construction sites. Workers were often responsible for bringing their own food and tools. In the case of very large repair jobs, however, the government often provided food, and sometimes even a daily wage.[91]

Specific examples of corvées from all regions of Egypt in the early nineteenth century further illustrate the magnitude and widespread nature of this system of forced labor. The maintenance of the Shibīn Canal, which ran past al-Maḥalla al-Kubrā in the central delta, demanded a yearly corvée of 50,000 to 60,000 men from villages throughout the delta.[92] The order for a corvée between 1817 and 1821 to rebuild the Ra's al-Wādī Canal, which flowed east from al-Zaqāzīq to Lake Timsāḥ, demanded the labor of 80,000 men. Within just eight days, this enormous number of peasants had been brought to the worksite from villages throughout the subprovince of al-Sharqiyya in the northeast of the delta.[93] In 1829, 32,300 men were brought to dig a new canal in the northwestern subprovince of al-Gharbiyya.[94] And in 1838, Mehmet ʿAli ordered 20,000 of his own troops to work on the Zaʿfarānī Canal, which fed water to Cairo.[95]

An order sent by Mehmet ʿAli to Upper Egypt in December 1835 for 24,000 men to clean and dredge canals and reinforce embankments makes clear the state's violently paternalistic treatment of the Egyptian peasantry used in corvée.[96] "If you say it upsets the fellahin [peasants] when there is no need," he wrote to his functionaries, "then I say the boy does not willingly go to school but is forced by his parents until he grows older and knows the value of learning, so driving all the men to dykes and canals is difficult for them but is necessary."[97] In this period, soldiers were instructed to forcibly (*qahran wa jabran*) move unwilling peasants to canal sites. Mehmet ʿAli continued, "If land is *sharaqi* [unwatered] because Jisr [the canal of] Banu Khalid is not well shored up there is no punishment but death." He added, "If we see one *qirat* of land unwatered we will bury you in it."[98]

Perhaps the most famous and deadly instance of corvée in the early nineteenth century was the reconstruction in 1817–20 of the Maḥmūdiyya Canal between Alexandria and the Rosetta branch of the Nile.[99] As discussed in chapter 4, an estimated 315,000 to 360,000 individual workers were forcibly moved for this project.[100] So many men were put to work on the canal that it caused labor shortages elsewhere in Egypt. In August 1818, for example, the *muḥtasib* (market overseer) of Cairo, Muṣṭafā Aghā, ordered certain city lanes, alleys, and dead-end streets cleaned of accumulated dirt and debris that was making them impassable.[101] The men who would normally have undertaken this task were away at the Maḥmūdiyya construction site, so shop owners and other urban residents were forced to do the work themselves.[102] This example illustrates one of the fragilities of a system based almost entirely on only one form of labor. Devoting hundreds of thousands of human hands to the Maḥmūdiyya Canal clearly meant there would be shortages elsewhere.

By all accounts, work conditions on the canal were very harsh. Of those more than 300,000 workers who were dispatched to dig and reinforce the canal, 100,000—nearly a third of the labor force—died.[103] Indeed, Mehmet ʿAli's earlier promise to bury unproductive peasant laborers in the dirt seems to have been fulfilled. According to al-Jabartī's account of forced labor on the canal in August 1819, "Dirt from the excavation was dumped on every peasant who fell, even if he were still alive."[104]

All of these instances of the corvée of human life in the first third of the nineteenth century were the end result of a transition in the energy regime of Ottoman Egypt from animal to human labor. They illustrate the massively transformative social power of the changing political economy of animals in the decades around the turn of the century. An enormous reduction in the number of animals in the countryside led to the emergence of land as the primary source of wealth in rural Egypt.

As small-scale rural cultivators lost their animals and were unable to acquire other forms of property, they found themselves increasingly left out of an economy now shifting away from animals to land. Large property owners too were severely impacted by animal deaths at the end of the eighteenth century, but unlike poorer farmers, they were able to replace the animals they had lost with some of the growing number of cheap peasant laborers. Thus, as the energy regime shifted on balance more toward humans, animals lost the centrality they had enjoyed in rural Egyptian society for millennia, and humans emerged as the primary form of labor in the new rural economy.

Conclusion

Animals played a vital role in the hugely transformative economic, social, and energetic changes that Egypt experienced between 1780 and 1810. These nonhumans' histories help explain the formation of large landed estates, shifts in the character and makeup of rural labor, and the administrative restructuring of Ottoman Egypt at the beginning of the nineteenth century. It was precisely because domesticated animals were so enmeshed in all facets of Egypt's rural economy, rural society, and rural energy regime that the massive reduction in their numbers at the end of the eighteenth century led to changes in all aspects of rural life. The waning of animals' centrality in the rural economy was thus one of the most crucial factors in Egypt's transition from an early modern agrarian subsistence economy to one based on commercial agriculture, large landholdings, and intensive human labor. We cannot fully understand this period without an understanding of the changing political economy of animals during these years—years that saw the establishment of Mehmet ʿAli's familial dynasty as the rulers of Egypt for nearly a century and a half and witnessed the unleashing of the massive force and power of the nineteenth-century Egyptian bureaucracy.

What happened in rural Ottoman Egypt as the eighteenth century gave way to the nineteenth also occurred in other early modern agrarian economies between the sixteenth and the nineteenth centuries. From Qing China to southern England, as land became concentrated in fewer and fewer hands, rural economies moved from subsistence to commercial agriculture.[105] One of the factors that made this global transition possible was a reconstitution of the energy regime of rural labor. Animal power became increasingly marginal as massive numbers of human laborers, and ultimately nonhuman machines, took over as the central motors of rural economies.

Imagine what will happen when our world suddenly finds itself without the fossil fuels that make so much of our lives possible today. Cities will take a different shape; the way we communicate and move will change; how we eat will be radically altered; we will have to seek out other forms of energy to sustain the economy. What happened in Egypt at the end of the eighteenth century was just such a process—a wholesale reconfiguration of the rural world precipitated by the loss of a historic source of energy.[106] The social, cultural, political, and special consequences of this shifting animal economy and energy regime were clearly enormous and deserve further study, both in Egypt and elsewhere. Our biases toward humans as the most important historical agents perhaps predispose us to miss some of these animal histories. We ignore them at our peril.

Part Four: Elemental

8

Food and Wood

The Ottoman Empire captured Egypt and much of the Arab Middle East from the Mamluks in 1517.[1] With this conquest came many spoils: a near doubling of the empire's territory; the inclusion of Islam's holiest sites into Ottoman domains; access to the Red Sea and Indian Ocean; strategic control of most of the eastern Mediterranean; sovereignty over some of the largest cities in the Middle East (Cairo, Aleppo, Jerusalem); and a massive influx of money, people, and resources from these newly conquered lands. All this notwithstanding, the conquest also presented the Ottomans with many logistical and administrative challenges. The most pressing from an imperial perspective was how to rule and collect taxes over such a large and widespread area, one with many already-ensconced bureaucratic and legal traditions.[2] This was a problem that the Ottomans—like all imperial states—regularly faced after the conquest and attempted absorption of new territories.

These rather conventional and routine administrative challenges aside, the Ottomans' territorial expansion of the years around 1517 also brought them face to face with many novel challenges they had never before encountered. One of the most important of these was a logistical problem involving the movement of two strategic goods that would prove crucial in shaping much of the empire's rule after 1517—wood and grain.[3] Concerns surrounding these goods, especially the wood, were largely a by-product of the Ottomans' entrance into the Red Sea and Indian Ocean worlds.[4] To benefit from—never mind to attempt to control—the lucrative commerce of the Red Sea, to challenge Portuguese power in the Indian Ocean, and to provision the yearly pilgrimage to Mecca and Medina, the Ottomans needed ships in the Red Sea. To build these ships, the Ottomans needed wood.[5] Herein lies the main problem. In and around the major Ottoman Red Sea port of Suez, there was a vital lack of usable wood supplies. For all its agricultural wealth and rich soils, Egypt simply did not have adequate domestic wood supplies to feed the growing Ottoman need for shipbuilding timbers in Suez.[6] This wood thus had to be brought from elsewhere.[7]

This chapter tells the story of the enormous logistical and bureaucratic effort and organization the Ottomans undertook to overcome the problem of wood supply in Ottoman Suez by bringing lumber from Anatolia. To trace this story, we must follow the wood. The wood tracked in this chapter is a group of timbers that were first harvested in the forests of southwestern Anatolia and ended up being bent to shape the hulls of three ships in Suez in 1725. These vessels would eventually sail from Egypt to the Hijaz (again, the region on the western coast of what is today Saudi Arabia housing the cities of Mecca and Medina) carrying massive quantities of grain to feed people across the Red Sea.[8] As we saw in chapter 3, Egypt and the Hijaz were intimately connected through Egypt's export of food across the sea. This relationship between the two regions is important for reminding us of the connections of empire that bypassed Istanbul and because it brings to the fore Egypt's eastern links.

At its heart, the story of this chapter is a story of provisioning—of the lengths to which the Ottomans were forced to go to make possible the movement of grain from Egypt to people in the Hijaz. The path taken by this amount of wood is important for what it illuminates about the economic history of the Ottoman Empire. The domestic markets of Egypt, Anatolia, Istanbul, or Suez alone were unable either to meet the demands for certain raw materials or to support an enormously complex and expensive logistical project like moving parts of a forest across the Mediterranean to Egypt and then overland to Suez. This sort of work could only be done by a political and organizational entity like the imperial administration of the Ottoman Empire. The example of these ships' construction thus illuminates how the early modern Ottoman Empire occasionally intervened in economic affairs and market relations in different parts of the empire to effect a desired outcome such as the undertaking of a massive infrastructural or construction project.

Grain Needs Ships

Most of the current scholarship on Ottoman shipbuilding and timber provisioning in the Red Sea focuses on the sixteenth century.[9] This was the period when the Ottomans first expanded into the Red Sea and captured parts of Yemen, Bahrain, and other sites on the Arabian Peninsula.[10] This was also, and perhaps more important from the perspective of modern scholarship, the heyday of Ottoman-Portuguese rivalry in the Indian Ocean. Our story of wood supplying, however, comes from the first half of the eighteenth century, a full 150 years after the Ottomans supposedly lost interest in the Red Sea. As we will see, Ottoman stakes in the Red Sea remained quite high into the eighteenth century and focused mostly on commerce and provisioning between Egypt and the Hijaz.[11]

Despite Ottoman-Portuguese high-seas imperial rivalries in the sixteenth century, the most consistent, longer-lasting, and historically more significant reason the Ottomans brought wood to Suez to build ships in the early modern period was to feed people in the Hijaz and to support transport and commerce in the region. The Hijaz was of symbolic value to the Ottomans because custodianship of the holy cities allowed them to make universalistic claims of authority, leadership, and sovereignty in the Muslim world.[12] With this symbolic power and religious status also came responsibilities. The yearly Muslim pilgrimage to Mecca and Medina was surely the largest annual gathering of people anywhere in the early modern world. It was an enormous undertaking in terms of logistics, transportation, and provisioning, which had to function smoothly and safely if the Ottomans wanted to ensure respect and pliancy from the thousands of pilgrims who came to the Hijaz every year and who would then return to their homes in Hyderabad, Tehran, and Sofia with accounts of their experiences.

A crucial aspect of this maintenance of the yearly pilgrimage was providing pilgrims with adequate food supplies. Such provisioning was the historic duty of a pious and proper Muslim sovereign and was also in the practical interests of the Ottoman state so as to ensure the health and well-being of its visitors.[13] Egypt came to play a central role in this system of Ottoman provisioning.[14] Not only was it the largest grain-producing region of the entire Ottoman Empire, but it was also, quite conveniently, right across the slender Red Sea from the Hijaz. Thus, throughout the imperial record of the Ottoman state, we have copious materials evidencing imperial interests in maintaining food supplies from Egypt for the yearly pilgrimage.[15] Indeed, after Istanbul, the Hijaz was the most common destination for Egyptian grains in the Ottoman period.[16] Ottoman concerns over food production in Egypt often took the form of imperial orders to maintain and repair rural irrigation works, since water was obviously the key to food production in the province.[17] For example, as we saw in chapter 3, a series of orders sent to Egypt between 1709 and 1711 about the repair of a very important set of dams and dikes in the region of Fayyum made the point over and over again that grains grown in Egypt were to be sent to feed pilgrims in the Hijaz and that it was therefore imperative that these irrigation works function properly to grow the needed amounts of food.[18] Like other regions of Egypt, almost all of Fayyum's surplus grain supplies went to the Hijaz. Additionally, Fayyum also maintained its status as a major exporter of grain to the Hijaz because of the many pious endowments (*awqāf*) for the holy cities established in the region and throughout Upper Egypt more generally. The chief function of these Upper Egyptian endowments attached to the Hijaz was to provide grains for both the pilgrimage caravan and the people of Mecca and Medina.[19]

Food grown in Fayyum and in numerous other regions in Egypt would eventually make its way to Suez to await shipment to the Hijaz. While Fayyum had always sent the majority of its export grain to Suez over other ports, during the fifteenth century, just a few years before the Ottomans conquered Egypt, Suez became even more important than Egypt's southern ports (Qusayr and Aydab most prominently) as the main hub of export from the province to the Hijaz. This was chiefly due to shifts in agricultural cultivation in the period that saw the delta (which is closer to Suez than to the southern ports) emerge as Egypt's richest area of food production.[20] Whether grown in soils in Fayyum or elsewhere in rural Egypt, once food was in Suez, it could only, quite obviously, continue its journey across the sea by ship. And for much of the Ottoman period, ships were readily available. Quite often, though, there were none to be had in Suez, because of either disrepair, shipwreck, or needs elsewhere.[21] Such was the case in the spring of 1725, when the Ottomans had plenty of food in Suez to send to the Hijaz but no ships to get it there.

Ships Need Wood

A series of cases from the archival record of Ottoman Egypt brings to life the complicated procedures involved in delivering wood to Egypt from forests in southwestern Anatolia to build three *galettas* (*kalite*) in Suez.[22] Wood was a strategic asset for the empire.[23] This was partly a function of the fact that it was available only in a few specific regions within the empire's borders (on parts of the southern Anatolian coast, around sections of the Black Sea littoral, and in Greater Syria).[24] The Ottoman Empire therefore came to manage wood supplies and their distribution and movement very closely.[25] Trees in the Ottoman Empire thus came to be controlled by the logic of the state rather than the market. Wood—and for that matter food as well—entered into an imperial chain of demand, need, and availability in which the deficiencies of one region were met by the excesses of others. In the same way that the Hijaz relied on Egypt for food, Egypt relied on other parts of the Ottoman Empire for wood. The rather complicated procedures for bringing wood to Egypt were thus Ottoman attempts to project imperial sovereignty through the management of an essential resource needed to move excess amounts of caloric energy stored in grain to be consumed elsewhere by —in this case—pilgrims from all over the Muslim world. Integral to this resource management was the fact that within the Ottoman world, it was only the imperial state itself that could undertake a project of such scale, intricacy, complexity, and expense.

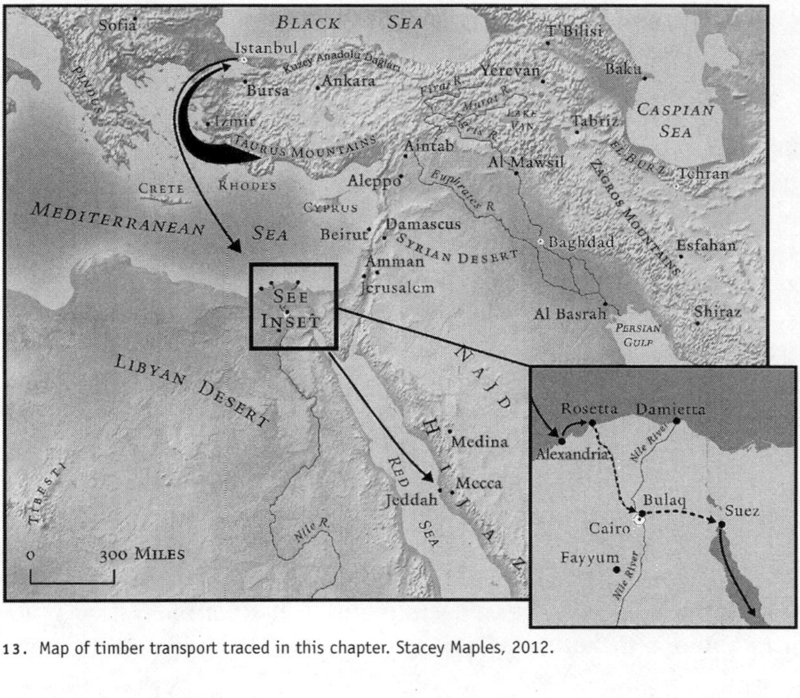

13. Map of timber transport traced in this chapter. Stacey Maples, 2012.

Wood in the Ottoman Empire was harvested in the forests of southwestern Anatolia and parts of the southern Black Sea coast every three to four years by peasants in those areas who were hired by the Ottoman state as temporary laborers. They worked for an entire season to cut trees and to transport them to imperial storage facilities in Istanbul.[26] The organization of this labor was overseen by the *kereste emîni* (timber superintendent), who was an official in the department of the Imperial Dockyards (*Tersâne-i Âmire*), the official body responsible for the collection of wood in the empire and the institution that would store the timber for later use and distribution.[27] The *kereste emîni* managed a veritable army of laborers (*amele*) in the work of cutting and moving trees.[28] Various military cadres (*yaya*, *müsellem*, *yörük*, *canbaz*) and specialized craftsmen (*neccar*, *teksinarcı*, *kalafatçı*) worked to turn trees into usable wood supplies.[29] Both southwestern Anatolia and the Black Sea coast were particularly good regions for timber harvest because of their extensive forest cover and proximity to coastlines.[30] Since these trees were primarily harvested for the construction of Ottoman naval ships, it made sense that the administration of the Ottoman Imperial Dockyards oversaw forest management in the empire.

Often these trees were moved to Istanbul on merchant ships rented by the state. In the case of southwestern Anatolia, for example, numerous orders were sent to the imperial governor of the island of Rhodes to organize the renting of these ships, as this region was under his administrative purview. Given the relatively high number of orders sent to this imperial governor in the late 1710s and early 1720s, it seems likely that most of the trees going to Istanbul's timber storage facilities in those years originated in southwestern Anatolia.[31] The important point about the empire's timber storage facilities is that they represented an attempt to monopolize the supply and control of wood as a strategic good. Other efforts at forest management included designating the vast majority of forests in the Ottoman Empire as having official (*miri*) status and promulgating numerous regulations forbidding the cutting of their trees and grazing, building, or hunting in imperial woodlands without the proper permissions.[32] By attempting to control the use of forests and by centralizing the distribution of the empire's trees in the Imperial Dockyards, an institution literally at the base of the palace, the sultan could control how his empire's wood was used and what projects it supported.[33]

In 1725, the palace directed that wood—again, wood mostly likely originating in southwestern Anatolia—be sent from the central Ottoman timber distribution facility in Istanbul to Alexandria on the Egyptian Mediterranean coast.[34] As in many instances of moving wood from parts of the Anatolian coast to Istanbul, merchant galleys (*tüccar sefineleri*) were rented to assist a group of imperial (*miri*) ships in moving this wood from the capital to Alexandria.[35] Alexandria was central to Ottoman interests in the early modern eastern Mediterranean.[36] It was used as an Ottoman naval base for various operations around the sea. Ships from Alexandria, for example, supported Ottoman military expeditions to Chios in 1566, Malta in 1575, and Crete in 1666 and 1715. The port was also crucial as a controlling hinge of trade between the Mediterranean and Red Sea and Indian Ocean.[37] An enormous amount of the goods coming from the Indian Ocean via ship eventually made their way to Alexandria overland from Suez to be put on ships sailing to points across the Mediterranean world (essentially the opposite direction of the wood in our case).[38] Because of Alexandria's military and economic importance, the Ottomans paid particularly close attention to the administration of this port city through a customs regime, legal and economic regulations, and a military presence.

The leg of the wood's journey from Istanbul to Alexandria was the longest (in terms of distance) of its entire itinerary from seed to ship hull.[39] It was also the most dangerous because of the Mediterranean's rough waters, threats of piracy, and various other possibilities for damage to the precious cargo. Piracy was a foremost concern of the Ottomans in the Mediterranean, and they undertook various measures over the years to try to combat it.[40] However difficult it might be for us to imagine, we must not underestimate the scale of moving such an amount of wood across the Mediterranean. Likely hundreds of very large logs that each required a dozen or so men to move were put on enormous ships and then sailed for hundreds of miles over the course of a fortnight or so, only to be taken off of those ships in another gigantic operation. There is indeed evidence of a floating crane in use in the port of Alexandria in the first half of the eighteenth century to aid in the loading and unloading of cargo and also to help in the repair of ships in port.[41] The merchant galleys that were rented to move this wood across the Mediterranean were obviously very large ships and had to be stable enough to deal with the treacheries of open sea. These ships and their captains were accustomed to this journey, as they were the ones who most often moved grains, textiles, finished goods, foodstuffs, and other products between the empire's most lucrative province and its capital. And like these other items, the wood in 1725 was recorded in the customs registers of Alexandria, a requirement of all cargo entering Egypt from the Mediterranean.[42]

Once this wood had made it safely to Alexandria and had been adequately registered by the state, it had to be moved to yet another set of ships. This transfer was necessitated by concerns of both geography and technology. Getting to Suez from Alexandria—there was of course no Suez Canal in 1725—required sailing down the Nile into the interior of Egypt.[43] In the early eighteenth century, however, there was no internal waterway connecting Alexandria to the Nile.[44] Thus, from Alexandria ships had to sail east along Egypt's Mediterranean coast to enter the Nile system either at Rosetta or Damietta, the two branches of the river that form the Nile delta. Unless prevented by rough waters, a storm, shipwreck, or some other impediment, almost all ships from Alexandria entered the Nile through the Rosetta branch because of its proximity to Egypt's second city. Although the enormous galleys that brought the massive load of wood from Istanbul to Egypt were extremely good at navigating the Mediterranean's rough seas, they were less well suited to sailing on smaller, narrower, and curvier bodies of water like the Nile. Thus, to overcome this navigational limitation, the wood in our case, like all cargo following this path, was transferred to a set of smaller, more compact, and more nimble ships known as *cerîm*.[45] As with each of the transfers in this wood's journey, this one required great care, patience, and effort to move the wood and to protect it against damage and theft. Furthermore, as with some of the galleys that crossed the Mediterranean, these smaller ships were also rented by the Ottoman administration from merchants in the area.[46]

14. Sailing the Rosetta branch of the Nile. Commission des sciences et arts d'Egypte, *État moderne*, vol. 1, pt. 2 of *Description de l'Égypte*, Basse Égypte, plate 79.

Now on the appropriate type of vessel, the wood sailed east, hugging the Egyptian coast, toward the mouth of the Rosetta branch of the Nile. Late summer and early fall were especially difficult times to enter the Rosetta mouth because of prevailing winds blowing out to sea and because of the force of the Nile's water near the end of the flood season pushing out to sea.[47] The *cerîm* ships in our case were, however, sailing in spring, so they had little trouble entering the mouth of the river's western branch. The wood's next stop was, perhaps not surprisingly, Cairo, specifically the area just to the north of the city known as Bulaq, Cairo's main economic port and commercial district.[48] Because of the unique economic status of wood as a commodity in the Ottoman Empire and because the timber in our case had already been earmarked by the state for a specific purpose, it did not enter into the business transactions that made Bulaq a hub of economic life in Egypt and the Mediterranean. The wood was noted but not traded. Instead, as before, it was transferred one last time to complete its journey to Suez.

This transfer was once again affected by Egypt's geography. For the entire journey until Cairo—from the forests to Istanbul, Istanbul to Alexandria, and Alexandria to Bulaq—the wood moved on water. Indeed, it was of the utmost importance to maximize the geographic distance traversed on water since this was clearly the easiest, cheapest, and most efficient means of moving such a heavy, unwieldy, and large quantity of wood. Between Bulaq and Suez, however, there was no waterway, only eighty miles of desert. Faced with no other option to get the wood to Suez, Ottoman authorities overseeing the project and the Egyptians they enlisted to help them arranged for the trickiest, most expensive, most complex, and most arduous part of the wood's journey. In Bulaq, they hired a convoy of camels to pull the wood through the desert.[49] Without the possibility of using ships to take advantage of water and wind power, camels were harnessed as the next best energy source affording the necessary power and stamina to move this load overland.[50]

The use of these camels, however, did not come cheap. The cost of renting the animals and of paying those who would load and unload the wood and lead the animals through the desert was the significant sum of 800 *niṣf faḍḍa* (or *para*)—450 *para* for the men and 350 *para* for the camels.[51] As further evidence of the cost of animal labor in transport, consider that in a list of forty-six expenses related to the pilgrimage of the year 1696 (the closest year to 1725 for which we have such figures), we find that animal labor represented nearly 10 percent of the total expended from the treasury of Egypt on the pilgrimage in that year—the impressive sum of 1 million *para*.[52] In the same way that historians have given much attention to the use of ships, navigation tools, and knowledge in the maritime commerce of the early modern period, so too must we recognize the importance of animals as means of transport, power, communication, commerce, and travel in the early modern world. As we have seen, the power, speed, and stamina provided by camels, water buffalo, donkeys, and other animals in the Ottoman Empire made possible commercial relations, imperial governance, and agricultural production. As the camels in this case show, animals—like wood—were vital commodities that allowed the state to accomplish and undertake tasks it could not do otherwise.[53] Historians estimate that camels in the Ottoman Empire could carry a quarter ton of weight for about fifteen miles a day, 20 percent more than horses and mules and over three times more than donkeys.[54] Accepting this estimate and giving some leeway to the enormous load of wood in our case, it likely took the camels and their handlers about a week to cross the desert from Cairo to Suez.

The wood had finally reached Suez. Now that it had moved from Istanbul across the Mediterranean to the Red Sea, the work of actually building the ships needed to move food from Egypt to the Hijaz—the ultimate goal of this project, lest we have forgotten—could finally begin.[55] The journey of this wood from southwestern Anatolia, to Istanbul, to Suez, through Alexandria, Rosetta, and Cairo was long, inefficient, and difficult for numerous reasons. Even if the construction of ships in Suez was an urgent matter, it would surely take at least several weeks as in this case (or perhaps longer) before the construction materials even arrived. And, of course, this was only the first step, since it would take anywhere from another six months to two years to complete the construction of the ships in port, depending on the size of the ships, the number of available laborers, unforeseen problems with the work, and other contingencies.

Another obvious problem with the wood's transport was the multiple transfers it required. Wood was packed from the royal dockyards onto ships in Istanbul, sailed to Alexandria, was then transferred to another set of ships, sailed to Cairo via Rosetta, was packed onto the backs of camels, dragged through the desert, and was only then finally unloaded in Suez to be used for its ultimate purpose of ship construction. All of these transfers, especially the overland leg between Bulaq and Suez, exposed the lumber to damage and theft. The wood could have been dropped, lost, chipped, stolen, or damaged in any number of ways. And, of course, the wood's transport involved great financial expense: the price of ships and sailors, customs duties, camels and camel drivers, food for sailors, and so on. Despite these difficulties and costs, however, the Ottoman imperial bureaucracy had few alternatives.[56] If the goal was the construction of ships in Suez, then wood was needed, and since Egypt had no wood, it had to be brought from elsewhere. These were the realities and costs of the absence of forests in Egypt.

Economy Needs State

The story of the construction of these three ships in Suez in 1725 reveals an important dimension of how certain kinds of economic resources were utilized in the Ottoman Empire. No individual, collective organization, or corporate entity in the empire could have built these ships; only the state was capable of undertaking such a project. When I use the word *state* in this context, I do not mean merely the sultan's imperial divan. Rather, I understand the state to be the entire bureaucratic apparatus of the imperial administration, which included merchants who rented their ships to the empire, temporary forest workers hired by the Imperial Dockyards, camels used to pull wood through the desert, and shipbuilders in Suez. A project of this case's scale and cost, employing so many different kinds of people in such disparate parts of the empire, could only be coordinated through imperial mechanisms—a centralized wood depot, a court system that facilitated communication across the Mediterranean, a bureaucracy with access to large amounts of cash and capital, and a network of merchants and craftsmen who knew that their goods and labor would be compensated.

By understanding what the empire could do that nothing else could, we gain a clearer understanding of what an often-vague notion of "the empire" actually was at the most basic level on the ground in a place like Egypt. The Ottoman Empire was an economic mechanism for coordinating resource management that provided the materials, finances, and organization to, among other things, build ships in Suez. Again, without this political and economic administration, no other entity in Egypt, or anywhere else in the empire, could have effected the construction of these ships. This fact stands in the face of several important assumptions about the empire in this period and thus helps us understand something of Ottoman imperial governance in the first half of the eighteenth century.

First, it has long been assumed that after the seventeenth century, the Ottoman imperial administration largely gave up any attempts to control economic affairs or to directly intervene in the economy.[57] Price ceilings (*narh*) are the classic example cited in this regard. The empire did away with them at the end of the sixteenth century, and they reemerged only much later in the last two decades of the eighteenth century. The present case of the three ships, however, shows the intense involvement of the state—again, in its most capacious meaning—in market relations, resource management, and transportation networks. This shipbuilding in 1725 was, in other words, an instance of Ottoman economic interventionism. The physical resources of the market, its transportation capacities, and the labor available to it could not provide the wood and muscle needed to build the three ships. Thus, the state had to intervene to provide and then manage these resources.[58]

This case also shows us that the Ottoman state in the first half of the eighteenth century was still able to influence, impact, and administer areas at some distance from the capital. Historians of Egypt and of elsewhere in the empire have traditionally assumed that Istanbul largely pulled back from the provinces in the eighteenth century and was not able to provision its army and other vital organs of the state.[59] In the present case, by contrast, the state clearly comes through as an actor of enormous economic and organizational wherewithal. Important, as well, much of the administration of resources that took place in this case was at the very edges of the empire—not in the empire's largest cities but, indeed, at some of the farthest possible points from imperial nodes of power like Istanbul and Cairo. According to much of the existing literature, southwestern Anatolia and Suez would have been the kinds of places where one might expect the imperial presence in the eighteenth century to have been rather weak, but, again, just the opposite seems to have been the case. Commodity histories thus open up new possibilities for understanding the geography and function of Ottoman imperial governance. As the cases of wood and grain show, commodity acquisition, transport, and utilization created multiple kinds of linkages and connections that are otherwise difficult to trace. Following a material object, in other words, usefully allows us to push beyond established ways of thinking about the inner workings of early modern polities like the Ottoman Empire.

Finally, the building of these three ships is a wonderful example of the involvement of local actors in the day-to-day governance and operation of the empire. At every point in the journeys of both the grain and the wood to Suez, and then eventually to the Hijaz, it was peasants, small-scale actors, local merchants, sailors, and laborers who directed the state how best to successfully meet its goals. Lumberjacks in the mountains of southwestern Anatolia knew the forests in their area better than anyone else and provided the expertise, knowledge, and experience the state needed to effectively manage and harvest wood. Likewise, sailors on the Mediterranean, who knew the best sea lanes and how to control their ships, brought the wood to Egypt. Camel drivers and their animals moved the wood across the desert, and established local shipbuilders in Suez were the ones who finally put the boats together. It was the unity and coordination of the Ottoman state that brought all these disparate actors together for the goal of building the ships. At no point in this process, however, were Ottoman imperial bureaucrats the ones actually carrying the wood, hammering the nails, and pulling the camels. Indeed, clearly the imperial administration could not have carried out this project of shipbuilding without the participation of local actors in the workings of the state.

This fact was in large part a function of the necessities and difficulties of managing and utilizing local environments and resources—forests, soils, the heat of the desert, and the flow of the river. Each of these particular environments and ecological forces demanded specific local knowledge, experience, and skill, and the Ottomans relied on locals to help them to operate in and use these natural environments, since this was the most expedient, sustainable, and efficient means of harnessing environmental resources and expertise. Thus, it was the collective knowledge and experience of local actors all across the empire that allowed the imperial state to function. At the same time, though, it was the connective administrative and economic links of the empire that tied the labor of lumberjacks in southwestern Anatolia to that of camel drivers in eastern Egypt. Neither the know-how and experience of local actors nor the administrative acumen and integrative powers of the Ottoman administration alone, without the participation of the other, could have moved wood to Suez or, more generally, managed natural resources in the empire. This imperial system was thus a commodity utilization chain of local labor and knowledge, linked together across the Mediterranean, from one continent to another, by the Ottoman administration.

State Needs Nature

Wood was unlike any other commodity in the empire. Its scarcity and strategic value dictated much of the way the state came to manage it and many of the economic relationships forged around it. The only usable supplies of forest in the empire were in areas of southern Anatolia, around the Black Sea coast, and in Greater Syria, and the state, as already mentioned, put in place a very sophisticated forestry management system to maintain these woodlands. As with almost all natural resources, much of wood's value in the Ottoman Empire came not from human labor but from the solar energy of the sun, nutrients in the soil, and copious amounts of water. Thus, in contrast to the traditional labor theory of value, which posits that the value of most goods is a reflection of the work humans do to *produce* objects for the market, much (or perhaps most) of the utility and value of the Ottoman Empire's wood came from nonhuman nature. Clearly, human labor and energy went into the cutting, transport, and readying of wood for use in Suez over five hundred miles away from its original growth site, but no amount of human labor, knowledge, or effort in Anatolia could *produce* the strong, durable, and desirable lumber of the trees needed in Egypt. As William Cronon writes in reference to the American West, "The fertility of the prairie soils and the abundance of the northern forests had far less to do with human labor than with autonomous ecological processes that people exploited on behalf of the human realm—a realm less of *production* than of *consumption*."[60] One could make similar arguments about other natural commodities—grains or animals, for example—but the case of wood nevertheless still stands out. Unlike, say, a field of wheat, a region of old-growth forests would take decades, if not centuries, to reproduce itself once harvested. This is a time scale unsuited to most human endeavors.

Thus, because the use of wood was essentially a process of the relatively irreversible consumption of natural resources, the Ottoman Empire was in many ways forced to centrally manage its forest supplies if it wanted to maintain them in any kind of long-term fashion. In other words, were market forces and personal interest allowed to have free rein in how forests were used, trees would be consumed very quickly, to the detriment of both populations around the empire and future populations alike. This is exactly what happened in the Great Lakes region of North America studied by Cronon. Other evidence from Ottoman Egypt further suggests the uniqueness of wood as a commodity. For example, an examination of estate inventories from the period shows that wood products were some of the most expensive items individuals owned in rural Egypt.[61] This high value of wood (and of its products) was again a reflection of its scarcity in Egypt.

Conclusion

Egyptian shipbuilders, camel drivers, and peasant cultivators, along with religious pilgrims from across the early modern Muslim world—all of whom had never seen Anatolia nor likely ever heard of the place

—affected its history in massively important ways. As forests were cut, ecosystems were altered or destroyed, soil fertilities depleted, and animal habitats forever transformed. What do these connections between Egyptians, pious pilgrims in the Hijaz, and Anatolian forests mean for our understanding both of the environmental history of the Ottoman Empire and of the ways that the specific properties of certain commodities shaped that history?

First, they suggest that the imperial calculus of the Ottoman state deemed certain natural resources and environments to be more important than others. Egypt, a place of great agricultural potential—one that helped maintain food supplies and Ottoman legitimacy in the Hijaz and elsewhere—was clearly, from both an Ottoman and a local Egyptian perspective, worth the alteration and consumption of other natural landscapes to provide the Nile valley with the materials needed to achieve this rich agricultural potential and to move the products of this potential to other parts of the empire. Thus, knowingly or not—most likely not—Egyptians participated in the consumption of large sections of Anatolian forest as they worked to construct ships to move food from Egypt to the Hijaz. These histories of Anatolian forests, of the Egyptian countryside, of the sustenance of pilgrims in the Hijaz, and of Ottoman imperial administration must therefore all be taken together as parts of a single process of the coordinated and connected consumption and use of nature.

Furthermore, from the imperial perspective of the Ottoman government in Istanbul and Cairo, connecting Anatolia to Egypt made perfect sense. These connections allowed the Ottoman Empire to shift a region's excess resources to places where that excess could fill a vital need that would eventually allow Egypt to grow food to feed people in yet other places. One can easily conceive of a process, then, whereby the Ottoman bureaucracy surveyed the empire, moving different pieces around to achieve an optimal configuration of rule. Lumber went to Egypt for both shipbuilding and irrigation purposes, thereby making possible the movement of grains and other foodstuffs to Mecca and Medina (and also Istanbul and other population centers in the empire). The imperial administration was thus essentially turning Anatolian trees into caloric energy for human stomachs in the Hijaz. By concentrating the labor, skill, and expertise found in certain areas of the empire on the production (or consumption) of natural resource commodities like lumber or grain, the Ottoman Empire was consequently able to increase its overall levels of agricultural and economic productivity. For this system to work most efficiently, transportation networks—as we have seen in some detail—had to move goods quickly and with a minimal amount of energy loss. The imperial system also had to rely upon and connect the actions and expertise of hundreds of actors across three continents and as many seas. Above all, this system of environmental comparative advantage and natural resource management was governed through an imperial administration that coordinated a vast system of local knowledge, autonomy, and action.

9

Plague Ecologies

Many historians have documented the intimate temporal and geographic coincidences of plague, famine, drought, flood, and price inflation in Ottoman Egypt.[1] Few, though, have shown how the connections between plague and these other phenomena fit into a recurring pattern of death and hardship during the sixteenth, seventeenth, and eighteenth centuries that came to inform Egyptians' experiences both of their natural environment and of plague itself. By examining one particular plague epidemic in Egypt—that of 1791—this chapter shows how plague was part and parcel of the pathology of the Egyptian environment at the end of the eighteenth century. The chapter thus highlights the multiple means through which plague functioned as a regular part of the Egyptian environment.

Plague in Egypt must be studied as one pathological element of the Egyptian environment that was known and understood by Egyptians at the end of the eighteenth century also to include floods, wind, drought, and famine. Like the annual Nile flood, increases in the prices of grain, famine, or other hardships, Egyptians considered plague an accepted and expected environmental reality. Thus, those who regularly dealt with plague at the end of the eighteenth century did not think of the disease as a kind of "foreign" invader, coming to Egypt in the hulls of ships from faraway ports. Indeed, although feared, plague did not cause Egyptians to flee; rather, it functioned and was thought of as a regular part of the Egyptian environment.[2]

While Ottoman armies conquered Egypt and made it a province of the empire in 1517, all evidence for the chronology and periodicity of plague indicates that 1517 did not represent a significant turning point in the frequency, severity, or treatment of plague in Egypt. Indeed, a new plague epidemic visited Egypt once every nine years on average for the entire period from 1347 to 1894.[3] This high incidence of plague suggests that the disease and its epidemics were regular and expected occurrences in the lives of most Egyptians throughout this period. It also suggests that historians should pay closer attention to the role of the disease in shaping Egyptian history and Middle Eastern history more generally.[4]

Despite the artificiality of 1517 as a dividing line in the history of plague, this date proves significant for the historiography of the disease because most work on plague in the Middle East focuses on its impact during the Mamluk or medieval periods, even though, as just noted, it was a consistent reality in Egypt (as elsewhere) throughout the entire Ottoman period.[5] This historiographic imbalance toward the earlier history of plague is commonly attributed to a greater number of sources for the study of the medieval disease than for comparable studies during the Ottoman period.[6] As much recent work has shown, however, such an explanation is difficult to maintain, since there are numerous kinds of sources for the history of plague after 1517: documentary records of Ottoman governmental bureaucracies, archives of Islamic courts, manuscript sources, and records of correspondence between the Ottoman center in Istanbul and various provinces of the empire.[7] In addition, the voluminous chronicles of the seventeenth and eighteenth centuries offer a great deal of information about the outbreaks, nature, and impact of plague during these two hundred years.[8]

The Plague of 1791

The year 1791 began on a thoroughly foreboding note for Egyptians. Based on the reports of numerous astrologers (*al-falakiyyīn*), they came to believe that at midnight on 1 February a great earthquake (*zalzala ʿaẓīma*) would strike Egypt and last for seven trembling hours.[9] Both the poor and the rich were convinced of the coming of this earthquake, and those who could manage to leave their cities and villages fled to broad open places like the desert or to one of Cairo's two main lakes—al-

15. Cairo's Lake Azbakiyya. Commission des sciences et arts d'Egypte, *État moderne*, vol. 1, pt. 2 of *Description de l'Égypte*, Le Kaire, plate 40.

Azbakiyya and al-Fīl—to ready themselves for the anticipated calamity. Egyptians braced for the event all night, but the quake never came, and all found themselves safe and sound the next morning.[10] Feeling thoroughly duped, people recited the following verse about their foolish naïveté: "And how many laughable things are in Cairo / but it is laughter like crying."[11]

However unfounded their fears of a colossal earthquake might have been, Egyptians had good reason to be afraid in February 1791. Later that month, plague struck with great force. One eyewitness reported that at the start of the outbreak, one thousand people died every day. Before long, that number rose to fifteen hundred per day.[12] Elsewhere, the same chronicler estimated that the plague killed two thousand people every day.[13] The disease did not discriminate between young and old, powerful and weak, pious and heathen.[14] An "uncounted number of babies, youths, maidservants, slaves, Mamluks, soldiers, inspectors, and amirs" all died in the spring of 1791.[15] Before attempting his planned escape to Istanbul, the leader Ismāʿīl Bey died, along with many of his followers.[16] Indeed, the plague of 1791 caused an enormous crisis of leadership in Ottoman Egypt since no appointed leader could stay alive long enough to rule effectively. An Ottoman firman sent to Egypt from Istanbul implored the surviving leadership of the province to do all it could to defend against the state of disorder (*perişanlık*) that was gripping Egypt due to the death of eight or nine of the province's most important beys. The decree further ordered this leadership to inform Istanbul of the names, physical characteristics, and notable attributes of those important men who had died and those who had replaced them.[17]

When both the *aghā*[18] (a high-ranking military official) and the *wālī*[19] (a police administrator who served under the *aghā*) died, successors immediately rose to power only to die themselves three days later. And those who replaced these plague victims themselves also died in the course of a few days. The Egyptian chronicler al-Jabartī wrote that "succession changed hands three times in one week."[20] Al-Khashshāb, another chronicler, described the situation slightly differently. He wrote that the appointment of an *aghā* and the need to replace that appointee because of his death from plague occurred three times in one day. Leaders came to power in the morning and died by late afternoon.[21] The pasha of Egypt, on the advice of his agents, left Cairo along with his amirs to seek refuge in the region of Ṭurā.[22] Many large Cairene families (*buyūt*) were decimated by the plague.[23] The speed of death after the onset of the disease was something noted throughout accounts of the 1791 plague.

So great a number of the soldiers and marines stationed in Old Cairo, Giza, and Bulaq died that mass graves were dug for their corpses, which were buried without ceremony or any final rites.[24] Funerals for those not connected to the military also had to be done en masse, with prayers being said for up to five people at a time. Indeed, the apparatuses charged with the management of death were stretched to their limits during this spring as the demand for undertakers (*al-ḥawānīt*) and corpse-washers (*al-mughassilīn*) far exceeded their available numbers. Most economic and social functions unrelated to the present grim circumstances ceased during the plague, since "there was not left for people any work except death and its attendant matters."[25] Accounts of the spring of 1791 devote significant attention to the sick, the dead, visitors of the sick, consolers, funeral-goers, those returning from funeral or burial prayers, those busy with preparing the dead, or those weeping in anticipation of their own death.

Later, in the summer of 1791, al-Jabartī tells us of another consequence of that year's plague: an absence of male heads of households.[26] When a group of amirs from southern Egypt attached to one Murād Bey came to Cairo in late July 1791, they found many houses without men, inhabited only by "women, maidservants, and slaves." Finding this situation agreeable, the amirs married these women, "replaced their bedding, and prepared their wedding feasts."[27] Any amir who did not have a house was free to enter any home he liked and to take it and everything in it without hindrance (*min ghayr mānīʿ*). In this way, these men took advantage of what plague had wrought to acquire land, houses, riches, and wives. From the imperial perspective in Istanbul, the outbreak of plague in Egypt created a worrisome opportunity for enemies of the state and rebellious bureaucrats to escape to Egypt undetected amid the mayhem and general disarray. Indeed, in very strong language, Ottoman authorities ordered those still present in Egypt to prevent any fugitives from entering the province to hide.[28]

The movements of peoples and goods were of the utmost concern for Ottoman authorities during plague outbreaks. It was, indeed, precisely because Egypt was so central to currents of trade and commerce within the Ottoman Empire and between the empire and elsewhere that plague was a constant presence in Egypt. Plague is permanently maintained among rodent populations in areas of Central Asia, Kurdistan, Central Africa, and northwestern India.[29] Plague came to Egypt from these and other places through the movement of merchants' wares, rats, fleas, and people.[30] As the American missionary John Antes wrote in Cairo at the end of the eighteenth century, "I could never find sufficient ground to ascertain that the plague ever broke out in Egypt, without being brought thither from other parts of Turkey [the Ottoman Empire]."[31] Indeed, the arrival of hundreds of ships and caravans every week from places like Istanbul, India, Yemen, the Sudan, China, Central Africa, and Iraq ensured a consistent flow and constantly replenished supply of goods, people, and vermin to Egypt. Plague therefore shows us yet again that the history of Egypt, the history of the Ottoman Empire, cannot be understood without a global perspective. The two main entry points of the disease were through the major trading areas of Egypt: its Mediterranean ports and the southern route from the Sudan.[32] Thus, plague either entered Egypt from the sea through the ports of Alexandria and Rosetta before making its way farther inland, or it traveled overland from Central Africa to the Sudan and then into Egypt. The plague of 1791 most likely entered Egypt through its Mediterranean ports, carried ashore from ships coming from Istanbul.[33]

Whether a particular epidemic's origins lay in Kurdistan or the Sudan did not affect the ultimate outcome of the disease on the Egyptian population. What mattered most for Egyptians' experiences of plague was that the disease functioned as though it were endemic to Egypt, given the consistency of its incidence and, more important, its regular appearance in the Egyptian environment as an ecological force alongside famine, flood, and drought.[34] For their part, many European observers considered Egypt to be the "cradle of the plague."[35] Interestingly, the Egyptian chronicler al-Jabartī seemed to concur with the notion that plague was somehow endemic to Egypt or, more specifically, that it was present in the Nile valley's soil. During the French occupation of

Egypt between 1798 and 1801, al-Jabartī subscribed to French views about the etiology of plague in Egypt. "They say that putridity or rottenness (*al-ʿufūna*) pollutes the depths of the ground. If winter arrives and the underground becomes cooled due to the flow of the Nile, rain, and humidity, the putrid vapors that had been trapped in the ground emerge and pollute the air, causing epidemic disease and the plague (*al-wabāʾ wa al-ṭāʿūn*)."[36] Al-Jabartī's explanations of the causes of plague in Egypt suggest both the persistence of the miasmatic theory of disease causation and the perception that plague was somehow natural to, endemic in, or constitutive of Egypt.

Despite the existence of multiple descriptions of the 1791 plague and of other outbreaks, determining the disease's demographic effects remains difficult.[37] There are very few studies of the Egyptian population at the end of the eighteenth century and during the period of Ottoman rule more generally.[38] Egyptian chroniclers' reports claiming that 1,000, 1,500, or 2,000 people died every day are therefore difficult to interpret and are in all likelihood incorrect.[39] Indeed, these descriptions seem to raise more questions than they answer.[40] How were these figures determined? For how many days did this same number of people die? What were the relative mortality rates in cities and the countryside? Were these symbolic figures meant to express the severity of a given outbreak rather than statistical numbers? Modern historian André Raymond estimates that the plague of 1623–26 killed close to 300,000 Egyptians.[41] He states that the same number died during the 1718 epidemic as well, representing close to an eighth of the entire Egyptian population at the time. For his part, historian Daniel Panzac reports that the plague of 1784–85 killed anywhere from 30,000 to 40,000 of Cairo's 300,000 inhabitants and that the plague of 1791 claimed the lives of a fifth of Cairo's population (still around 300,000 in that later year).[42]

The Flood before the Plague

During the fall of 1790, before the plague's attack in the spring of 1791, an unusually large amount of rainfall in Egypt caused many parts of Cairo to flood. In his characteristically hyperbolic style, al-Jabartī described this large rainfall as the result of one enormous deluge.[43] On the night of Thursday 14 October 1790, the skies above Cairo poured water over the city, "as if from the mouths of waterskins."[44] Accompanying this rain were continuous claps of thunder and lightning powerful enough to blind all those who saw it. The rains continued all night and for the entire next day, rushing down off the mountains and filling the desert outside the city's walls. The waters destroyed tombs and graves and caused houses to collapse, killing those trapped inside.

As if all this were not disastrous enough, this saturated Friday coincided with the return of pilgrims to Cairo from the annual pilgrimage to Mecca and Medina. Instead of enjoying the celebrations that normally marked their return, these pilgrims were cruelly welcomed back to the city with floods that carried away the pavilion of the Amīr al-Ḥajj.[45] The waters had by this time entered the city and flooded its numerous *wakāla*s (storage facilities for grains and other foodstuffs), caravansaries, and mosques. Businesses, residences, and even entire neighborhoods were destroyed. For example, more than half of the houses in the district of al-Ḥusayniyya were swept away.[46] A huge lake was created in and around the city. Al-Jabartī ends his account of this rainfall with the simple statement that "this was a most terrible affair."[47]

Whether this flooding occurred as the result of one deafening deluge or merely a steady rain is ancillary to the result: significant destruction of the city and its resources.[48] Similar to years in which the Nile flooded far beyond its banks, the waters of 1790 likely destroyed vast areas of agricultural land in and around Cairo and its hinterland.[49] This meant a significantly lower amount of food production for the coming harvest season. And more destructive than even an excessive flood season, the torrential rains of 1790 destroyed stored grain supplies in the markets and *wakāla*s of Cairo. This doubly magnified the danger for the coming year in and around Cairo, since both fields and grain supplies were washed away. The population thus likely experienced food shortages and even famine, weakening its resistance to and resolve against disease. That plague is preceded by famine is commonly observed throughout the history of plague epidemics in the Middle East and elsewhere.[50]

Another result of the floods in Cairo was the movement of thousands of rats seeking refuge. Although most species of rat do swim, these rodents—much like humans—seek out dry areas to escape rushing water.[51] In the fall of 1790, rats and humans were therefore competing for space in areas of Cairo not damaged by the floodwaters inundating the city. Rats sought refuge in the thatched roofs of homes, in homes themselves, and in other protected areas. In the countryside along the Nile, the flood caused rats and their fleas to escape from areas near the river, from places like fields and embankments, in search of higher ground.[52] When the Nile flooded and then receded quickly, the Egyptian countryside was especially susceptible to the dangers of rodent and insect infestation. As John Antes noted, "Should it [the Nile] happen to rise suddenly to a very great height, but not remain long enough to

16. Alexandrian rat at the end of the eighteenth century. Commission des sciences et arts d'Egypte, *Histoire naturelle*, vol. 1, pt. 3 of *Description de l'Égypte*, Mammifères, plate 5.

soak the fields sufficiently, it will not be a fertile year, and other bad consequences may likewise follow if it leaves the fields too soon, before the air begins to cool, for many sorts of vermin will breed in the ground which are pernicious to some kinds of vegetables."[53] Although rats and humans did often share the same dry, protected spaces, rats usually hid in dry places that humans could not or did not enter, thus allowing the rodents ample space in which to breed. As the waters receded and people returned to their dwellings, they encountered a large concentration of rats—as well as fleas and other insects—feasting on the contents of numerous unearthed graves and a great quantity of wet food left by the flood. Thus, the flood caused rats, fleas, and humans to come into much closer proximity than they normally would have.[54]

This proximity is a key factor in the etiology of plague, since the primary vectors of plague transmission in humans are rats and the fleas associated with them.[55] As long as the rat population in a particular area is large enough to support a sizable flea population, plague will survive.[56] Egyptian chroniclers do not directly address the epidemiological character of the 1791 plague. Nevertheless, enough evidence exists to suggest that this episode was bubonic rather than pneumonic or septicemic plague.[57] Although al-Jabartī states that plague victims were not feverish and that they died in a matter of two or three days, suggestive of pneumonic plague, he goes on to describe a situation in which the sick, the dead, and the healthy were all present together in very close proximity.[58] He writes of those going to visit the sick and of those caring for the sick in their own homes. He relates also that despite the close interaction between amirs and their wives, only the former died of plague while the women themselves did not fall ill.[59] If the 1791 outbreak was indeed pneumonic plague, then all of those not infected would have contracted the disease from their close interactions with the sick. Victims of pneumonic plague most often experience violent coughing spells, and a simple cough or sneeze would be enough to infect a healthy person in close proximity. Since al-Jabartī makes no mention of whole families or neighborhoods being wiped out by plague and because he discusses at some length how the healthy interacted with and cared for the sick, it seems likely that the 1791 plague was bubonic.

Likewise, John Antes described a similar situation of intimate proximity and contact between plague victims and their caretakers at the end of the eighteenth century. "It [plague] perhaps takes one or two only out of twelve, fifteen, or more, and those sometimes die in the arms of others, who, with all the rest, escape unhurt. There are instances of two people sleeping in one bed, one of whom shall be carried off by it, and the other remain unaffected."[60] Antes's descriptions of the behavior of some Europeans resident in Egypt at the end of the eighteenth century also suggest that this period's plague episodes were less contagious. For instance, the Friars de Propaganda Fide stationed in Cairo "always appoint two of their number to visit the sick, and to administer extreme unction to those of their persuasion who are dying: and it happens but seldom, that any of these visitors die of the plague, which constantly inclines them to make a miracle of it."[61] Antes also wrote of a Venetian doctor in Cairo who regularly visited plague victims but was never stricken with the disease himself.[62]

Offering even clearer evidence that Egypt's late eighteenth-century plagues were indeed of the bubonic variety, Antes writes that victims had "buboes in the arm-pits, or the soft part of the belly, with a few dark purple spots, or carbuncles, on the legs. When the buboes break, and discharge a great deal of matter, such patients may chance to recover. . . . The sick commonly complain of intolerable heat, and say they feel as if thrown into a fire."[63] Indeed, most historians of plague agree that bubonic plague was the most common form of the disease in Egypt.[64]

Nevertheless, one pocket of pneumonic plague was known to exist at the turn of the nineteenth century around the southern city of Asyūṭ.[65] Al-Jabartī includes in his chronicle a letter from his friend and associate Ḥasan al-ʿAṭṭār, who was present in Asyūṭ in May 1801 during an especially virulent plague epidemic.[66] Al-ʿAṭṭār's letter states that this unprecedented plague killed more than six hundred people every day in Asyūṭ alone and that it "exterminated most of the people of the region."[67] Although these statements are far from conclusive about whether or not this plague in May 1801 was pneumonic or bubonic, given the almost certain fatality of pneumonic plague and the stated severity of this outbreak, it was in all likelihood pneumonic. This seems all the more probable since even during the Black Death, "pneumonic plague was far more frequent in the south [of Egypt] than in the north."[68] Exceptionally mild temperatures in Asyūṭ, along with large rat and flea populations, contributed to the plague's historically endemic and likely pneumonic character in the city.[69]

The Famine after the Plague

If the causes of the 1791 plague are to be found in an excess of water, its effects in the form of continuing hardship for Egyptians at the end of the eighteenth century are to be found in a dearth of water. On 21 August 1791, the Nile crested.[70] Every year at the cresting of the river, the Ottoman government of Egypt held a series of celebrations that culminated in the breaking of a dam constructed at the mouth of an artificial canal in Cairo known as al-Khalīj.[71] This dam was built every year to prevent water from flowing into the canal except during the flood season. When the Nile's waters crested, the dam was broken and water then rushed through the canal into Cairo, symbolically inaugurating the annual flood and the beginning of the agricultural year.[72] Although usually an event of great joy, this ceremony could also be a harbinger of hard times ahead, for a lower- or higher-than-expected flood would mean famine and death for many Egyptians. This was exactly the case in 1791.

Similar to instances of excessive flooding (like that experienced in 1790), a poor inundation also meant food shortages, famine, and death. These effects of a low flood were even more acute in the late summer and fall of 1791 because of the ravages of plague earlier that year. Moreover, the populations of Cairo and other parts of Egypt were already suffering from lowered resistance and hence greater susceptibility to plague because of the floods of 1790. When the disease hit in 1791, the stage was thus set for a very bad epidemic. And by the end of 1791, drought and famine had fully set in.[73] The combination of flood in the fall of 1790, plague in the spring of 1791, and drought in the fall of 1791 resulted in widespread famine, severe price inflations, and a massive human toll.

Indeed, as al-Jabartī tells us, later in 1791 irrigation canals dried up and fields became parched because of a lack of water.[74]

With crops withering and dying, peasants became extremely anxious—"the people clamored," in al-Jabartī's words.[75] The poor harvest meant an increase in grain prices for that year and the corollary of revolts and agitations by peasants and the poor against increased food prices and against their rulers. In November and December of 1791, Egyptian authorities began seizing the property and land of merchants and peasants, ostensibly to relieve the economic pressures brought about by plague and famine.[76] Drought continued through these months and into January 1792. Al-Jabartī writes that "not one drop of water fell from heaven."[77] Some peasants did their best to farm land that seemed salvageable, but when they plowed, they found only worms and rats. These vermin competed among themselves and against their human rivals for fruits and the precious few crops that were grown in fields that year.[78] Many people had to make do with weeds, and cattle had no spring feed. There was thus an exceptionally large number of rats in rural Egypt in 1791, aiding the movement and tenacity of plague.[79]

This cycle of flood, plague, drought, famine, price inflations, and death in Egypt suggests that there was a kind of cyclical pathology to the economy that also functioned alongside the ecological pathology of plague in Egypt.[80] As foodstuffs decreased, prices and the severity of official measures to secure adequate supplies for the powerful and the military increased.[81] In response, there were numerous instances of agitations, complaints, and small-scale revolts by peasants and merchants aimed at these official actions. For example, when Mūrad Bey and his amirs—the same group of soldiers who married the wives of men who died during the plague—entered Cairo in July 1791, grain prices began to soar.[82] Taking advantage of the low Nile and the weakness of the population (*ḍaʿf al-nās*) from plague and other hardships, these military men began seizing grain for themselves and their entourage. Their cruelty was proved when one of them attempted to extract an unjustly large quantity of grain from a village outside of Cairo. A revolt broke out in Cairo and in the village itself in response to this draconian move on the part of one of Mūrad Bey's amirs. The villagers refused to hand over the foodstuffs, and the *ʿulamāʾ* (members of the elite classes of Muslim religious scholars) denounced the man's illegal actions. A violent struggle between the amir and the villagers soon ensued, but Mūrad Bey, fearful that unrest might spread, reined in his underling and apologized to the *ʿulamāʾ* and the villagers.

The relationship between the city of Cairo and rural villagers during plague epidemics serves to highlight yet again the importance of urban grain reserves during times of want. Rural depopulation was a common phenomenon during plague epidemics and the famines that usually accompanied them, as peasants from the Egyptian countryside fled their lands in search of food and work.[83] The flight of rural labor to cities further exacerbated the many hardships brought on by food shortages—the very problem peasants were escaping. Thus, ironically, "In time of famine, peasants actually came to Cairo in search of food rather than the reverse."[84] In addition to food, cities also offered peasants access to physicians, healers, and religious institutions.[85]

The Season of Plague

In describing the death of the amir Riḍwān Bey, al-Jabartī writes that the man's candle was extinguished by a plague that came like "the icy gale of death" (*ṣarṣar al-mawt*).[86] The trope of likening plague to wind further suggests that the disease was considered part of the natural world in Egypt. Plague is elsewhere described as something that scatters the lives and possessions of its victims.[87] In the only eyewitness account of the Black Death in the Middle East, Ibn al-Wardī (who died of plague in Aleppo in 1349) likened plague to a cloud: "It eclipsed totally the sun of Shemsin and sprinkled its rain upon al-Jubbah. In al-Zababani the city foamed with coffins."[88] Later, Ibn al-Wardī wrote that "the air's corruption kills," an obvious reference to the dominant miasmatic theory of disease causation.[89] A later plague was furthermore compared to winds that came and scattered the dried foods of fields in southern Egypt.[90]

Perhaps the most important reason for the association of plague with wind is that, as in 1791, plague usually afflicted Egypt contemporaneously with the *khamāsīn*—the warm southerly winds that blew into Cairo every year in late spring and early summer.[91] These winds covered the city with sand and dust from the deserts south of Cairo. John Antes described the *khamāsīn* as follows: "In spring it [the wind] often changes to south-east, and then it is of a whirling nature, filling the atmosphere with such quantities of sand and dust as to make it almost totally dark. I once remember being obliged to light a candle at noon on such a day, as the sky was at the same time covered with thick clouds."[92] The often contemporaneousness of the *khamāsīn* and plague caused many to think that plague, like dust and sand, was brought in these annual winds. For instance, Aḥmad al-Damurdāshī Katkhudā ʿAzabān drew a connection between the plague of 1690–91 and the *khamāsīn* that occurred at the same time. He wrote that the plague swept into Cairo like the winds of the *khamāsīn*, filling the city's quarters and alleyways with dead bodies.[93]

The coincidence of plague and the annual *khamāsīn* also helps to underscore the timing of the so-called "season of plague" in Egypt—the period of greatest recurrence of the disease in any given yearly cycle.[94] In 1791, plague began in late winter, seems to have been most deadly in the spring, and waned toward the middle of the summer. This pattern is the one most commonly observed for the course of plague in Cairo, which Panzac identifies as beginning in February, peaking in May, and dissipating toward its eventual end in July or August.[95] A more nuanced study of the season of plague in Egypt suggests a gradual movement of the disease from the south toward the Mediterranean.[96] Those plague epidemics that traveled from Upper Egypt (as the south is known) to Cairo seem to have occasioned much more fear among the population than those that originated in Egypt's Mediterranean ports. John Antes observed, "There is a saying among the people, that the plague, which was brought from Upper Egypt, was the most violent."[97] In Upper Egypt, plague begins in March and ends in May. In the middle of the country, it begins in April and ends in June. In the Egyptian delta above Cairo, plague begins in April and ends in July. And, in Egypt's Mediterranean ports, the disease begins in May and ends in October. The reasons for this slow movement of plague from the south to the north are climatic, since temperatures rise first in the south and then move north as the country warms through spring into summer. The ideal meteorological conditions for plague are temperatures between 20°C and 25°C with mild humidity, precisely the conditions that move up the Nile valley from spring into summer, taking mild temperatures—along with plague—up the country. Moreover, fleas—the primary agents of plague transmission—are most active during the warm months of the spring and early summer.[98] The dry heat of the summer eventually makes Egypt unfavorable to plague and fleas, as temperatures rise well above 27°C and humidity remains lower than 40 percent.[99]

For his part, Antes also ascribed the disappearance of plague in Egypt to the increasing temperatures of the summer months.[100] To illustrate this point, he related the following:

> In the year 1781, the plague broke out about the middle of April, and increased with such rapidity and virulence, that sometimes one thousand people died of it in one day at Grand Cairo; but, about the middle of May, the wind shifted to the east, which

> occasioned a few days violent heat, in consequence of which it immediately diminished; and though, as the weather became again cooler, the plague did not leave the country before the end of June, yet it never encreased to the same degree as before, but continued dwindling away, till it ceased entirely when the summer heat became regular. It has always been observed in Egypt, that a great degree of heat, if even but for a few days, had this effect; but this time it was very remarkable.[101]

According to Antes, heat was also effective in curing those already afflicted with plague. He often observed that the sick arriving in Egypt from other parts of the Ottoman Empire during the summer months would soon recover from the disease after only a few days in the province. In cities like Istanbul and Izmir, he continues, plague was much more virulent since heat was never as regularly intense in those places as it was in Cairo.[102]

Egyptians took 24 June as the date past which no plague could exist in Egypt due to the summer's consistently high temperatures.[103] This tradition was connected to several events celebrated annually by Egyptians, especially by Copts, at the end of June. The first was the celebration of al-Nuqṭa (the drop) on 17 June, which commemorated the beginning of the yearly flood season; it was also the Coptic festival of the Archangel Michael. It was believed that on this day the archangel threw into the Nile one drop of holy water of such fermenting power that it made the river overflow its banks, thereby flooding the entire land of Egypt. On this day also, the archangel commanded all other angels to cease striking the people of Egypt with plague, since it was believed that angels were sent by God to infect those intended as sacrifice. If by 24 June, the festival of Saint John, any angels were found to be still lurking about in search of humans to strike down with plague, they would have to face the heavenly consequences.[104]

Conclusion

This chapter argues for the inclusion of plague in the study of the Egyptian environment alongside natural elements and forces such as flood, famine, wind, animals, and drought. In doing so, it points to questions about the relationships between plague and Egyptian society and culture. If plague was indeed a regular feature of the Egyptian environment, one that connected Egypt to regions around the globe (a theme we will pick up in the next chapter as well), then how did this regularity affect the ways Egyptians interacted with and thought about other parts of the natural world and other parts of the globe? How, for example, did social institutions like the family and religion adapt to, explain, and employ the constant presence of plague in Egypt? How did plague enable the deployment of sets of practices and institutions related to sickness, disease, and death? Did Egyptians associate plague with global trade and the movements of peoples and goods? Answers to these and other questions show us that to properly understand the history of plague, we must consider it in all its social, cultural, economic, political, epidemiological, and environmental contexts and complexities.

10

Egypt, Iceland, SO_2

A volcano erupted in Iceland; people starved in Ottoman Egypt. This chapter tells the how, why, and when of the connections between Iceland and Egypt through an examination of the environmental impacts of the explosion of the Laki volcanic fissure in 1783 and 1784.[1] One of the largest volcanic eruptions in recorded history, it led to two years of cold summers and extreme winters in Europe, North America, the Mediterranean, Central Asia, and beyond. The story of its impacts on Egypt and the Ottoman Empire has not been told.[2] In telling it, this chapter undertakes the examination of the material and political effects of a specific small-scale instance of climate change.

Climate is a big story. In turn, environmental historians of the climate's past tend to tell big stories about that past.[3] Climate events are transcontinental or sometimes global; the time scales historians use are often on the order of millennia; the effects of climate change include rising seas, atmospheric flux, and species extinctions. Against the backdrop of this dominant mode of doing climate history, the story of Laki and Ottoman Egypt affords the chance to think about climate on what is comparatively a very small scale.[4] The climatological data offer an exceptionally clear and direct cause and effect—Laki's explosion reduced the Nile's flood.[5] And the volcano's ecological effects on Egypt were, moreover, only temporary—just a few years in the middle of the 1780s. As I will discuss below, Laki impacted many parts of the globe, and its political and environmental effects on Ottoman Egypt were clearly real and extremely deadly. Still, this one small slice of the history of just a single nonanthropogenic climate event allows us to think about climate change's individual causes and individual effects rather than seeing the phenomenon as only an abstract all-encompassing problem of planetary proportions.

To put it differently, climate change has been, for obvious reasons, one of the dominant and most productive modes of doing global history.[6] Iceland and Egypt are two places that have so far been peripheral to these discussions. However, they were not peripheral to each other. Connecting them through the remarkably linear story of Laki's effects on Egypt thus allows us to avoid the kinds of universal accounts of climate change that do not address the one-to-one influences of individual places on one another. Putting climatological data together with narrative accounts makes it possible for us to trace the earth's energy up through Iceland's substrata into the stratosphere and back down to individual villages in Egypt. By following this ecological story from Iceland to Egypt, we gain a better understanding of the many complicated political and material changes that occurred in the Ottoman Empire at the end of the eighteenth century. The tools of environmental history thus allow us to center these peripheries through a story of climate.

Indeed, with this directly linked climate history of Iceland's environmental and political impacts on Ottoman Egypt, not only do I seek to overcome conventional conceptions of political geography (I am not aware of any histories that integrate the two places), but I also want to put in dialogue the history of Laki with the history of the Middle East.[7] Current scholarship on the global history of the Laki explosion by historians, climatologists, geochemists, and others focuses mostly on the volcano's impacts on Western Europe; one could make a similar observation about climate history more generally. Adding the Egyptian story helps to paint a more complete picture of Laki's global effects—explaining the volcano's impacts on Indian Ocean monsoons, on the longest river in the world, and on one of the most agriculturally rich areas in all of Africa and the Mediterranean. Similarly, Laki is crucial to the study of Middle Eastern history. The volcano's environmental history helps to explain some of the political, economic, and social history of Egypt at the end of the eighteenth century, a history that indeed remains incomplete without an understanding of Laki.

Iceland Erupts

On the morning of 1 June 1783, Iceland began shaking.[8] After a week of earthquakes, in the south of the island, a landscape of volcanoes and glaciers, "a black haze of sand" appeared over the twenty-seven-kilometer-long Laki fissure.[9] For the next eight months, this seam in the island's rock would spew tephra, dust, and smoke into the air and rattle Iceland's residents with persistent earthquakes and tremors. When all was said and done, the Laki eruption produced more basaltic lava flow than all but one other volcano in recorded history.[10] It shot tephra as high as fourteen hundred meters into the air, affected temperatures from Greenland to China, and caused famine and devastation that killed about a fifth of Iceland's population.[11]

The land of fire and ice was, of course, no stranger to volcanoes.[12] Throughout its history, Iceland had been forged and reforged through the awesome power of earth's fire.[13] The eruptions of 1783 and 1784 were, however, more colossal than nearly anything else the island had ever seen.[14] Jón Steingrímsson, pastor of the church of Kirkjubæjarklaustur, a town in south-central Iceland, was an eyewitness to the Laki explosions and the destruction that ensued. Known to history as the "fire priest," because of a "miracle" in which flowing lava destroyed everything around his church but not the church itself, Steingrímsson understood the volcano, not surprisingly, through a Christian idiom—"as a sort of reflection and reminder of the end of the world and the Doomsday fires."[15] The entries in his journal, set down on paper only in 1788, provide a near play-by-play account of the Laki eruption event. Combining Steingrímsson's on-the-ground descriptions with modern climatological studies offers a detailed picture of the course of Laki's explosions from June 1783 to the winter of 1784.

The eruptions were most intense at the beginning of this eight-month period; ten occurred in the first five months.[16] Laki's magma volume in June 1783 was roughly 7.67 cubic kilometers; around 1.5 cubic kilometers in July; 2 cubic kilometers in August; 1 cubic kilometer in September; and then tapered off precipitously until the final explosions in February 1784.[17] About 93 percent of the magma volume had discharged by October 1783.[18] Observing the most powerful early weeks of the lava flow, Steingrímsson writes that on 19 June, "the great flood of fire which poured forth . . . quickly filled up the course of the river [Steinsmýri]."[19] Just a few days later, on the twenty-fourth, "The new lava had piled up so high that when I stood on a cliff, just above the upper farm, Efri-Steinsmýri, and looked westward across it, I could only see the top of Hafursey, a single peak which stands far away in the wastelands of Mýrdalssandur."[20] Into July and August, "The liquid fire poured forth over the land so that everything became mixed together."[21] The enormous amounts of sulfuric ash released into the air that summer meant that "neither sun nor sky could be seen for the thick clouds of fumes and smoke which blanketed the area."[22] Things improved at the end of August, when "from the 17th to the 23rd the flow of the fire forward slowed considerably."[23] Laki's lava may have begun to creep rather than flow, but earthquakes were still frequent and clouds of ash continued to envelop the sky into the fall of 1783.[24] For Steingrímsson the occasion of the New Year was an opportunity for optimism, as he observed that "this sorrowful era came to an end, with God's help, and a new and different one began."[25]

Things were not so simple. Iceland in early 1784 continued to experience lava flows, earthquakes, and ash cloud cover.[26] The last of Laki's explosions was on 7 February 1784.[27] After eight months of eruptions, lava had collected in valleys, riverbeds, and streams. Even as the earth no longer exploded, shook, and spewed into the air, the people of Iceland were only just beginning to take stock of all the destruction of the last few months.[28]

Agricultural land had been destroyed. Approximately 565 square kilometers of the island were covered by lava extrusion of a depth of about twenty meters.[29] "There is no more than a quarter of the former upland pasture area of the Síða region [the primary area affected by the Laki eruption]," Steingrímsson wrote of early 1784, "remaining un-

Table 6 Laki magma volume discharge and SO_2 release, 1783–84

Date	Magma volume discharged (km^3)	SO_2 released (megatons)
8–10 June 1783	1.25	10.3
11–13 June 1783	2.01	16.7
14–21 June 1783	2.80	23.2
25 June–1 July 1783	1.61	13.4
9–21 July 1783	1.33	11.0
29 July–9 Aug. 1783	1.97	16.3
31 Aug.–4 Sept. 1783	1.15	9.6
7–14 Sept. 1783	0.87	7.3
24–29 Sept. 1783	0.66	5.4
25–30 Oct. 1783	0.44	3.6
1 Nov. 1783–7 Feb. 1784	0.61	5.1
	14.70	121.9

Source: Adapted from Thorvaldur Thordarson and Stephen Self, "Atmospheric and Environmental Effects of the 1783–1784 Laki Eruption: A Review and Reassessment," *Journal of Geophysical Research* 108 (2003): 5.

covered by hardened lava, pumice and sand, and much of what is left is so badly choked that it is not certain that grass will ever grow there as before."[30] The destruction of pasture and agricultural land soon led to food and fodder shortages, causing "famine among men and animals alike."[31] Not only was land covered in ash and sand, but the scourge of sulfuric air caused plants, animals, and indeed humans to wither away. "The poisoning effects of the fires" killed off countless sheep, horses, and cattle, as well as smaller mammals, birds, and fish.

Eighty percent of the island's sheep, 75 percent of its horses, and 50 percent of its cattle died.[32] Steingrímsson relates that "horses lost all their flesh, the skin began to rot off along the spines of some of them, the hair of the tail and mane rotted and came off if pulled sharply. Hard, swollen lumps appeared at joints, especially the fetlocks. Their heads became swollen and disfigured, and their jaws so weak they could hardly bite off or eat grass, as what little they could chew fell out of their mouths again. Their innards decayed, the bones shrank and lost all marrow."[33] This description was eerily similar to what happened to many hapless humans after the volcano:

> Ridges, growths and bristle appeared on their rib joins, ribs, the backs of their hands, their feet, legs and joints. Their bodies became bloated, the insides of their mouths and their gums swelled and cracked, causing excruciating pains and toothaches. Sinews contracted, especially at the back of the knee. . . . The inner functions and organs were affected by feebleness, shortness of breath, rapid heartbeat, excessive urination and lack of control of those parts. This caused diarrhoea, dysentery, worms and sore growths on necks and thighs, and both young and old were especially plagued by loss of hair.[34]

Thousands of "eruption-people," as they came to be known, fled this poisoned air and damaged land in search of food and shelter elsewhere on the island.[35] From 1783 to 1786, Iceland was gripped by famine and what are generally termed the "Haze Hardships" (Móðuharðindin).[36] A smallpox epidemic broke out in these years as well.[37] In the end, it was the volcano's aftermath of famine and disease—not the force of the explosions themselves—that killed nearly a fifth of the island's humans: thirteen thousand people, reducing Iceland's population to thirty-eight thousand, roughly what it was after the island was first settled at the end of the ninth century.[38]

Not only Icelanders had to deal with the consequences of Laki. From

Alaska in the west to Central Asia in the east, people all over the world were affected by the explosions of 1783 and 1784. The primary means through which Laki impacted the globe was sulfur dioxide (SO_2).[39] The volcano released approximately 122 megatons of the gas into the atmosphere (see Table 6).[40] Within a few short days of the first eruption on 8 June 1783—in some cases within forty-eight hours—Laki's sulfur dioxide had impacted Greenland, mainland Europe, and the Mediterranean. Information gleaned mostly from weather logs and contemporary accounts from the summer of 1783—by the likes of Benjamin Franklin, no less—allow for a reconstruction of the chronological course of ash and acid rain that came with the Laki haze.[41] Just a day after the initial eruption, a blue sulfuric fog appeared over eastern Iceland.[42] By 16 June, a thick haze covered western and northern Iceland. At the end of the month, the Laki haze was moving over Nuuk in western Greenland. It was seen to the south in the Azores a few weeks later. To the east, acid rain and ash fell on the Faeroe Islands, the western coast of Norway, and possibly northern Scotland as early as 10 June.[43] In other parts of Britain, the haze was first seen on 16 June; in Denmark on 24 June; St. Petersburg on 26 June; Moscow on the thirtieth; Hungary on 23 June; Munich, Berlin, Middelburg (Holland), and Padua on 17 June; Paris on 18 June; Rome on 16 June;[44] Tripoli (Lebanon) on 30 June; and over the Altai Mountains in Mongolia, about seven thou-

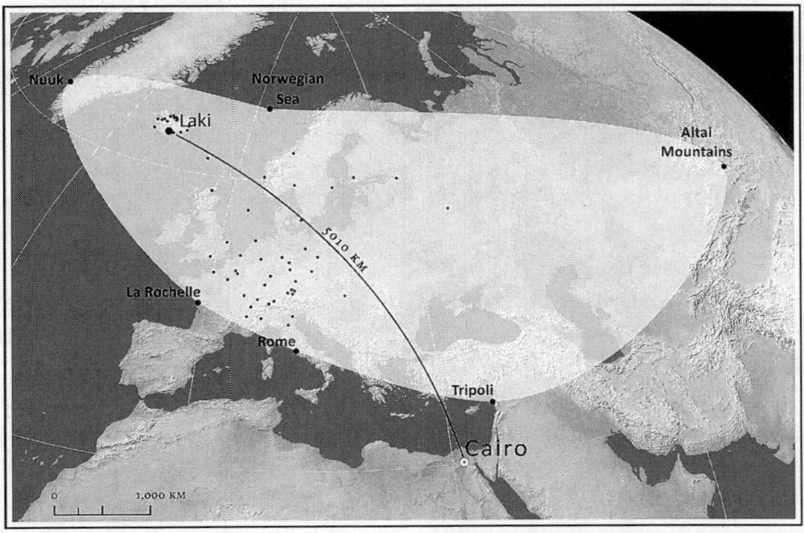

17. Visible Laki haze. Dots represent locations of eyewitness reports of the haze from the summer of 1783. Stacey Maples, 2014.

sand kilometers away from Iceland, on 1 July. Other evidence points to the Laki haze in China and Alaska as well.[45] In just three short weeks, Laki had thus spewed its ash all across the Northern Hemisphere, from 35°N to the North Pole.

As it did in Iceland, Laki immediately affected plants, animals, and humans in continental Europe and beyond. Acid rain damaged vegetation in Norway, Denmark, and England.[46] In Holland at the end of June 1783, many people reported headaches, respiratory problems, and asthma resulting from the sulfuric haze hanging over the continent. Acid precipitation and sulfuric haze were most pronounced in northwest Europe in June and July of 1783, but their effects were felt all the way to the northern Mediterranean coast that summer. The initial effects of the visible Laki haze lasted through the summer of 1783 before beginning to dissipate in the fall.

After the haze, there were many much longer and more impactful consequences of the Laki eruptions. These had to do with the volcano's unique climate effects.[47] Laki was exceptional for several reasons. First was the sheer volume of sulfur dioxide it produced—again, nearly 122 megatons.[48] Most important for its climate impacts, Laki's eruption columns extended from nine to fourteen kilometers into the air, much higher than those of most other volcanoes.[49] This meant that an extremely high percentage of Laki's 122 megatons of sulfur dioxide was released into the atmosphere's upper levels. It is estimated that nearly 95 megatons of the volcano's sulfur dioxide made it into the upper troposphere and lower stratosphere—the polar jet stream. Moreover, because Laki sent sulfur dioxide to such high elevations consistently over the course of several months, concentrations of the gas were continually replenished, remaining elevated over a long period. Once in the air, the sulfur dioxide aerosolized with atmospheric moisture to produce an estimated 200 megatons of sulfuric acid (H_2SO_4).[50] The polar jet stream distributed these aerosols far to the east and west of Iceland. It was this global aerosol distribution that ensured that the effects of Laki would linger for much longer and far beyond the initial visible haze. Around the globe, the summers of 1783 and 1784 and the intervening winter exhibited strong Laki effects.

Although the climate anomalies Laki produced varied somewhat by region, the largest statistically significant impact was a cooling effect.[51] Surface temperatures in Europe and North America in the three years after the Laki eruptions were 1.5°C below average; indeed, the years 1784, 1785, and 1786 were the coldest of the second half of the eighteenth century.[52] Temperature series readings at the New Brunswick station, for example, show that the winter of 1783–84 was the coldest there in 250 years.[53] Ice in the Chesapeake Bay's harbors and channels led to its longest closing to boat traffic ever in its history.[54] There were even ice blocks in the Mississippi River at New Orleans. In Iceland itself, the winter of 1783–84 began extremely early, in September, with unusually thick ice cover in its lowlands and fjords and severe sea ice.[55] While winter mean temperatures in twentieth-century west and north Iceland were −0.9°C and −1.7°C, respectively, in the winter of 1783–84 these temperatures stayed below -15°C for most of the season. Boats could not cross between islands in the Danish straits because of ice cover; people in Amsterdam drove wagons over frozen canals; Paris froze in January and February, with temperatures of −4°C; Vienna could not get firewood because the Danube was solid ice; and severe temperatures in Italy, Munich, Prague, and Moldavia led to food shortages and great hardship.[56] Farther east, there was frost in Mongolia in July 1783, and China suffered intense cold that summer too.[57] Exceptionally low summer temperatures in Japan led to one of the archipelago's worst famines ever.[58] In Lebanon, "The winds blew as in wintertime."[59] In northwest Alaska, 1783 was the coldest summer of the past four hundred years, and probably of the past nine hundred years.[60]

Temperatures dropped globally in 1783 and 1784 for two reasons. First, the enormous amount of sulfur dioxide in the atmosphere increased the earth's albedo, thereby reflecting more solar energy back into space.[61] Second, the heavy H_2SO_4 aerosol burden in the Arctic north led to "substantial heating of the Arctic atmosphere and subsequent reduction of the equator-pole thermal gradient."[62] The result was a weaker westerly jet stream of warm air. The cooler temperatures of a weak westerly jet stream also contributed to a strong dynamical effect of lessening the African and Indian Ocean monsoon circulations, a phenomenon that would have significant implications for Egypt.[63]

Egypt Suffers

The Indian Ocean monsoons fed the Nile.[64] Moving over the Ethiopian highlands in early summer, these rains swelled the upper reaches of the Nile system, eventually flowing into Egypt in June. The river rose in the south at Aswan in June and in Cairo by July, peaking in the capital in late August or early September. Laki erupted in June—just in time to interrupt the Indian Ocean summer monsoons. Climatological studies make clear that Laki led to reduced Nile floods in 1783 and 1784.[65] Estimates are that the Nile's flow decreased by as much as 18 percent in these years.[66] Of all the annual floods between 1737 and 1800, 1783 saw the lowest flood and 1784 the third lowest of the entire period.[67] Given that the Nile was Egypt's ultimate source of food, revenue, and power, a reduction of nearly a fifth of its waters obviously had devastating consequences for the social, economic, and political structures built by the wealth the river produced. Although likely no rural Egyptians in the 1780s had ever heard of a place called Iceland, Laki was a big part of the reason they suffered.

To fully understand the powerful effects of Laki on Ottoman Egypt, we must first understand what was going on in Egypt during that period. The volcano came at a particularly bad time for the Ottoman Empire in Egypt, both politically and ecologically. As we have seen, beginning in the 1760s, local elites in the Egyptian countryside had started to pull away from the central power of the Ottoman state.[68] While this growing autonomy of course served these men personally, it had deleterious effects on local populations and environments. As these emerging rural leaders took to carving out for themselves zones of influence and bases of rural capital, they monopolized rural resources, forcibly seized large swaths of land, adopted modes of extractive agricultural commercial production, forcibly moved peasants to work fields, withheld tax revenue from the Ottoman state, and raised small armies to protect their growing economic and political interests. Theirs was, in short, a project of local political centralization forged through resource monopolization.

Needless to say, these locals' attempts to seize control of rural human and environmental capital were a cause of great concern for the Ottoman Empire. As never before, at the end of the eighteenth century provincial elites in Egypt, and elsewhere in the empire, exercised nearly autonomous control over land and tax revenue.[69] The most prominent of these magnates in Egypt during the second half of the eighteenth century were the provincial governors ʿAlī Bey al-Kabīr and Muḥammad Bey Abū al-Dhahab.[70] In the 1760s and 1770s, these two men seized wide tracts of land and used these properties to raise revenue to pay soldiers and patronize political and economic partners in the province and elsewhere. These provincial elites reached the height of their power in the 1770s, when their private armies carried out several invasions of Ottoman Greater Syria.[71] They held some territory for a few years before being pushed back to Egypt by Ottoman imperial forces. The Russo-Ottoman wars of the last half of the eighteenth century, the expanding corporatization of the Ottoman military, and inflationary pressures on the Ottoman economy all contributed to the realignment of power in various Ottoman provinces at the end of the eighteenth century.[72] Local strongmen in Egypt took advantage of the opportunities created by these imperial changes to gain more power and control over rural resources at the expense of the empire.

These political developments at the end of the eighteenth century worked in concert with and in many ways depended on the period's extraordinary ecological stresses. The 1780s and 1790s were especially trying decades for rural Egyptians for several reasons. First, a number of major plague epidemics and epizootics hit the countryside in these twenty years. Foremost among them were in 1784, 1785, 1787, 1788, 1791, 1792, and 1799.[73] Working in tandem with these disease outbreaks to weaken populations and cause economic and political chaos in the countryside were years of drought, famine, and poor yields. Low Nile floods were the ultimate cause of agricultural want in Egypt; in all but two years of the last two decades of the eighteenth century, flood levels were below average, and, again, the lowest flood of the period from 1737 to 1800 was in 1783.[74] Taken together, famine, disease, drought, low agricultural yields, and widespread human and animal death created a power vacuum in the Egyptian countryside that allowed local elites to seize land and resources through force and theft, to consolidate and extend their political power, and to use these gains to chip away at Ottoman central authority in what was the empire's most lucrative province.

In many ways Laki was therefore the straw that broke the camel's back. Its effects on Egypt came at an opportune moment for those seeking to pull away from the empire. The drought, famine, and rural economic chaos it caused allowed local elites to further their project of rural resource consolidation—a phenomenon that eventually demanded a response from Istanbul. Had Laki not erupted, Egypt's growing autonomy from the empire that began in the 1760s likely would have continued, but clearly along a rather different path with different causes and different effects. Thus, Laki was a major factor in Egypt's late eighteenth-century transition toward semi-independence driven by the centralizing military and commercial interests of local elites.

The sources from this period make clear that Laki's effects on Egypt precipitated a massive crisis in the countryside. Documenting the early fall of 1783, the Egyptian chronicler al-Jabartī wrote of the Nile's dearth that year and the food shortages that followed. "The Nile did not rise sufficiently, and it fell rapidly. . . . The ground remained dry in the South as well as the north. Grain became scarce. . . . The price of wheat was on the loose . . . and the poor suffered greatly from hunger."[75] Almost a year later, another lack of summer floods exacted a similar toll on Egyptians, leading to great "scarcity and dearth" (*kaht ü galâ*).[76] Al-Jabartī wrote that the fall of 1784 was "like the preceding one with distress, rising prices, an inadequate rise of the Nile, and continual internal strife."[77] Two consecutive years of bad floods ravaged the countryside, Egypt's economy, and its rural social structure. Land became so progressively unproductive that the taxes garnered from rural Egypt in 1785 were the second lowest total in over sixty years.[78] "The land turned to waste," "peasants abandoned their villages because of a lack of irrigation," and "many of the poor starved to death."[79] Moreover, "store-houses on the river stayed empty of grain for a whole year and the granaries also remained closed. People's daily bread and subsistence were cut off, and they perished regardless of whether they compromised or cheated."[80] Traveling in Egypt in those years, the French philosopher and Orientalist C.F. Volney corroborated that "the inundation of 1783 was not sufficient, great part of the lands therefore could not be sown for want of being watered, and another part was in the same predicament for want of feed. In 1784, the Nile again did not rise to the favourable height, and the dearth immediately became excessive."[81] By the end of 1784, "Many men and animals had perished from hunger."[82] As evidence of just how hungry people had become, Volney reported seeing two men "sitting on the dead carcase of a camel, and disputing its putrid fragments with the dogs."[83]

In Egypt (as in Iceland), drought and hunger in 1783 and 1784 made people more susceptible to plague and other diseases. Volney guessed that these years' "famine carried off, at Cairo, nearly as many as the plague."[84] The plague began in the winter of 1783–84, with "not less than fifteen hundred dead bodies" taken out of Cairo each day.[85] It increased its deadly intensity in the summer and fall of 1784, likely because the previous years' food shortages had weakened rural people's immunities.[86] It continued into 1785.[87] The combined forces of drought, famine, and disease decimated rural populations through both death and flight. Volney estimated, citing "received opinion," that Egypt lost one-sixth of its total population between 1783 and 1785.[88]

The political possibilities produced by Laki's environmental impacts immediately contributed to the economic, political, and social transformation of rural Ottoman Egypt. Local power brokers throughout the countryside took the combined growing rural stresses of drought, famine, depopulation, and disease as an opportunity for theft and a chance to tighten and extend their authority over territories and communities. Banditry, plundering, and violence thus gripped Egypt in the middle of the 1780s.[89] "During this period," al-Jabartī wrote, "lawlessness increased."[90] Local amirs and their henchmen looted cargo from ships on the Nile and from transport caravans on roads; exacted protection money from local communities; stole grain, animals, and cash; and destroyed crops.[91] This violence, theft, and turmoil further encouraged rural depopulation as countless people fled these dreadful circumstances. "Extortions and acts of tyranny committed by the amirs followed one another, and their followers spread through the country to levy money from the villages and towns and invented illegal contributions . . . until they ruined the peasants, who became unable to bear the burden and abandoned their villages."[92] Again, the consequences of the ecological stress Laki's eruption helped produce were a major component of the political and economic story of Egypt in the 1780s and 1790s.

The most important and direct outcome of Laki on Egypt's political history was the strengthening of the general decentralization of power that had been occurring in Egypt since the 1760s, when rural elites first began consolidating their influence in the countryside.[93] Attempting to take control of as much of the countryside's dwindling human, agricultural, and financial resources as possible, these men carved out for themselves various zones of autonomy away from the direct political and military grasp of the Ottoman Empire.[94] In the 1780s, they furthered their monopolization of rural tax revenue earmarked for the state, raised small armies, and fought wars against one another for control of Egypt. As a result, in June 1785, a special meeting of the imperial council (divan) was held in the Ottoman palace in Istanbul to determine how the empire could restore a strong presence in its most lucrative province. Sultan Abdülhamid I and his retinue determined that the only plausible course of effective action was a full-scale military operation by land and sea to drive out the rebellious leaders. To help with this venture, the sultan commissioned a secret report on the current conditions maintaining in Egypt and began preparations in Istanbul for the invasion.[95] Ottoman forces led by the grand admiral Gâzî Ḥasan Paşa arrived in Egypt in July 1786 and made quick work of their enemies. Within two months, most of the province's upstarts had been driven into hiding. Ottoman soldiers continued to pursue them but were forced to retreat from Egypt in October 1787 to join the empire's more pressing military efforts against Russia.[96] Once Ottoman forces had withdrawn, Egypt's defiant rural leaders returned and quickly picked up where they had left off.

The Ottoman offensive of 1786–87 was an attempt to prevent Egypt from falling into the hands of an emerging local elite. It was also an attempt to fight against the ecological consequences of the Laki eruption. Laki contributed to the processes that helped push the Egyptian countryside into the drought, famine, and rural economic and political mayhem that allowed local power brokers to gain an advantage at the end of the eighteenth century. In the final analysis, the Ottomans lost to the volcano. The brief reestablishment of central Ottoman control in Egypt in the middle of the 1780s could not stem the tide of the local autonomy and strongman politics that were developing in this period and that Laki helped to further, a phenomenon we have touched on in other chapters. Indeed, what emerged in the 1780s was a kind of crony politics that would shape Egypt for at least the next 150 years. As we have seen, this is a history that Egyptian and Ottoman historians have long tried to explain through an examination of economic and political factors as varied as the emergence of a particular brand of eighteenth-century rural Egyptian capitalism, Napoleon's invasion of 1798, and the creation of bureaucratic state institutions in the early nineteenth century.[97] Acknowledging the importance of this historiography, an element we need to understand better is how the rural flux of the 1780s underpinned various political and economic efforts to remake the capital regime and political landscape of Egypt. Laki is one part, a central part, of the story of how Egypt transformed at the end of the eighteenth century. Its climatic effects were a motor force of the massive environmental and political distress—distress that included famine, disease, drought, death, economic crisis, military invasion, and forced labor—that gripped Egypt in the last decades of the century.

Conclusion

This chapter's connected climate history of Laki and Ottoman Egypt has sought to explain the effects of an Icelandic volcano on the Ottoman Empire. The story of Laki's impacts on Ottoman Egypt is one we miss if we only see Egypt as a geographically bounded political space in northeast Africa. At one level, it of course was this, but it was also an integral part of a global ecosystem and an empire that stretched across the Mediterranean. In the middle of the 1780s, ecosystem and empire clashed in Egypt. The faraway Ottoman capital fought to prevent a farther away volcano from pushing Egypt into the hands of local upstarts who sought territory and rural capital for themselves. Overcoming normative geographic and political boundaries, climate history thus allows us to bring both seemingly disparate historical agents (peasants, local elites, wind currents, sheep, the Ottoman palace, Laki, the Nile) and distant parts of the globe (Iceland, Egypt, Istanbul, Alaska) together into a single interpretative framework.

Furthermore, like the rest of this book, this chapter's climate history of Laki also contributes to the effort to bring together the fields of environmental history and Middle Eastern history. Global environmental history has, for the most part, ignored the Middle East. However, the history of Laki shows the importance of including the Middle East. Not only does it add empirically to the global history of Laki—explaining its effects on Indian Ocean monsoons, the Nile, and agriculture in one of Africa's most fertile ecological zones—but it also reveals something of the specific political outcomes of the volcano, namely how it contributed to the growing political and economic autonomy and power of rural elites in Egypt against the interests of the Ottoman Empire. Thus, the history of the Middle East helps us to understand Laki. The converse is undeniably also true: Laki helps us to understand the history of the Middle East—specifically, the history of an agricultural crisis with deep political and economic implications and outcomes for the Ottoman Empire's most lucrative and strategic province. A history of late eighteenth-century Ottoman Egypt that ignores Laki cannot fully explain the causal mechanisms of that period's epochal transformations.

Finally, this chapter has attempted to show the utility of analyzing small-scale instances of global climate change. Examining how specific examples of climate change—such as the one Laki produced—intersect with larger, more general narratives and interpretations of climate history offers us the opportunity to test global descriptions of environmental change against particular case studies. The story of Laki is indeed noteworthy because it allows us to follow an unusually clear instance of a singular climate event's exact causes and precise effects. A volcano erupted in Iceland; people starved in Egypt; local elites took advantage of the resulting chaos; the Ottoman Empire reacted. This is the climate history of Laki's direct impact on the Ottoman Empire—the story of how Iceland became a part of Ottoman history.

Conclusion: Empire as Ecosystem

The Ottoman Empire was an ecosystem. The late fifteenth-century Ottoman historians who first wrote down Osman's dream understood this. We early twenty-first-century historians are beginning to as well.

Analyzing the Ottoman Empire as an ecosystem opens up Ottoman historical realities in all their complexities to reveal sets of relationships among resources, peoples, ideas, animals, and places in which all the elements of the system are connected to and dependent upon one another. A change or perturbation in any one part of the environment affects all others. The idea of the Ottoman Empire as an ecosystem of collective dependence and determination foregrounds how the smallest and largest of imperial actors were connected, across time and space, through means of exchange, administration, and mutual reliance. The examples of irrigation in Ottoman Egypt we have examined in this book clearly show how farmers in sometimes very remote parts of the empire were in constant dialogue with the palace in Istanbul and how the two worked together to make the countryside productive. Peasants used the empire and the empire used peasants. A volcano in Iceland, rats in northern India, timber stocks across the Mediterranean, water buffalo in villages throughout Egypt—all of these impacted the Ottoman Empire and must be brought into our analytical frame to properly understand the empire's history.

Whether the focus is productivity, tension, struggle, or violence, viewing the Ottoman Empire and other polities as ecosystems allows for a nuanced and capácious understanding of politics, economy, and society.[1] The likening of the empire to an ecosystem, let me stress, extends far beyond considerations of only nature or environment. From the perspective of empire-as-ecosystem, almost everything within and around an imperial realm comes to be connected, which then allows for analyses of how these connections formed, functioned, were challenged, reconstituted, and ultimately defined. This kind of analysis could operate at many registers. A study of Ottoman ecologies could focus on pushes and pulls within the relatively small confines of the palace and its internal dynamics, or could work to elucidate how Ottoman merchants in the Mughal Empire affected economic networks in Ottoman Salonica. More than just a focus on nature or environment, the eco-approach to empire is about analyzing and understanding connections, relationships, and impacts far beyond the immediately apparent or geographically proximal.

Viewing the Ottoman Empire as an ecosystem means that Egyptian (or Turkish or Arab or Balkan) history cannot be understood as somehow separate from Ottoman history, and vice versa. Ottoman history made Egyptian history, and Egyptian history made Ottoman history. An ecological perspective encompasses the broad Ottoman world that set the rough parameters for nearly everything that occurred in the eastern Mediterranean, Anatolia, and the Balkans for many centuries and that impacted regions far beyond. Thus, an ecological approach to the Ottoman Empire reveals how the empire's variegated geographies, overlapping chronologies, and connected histories functioned across space and time and how small changes in one part of the empire affected places, ideas, and peoples across the imperium and beyond.[2]

The idea of the empire as an ecosystem therefore opens up new geographies for Ottoman history. Indeed, the very practice of environmental history is necessarily transterritorial, as water currents, disease vectors, climate patterns, and migratory animals do not recognize nor (until perhaps very recently) are they controlled by politico-territorial demarcations of geographic space. To study any one of these subjects—as well as trade, religious pilgrimages, intellectual networks, and so on—one cannot simply take the space of the imperial province or the nation-state as a bounded and sealed unit of analysis. While clearly the artificiality of these spatial distinctions has long been recognized, an ecological approach to empire centers the crossing of borders and the transterritoriality of movement, influence, and cause and effect as foundational rather than innovative. There is really no way of getting around the fact that the Nile flows through most of the countries of East Africa or that plague followed commerce around the globe or that the Little Ice Age affected all of the Ottoman Empire (let alone much of the world) and not simply one or another province. Thus, to do a proper analysis of the ecology of the Nile, the history of plague, or the effects of climate change on the Ottoman Empire, one must consider the impacts of the river, of pathogens, and of temperature in a multitude of places with all of their political, cultural, and geographic nuance, complexity, and difference. Similarly, as we saw with the examples of food and wood and the story of Laki's sulfur dioxide, one of the most productive ways of getting past any static notion of territory or movement is to follow the path of a particular set of natural resources, processes, or even molecules. By making the history of a material object, chemical compound, or natural resource the focus of our inquiries, we can bypass many of the assumptions and preconditions already embedded in much historical work that accepts the territorial unity of the imperial province or the nation-state as a given. For these and related reasons, Middle East environmental history has indeed proved itself one of the most effective and exciting ways of connecting the Middle East to world history, of doing a kind of global history that encompasses the Middle East, a region that has up to now been ignored by such analyses. From an ecological perspective, the Ottoman Empire's geography looks very different indeed.

As does its chronology. Ottoman history organized around climate change events, for example, tells a very different story than does one organized around politics. What if we periodized imperial history based on the rise and fall of a single canal in the Egyptian countryside? Would the history of labor on the canal, imperial interest in the waterway, taxation, productivity levels, and food export open up a different window on the empire's fortunes over time? How did an ox in the Egyptian countryside experience the empire? From the bovine perspective, was the sixteenth century different from the nineteenth? These sorts of questions all point to different ways of conceiving of and analyzing chronologies of Ottoman history that move us past more standard periodizations organized around the reigns of sultans or wars. In this book, I have tried to use environmental history to open up new ways of conceiving of the pace of Ottoman history, its turning points, its periods. Most specifically, I have tried to offer a new interpretation of the turn of the nineteenth century—one that extends that hinge period's chronology further back in time to encompass a novel set of economic, ecological, political, and social causes and effects.

The ecological approach to empire also helps to integrate multiple kinds of actors—many of whom have remained in the empire's historiographical shadows—into the study of Ottoman history. Peasants, water buffalo, silt, fleas, dirt, salt, microbes, trees, volcanoes, and water currents have all had a role to play in this book. Of all these agents, peasants have obviously and rightly received the most attention from scholars, and, needless to say, they have figured prominently in this book as well. A generation of scholars using court records from various locations throughout the Ottoman Middle East has given us a quite detailed picture of how peasants interacted and dealt with the various state polities under which they lived.[3] Most of these studies have attempted to insert peasants into larger frameworks—imperial systems of governance, global commercial networks, and wider religious or cultural communities. Building on these analyses, an environmental perspective shows how peasants were not merely participants within these larger webs of power, but were often those actually holding the reins of power. When it came to understanding the intimate details of a village's environment, the cultivation history of a field, or irrigation in rural Ottoman Egypt, only Egyptian peasants, who had lived on that land for generations and who knew its topographies and idiosyncrasies, could direct the management and usage of that land. As we have seen, they were the ecological experts on whom the Ottoman state had to rely.

Nonhuman animals were also key historical actors in the agrarian economy of the Ottoman Empire. Water buffalo, camels, oxen, donkeys, and various other domesticated ungulates transported heavy loads across vast distances, turned waterwheels, and dredged canals. They were significant forms of property, sources of food and heat, stores of economic value, hedges against crop failure, and status symbols. No history of the Ottoman Empire, a vast rural agrarian society, can therefore afford to ignore how the ubiquity of domesticated work animals shaped almost everything that undergirded the empire's social and economic life.

Much of the work of environmental history has gone toward arguing for the role of nonhuman nature as a historical actor. This book has focused on the water flowing through Egypt's canals, the dirt that clogged these waterways, the salt that saturated much of the Egyptian delta, disease microbes, trees, wind patterns, camels, and much more. All affected the history of Ottoman Egypt in ways as important as—and often more important than—imperial bureaucrats, Egyptian peasants, wars, sultans, or global commodity prices. Indeed, it would not be an overexaggeration to say that the rainfall in the Ethiopian highlands that created the annual flood had more of an impact on Egypt and its past than any political entity that controlled the territory over its millennia of documented history.

Across these millennia, countless historians, poets, scholars, politicians, pundits, artists, activists, and others have offered the entire spectrum of ideas from grand narratives to pithy platitudes about the role of the environment in the history of the Middle East. These include Egypt being the gift of the Nile; ideas connecting irrigation to Oriental despotism; competing notions that the scarcity of water or the excess of oil create certain modes of authoritarian politics; tensions between desert and sown; and images of the fertile crescent or the granary of Rome. Slices of more general Orientalist notions of difference between east and west, Asia and Europe, these portrayals of the Middle Eastern environment take it to be uniquely and dangerously fragile, flawed, degraded, fallen, contradictory, excessive, out of balance, always on the brink of collapse. "An unnatural nature," in Timothy Mitchell's words, "appears to determine Middle Eastern history."[4]

Understanding the Ottoman Empire—the longest-lasting empire in the Middle East since antiquity—to be an ecosystem helps to overcome these timeless colonially tinged notions of an aberrant Middle Eastern environment. It brings us down to earth, forces us to slog through the muddy canals of the countryside, and focuses our attention on ecological processes. How did peoples, animals, trees, diseases, soils function together within the contexts of ecology, empire, and geography? Empire-as-ecosystem puts process before idea, function before theory, archive before ideology. Overarching narratives and interpretative frameworks—whether about the Middle East, climate, or politics—must be built up from collections of empirical case studies and specific stories. Very often, though, they are not. This book has tried to be different. It has sought to argue that a new understanding of the Middle East and of the history of the environment can only come from digging our feet deep in the soil, feeling the water up to our shins, hearing the donkeys bray, all while reaching our hands up to the sky.

Acknowledgments

I thank the following venues for providing me the space and support to think through many of the ideas undergirding this book: "Global Implications of the Middle Eastern Environment," *History Compass* 9 (2011): 952–70; "Oriental Democracy," *Global Environment* 7 (2014): 381–404; "From the Bottom Up: The Nile, Silt, and Humans in Ottoman Egypt," in *Environmental Imaginaries of the Middle East and North Africa*, ed. Diana K. Davis and Edmund Burke III, 113–35 (Athens: Ohio University Press, 2011); "An Irrigated Empire: The View from Ottoman Fayyum," *International Journal of Middle East Studies* 42 (2010): 569–90; "Labor and Environment in Egypt since 1500," *International Labor and Working-Class History* 85 (2014): 10–32; "Engineering the Ottoman Empire: Irrigation and the Persistence of Early Modern Expertise," in *Ottoman Rural Societies and Economies*, ed. Elias Kolovos, 399–413 (Rethymno: Crete University Press, 2015); "Animals as Property in Early Modern Ottoman Egypt," *Journal of the Economic and Social History of the Orient* 53 (2010): 621–52; "Unleashing the Beast: Animals, Energy, and the Economy of Labor in Ottoman Egypt," *American Historical Review* 118 (2013): 317–48; "Anatolian Timber and Egyptian Grain: Things That Made the Ottoman Empire," in *Early Modern Things: Objects and Their Histories, 1500–1800*, ed. Paula Findlen, 274–93 (New York: Routledge, 2013); "The Nature of Plague in Late Eighteenth-Century Egypt," *Bulletin of the History of Medicine* 82 (2008): 249–75; and "Ottoman Iceland: A Climate History," *Environmental History* 20 (2015): 262–84. The portions of these articles and chapters I used in this book were all significantly revised. I have removed any redundancies, reworked them to fit together as a new whole, and attempted to bring out the synergistic themes tying them together.

I relied heavily on Camille Cole to complete this book. She was the perfect editorial assistant. Her efficiency, meticulous attention to detail, and challenging ideas made this book both possible and better. My enormous thanks to this impressive scholar. Three anonymous reviewers offered their incisive critiques and very helpful suggestions on a draft of this book's manuscript. They nudged me in useful directions, saved me from many costly mistakes, and pushed me to make this book's audience even wider.

I hesitate to offer a long list of specific people deserving of acknowledgment. Not because there aren't any, but precisely because there are far too many! It is not an overexaggeration to say that every teacher, co-panelist, friend, editor, colleague, reviewer, reader, interlocutor, and student whom I have had the pleasure and good fortune to know has helped to shape this work. Many of their names can be found in the works listed in the paragraph before last. This book is a summation not only of a decade of scholarly work but of a decade of life. To all, my thanks immeasurable, unending.

Karen Merikangas Darling at the University of Chicago Press put this book in your hands. She supported it from the very beginning and made the publication process both enjoyable and productive. Evan White and Mary Corrado at the press oversaw the book's production with the perfect combination of masterful efficiency and detailed care. Susan J. Cohan's impeccable copyediting improved the book. For their mapmaking wizardry, I thank Stacey Maples and Kevin Quach. I owe the index to Derek Gottlieb.

For their patience and help over the years, I thank the staffs of Dār al-Wathāʾiq al-Qawmiyya in Cairo, the National Archives of the United Kingdom in Kew, and the Başbakanlık Osmanlı Arşivi and the Topkapı Sarayı Müzesi Arşivi in Istanbul.

The research and writing of this book were made possible by the financial support of Yale University; the Andrew W. Mellon Fellowship of Scholars in the Humanities at Stanford University; the Department of History at the University of California, Berkeley; the Institute of Turkish Studies; the Fulbright-Hays Doctoral Dissertation Research Abroad Fellowship; and the American Research Center in Egypt. This book would not exist without the generous support of these institutions, for which I am exceedingly grateful.

Notes

ABBREVIATIONS USED IN THE NOTES

ARCHIVES

BOA	Başbakanlık Osmanlı Arşivi, Istanbul		
	HAT	Hatt-i Hümayun	
	MM	Mühimme-i Mısır	
DWQ	Dār al-Wathā'iq al-Qawmiyya, Cairo		
TNA	The National Archives of the United Kingdom, Kew		
	FO	Foreign Office	
	PC	Privy Council	
TSMA	Topkapı Sarayı Müzesi Arşivi, Istanbul		
	E.	Evrak	

ISLAMIC MONTHS IN OTTOMAN TURKISH (ARABIC IN PARENTHESES)

M	Muharrem (Muḥarram)
S	Safer (Ṣafar)
Ra	Rebiülevvel (Rabīʿ al-Awwal)
R	Rebiülahir (Rabīʿ al-Thānī)
Ca	Cemazilevvel (Jumādā al-Ūlā)
C	Cemaziyel'ahır (Jumādā al-Ākhira)
B	Receb (Rajab)
Ş	Şaʿban (Shaʿbān)
N	Ramazan (Ramaḍān)
L	Şevval (Shawwāl)
Za	Zilkade (Dhū al-Qaʿda)
Z	Zilhicce (Dhū al-Ḥijja)

PREFACE

1. For a discussion of this dream story and other accounts of Ottoman origins, see Colin Imber, "The Ottoman Dynastic Myth," *Turcica* 19 (1987): 7–27.

INTRODUCTION

1. For works that discuss the Middle East as a Eurasian contact zone, see Janet L. Abu-Lughod, *Before European Hegemony: The World System A.D. 1250–1350* (New York: Oxford University Press, 1989); Richard W. Bulliet, *Cotton, Climate, and Camels in Early Islamic Iran: A Moment in World History* (New York: Columbia University Press, 2009); Richard M. Eaton, "Islamic History as Global History," in *Islamic and European Expansion: The Forging of a Global Order*, ed. Michael Adas, 1–36 (Philadelphia: Temple University Press, 1993); Gagan D.S. Sood, "Pluralism, Hegemony and Custom in Cosmopolitan Islamic Eurasia, ca. 1720–90, with Particular Reference to the Mercantile Arena" (PhD diss., Yale University, 2008).

2. For instructive treatments of pastoralism in the Middle East, see Reşat Kasaba, *A Moveable Empire: Ottoman Nomads, Migrants, and Refugees* (Seattle: University of Washington Press, 2009); Arash Khazeni, *Tribes and Empire on the Margins of Nineteenth-Century Iran* (Seattle: University of Washington Press, 2009); Andrew Gordon Gould, "Pashas and Brigands: Ottoman Provincial Reform and Its Impact on the Nomadic Tribes of Southern Anatolia, 1840–1885" (PhD diss., University of California, Los Angeles, 1973).

3. For a study of some of these connections in the early modern Muslim world, see Muzaffar Alam and Sanjay Subrahmanyam, *Indo-Persian Travels in the Age of Discoveries, 1400–1800* (Cambridge: Cambridge University Press, 2007).

4. The most ambitious work to tackle this subject is John F. Richards, *The Unending Frontier: An Environmental History of the Early Modern World* (Berkeley: University of California Press, 2003). In a telling indication of the absence of the Middle East in the global history of the environment, this magisterial work has very little to say about the region.

5. Andrew Watson, *Agricultural Innovation in the Early Islamic World: The Diffusion of Crops and Farming Techniques, 700–1100* (Cambridge: Cambridge University Press, 1983). For a critique of certain aspects of the "Islamic green revolution" thesis, see Michael Decker, "Plants and Progress: Rethinking the Islamic Agricultural Revolution," *Journal of World History* 20 (2009): 187–206.

6. Michael W. Dols, *The Black Death in the Middle East* (Princeton, NJ: Princeton University Press, 1977); William H. McNeill, *Plagues and Peoples* (Garden City, NY: Anchor Press / Doubleday, 1976); Michael W. Dols, "The Second Plague Pandemic and Its Recurrences in the Middle East: 1347–1894," *Journal of the Economic and Social History of the Orient* 22 (1979): 162–89.

7. Thomas F. Glick, *Irrigation and Hydraulic Technology: Medieval Spain and Its Legacy* (Brookfield, VT: Variorum, 1996); Thomas F. Glick, *Irrigation and Society in Medieval Valencia* (Cambridge, MA: Harvard University Press, 1970).

8. Ralph Hattox, *Coffee and Coffeehouses: The Origins of a Social Beverage in the Medieval Near East* (Seattle: University of Washington Press, 1985); Michel Tuchscherer, ed., *Le commerce du café avant l'ère des plantations coloniales: Espaces, réseaux, sociétés (XVe–XIXe siècle)* (Cairo: Institut français d'archéologie orientale, 2001); William Gervase Clarence-Smith and Steven Topik, eds., *The Global Coffee Economy in Africa, Asia, and Latin America, 1500–1989* (Cambridge: Cambridge University Press, 2003).

9. Ghislaine Lydon, *On Trans-Saharan Trails: Islamic Law, Trade Networks, and Cross-Cultural Exchange in Nineteenth-Century Western Africa* (Cambridge: Cambridge University Press, 2009); Ghislaine Lydon, "Writing Trans-Saharan History: Methods, Sources and Interpretations across the African Divide," *Journal of North African Studies* 10 (2005): 293–324.

10. On maize, see J.R. McNeill, *The Mountains of the Mediterranean World: An Environmental History* (Cambridge: Cambridge University Press, 1992), 89–90; Faruk Tabak, *The Waning of the Mediterranean, 1550–1870: A Geohistorical Approach* (Baltimore: Johns Hopkins University Press, 2008), 255–69. On the diffusion of maize from Egypt and North Africa to other parts of Africa, see James C. McCann, *Maize and Grace: Africa's Encounter with a New World Crop, 1500–2000* (Cambridge, MA: Harvard University Press, 2007).

11. In this regard, see, for example, Timothy Mitchell, "Carbon Democracy," *Economy and Society* 38 (2009): 399–432.

12. On the Little Ice Age and the Ottoman Empire, see Sam White, *The Climate of Rebellion in the Early Modern Ottoman Empire* (Cambridge: Cambridge University Press, 2011). On the Icelandic eruptions, see Luke Oman, Alan Robock, Georgiy L. Stenchikov, and Thorvaldur Thordarson, "High-Latitude Eruptions Cast Shadow over the African Monsoon and the Flow of the Nile," *Geophysical Research Letters* 33 (2006): L18711. On El Niño famines, see Mike Davis, *Late Victorian Holocausts: El Niño Famines and the Making of the Third World* (London: Verso, 2001). On the famines in Anatolia, see Mehmet Yavuz Erler, *Osmanlı Devleti'nde Kuraklık ve Kıtlık Olayları, 1800–1880* (Istanbul: Libra Kitap, 2010); Zozan Pehlivan, "Beyond 'The Desert and the Sown': Peasants, Pastoralists, and Climate Crises in Ottoman Diyarbekir, 1840–1890" (PhD diss., Queen's University, 2016).

13. For an instructive example in this regard, see Timothy Mitchell, "Can the Mosquito Speak?," in *Rule of Experts: Egypt, Techno-Politics, Modernity*, 19–53 (Berkeley: University of California Press, 2002).

14. The Hijaz is the name of the region containing Mecca and Medina on the western coast of the Arabian Peninsula.

15. Although I will make a few forays into earlier periods, my primary focus in this introduction is on the period after the rise of Islam in the seventh century and, even more so, on the period after 1500. I am thus regrettably not engaging the illuminating work done on the ancient Middle Eastern environment about such topics as Persian Gulf seafloor changes, desertification, irrigation, and salinization. For some of this literature, see, for example, P. Kassler, "The Structural and Geomorphic Evolution of the Persian Gulf," in *The Persian Gulf: Holocene Carbonate Sedimentation and Diagenesis in a*

Shallow Epicontinental Sea, ed. B.H. Purser, 11–32 (Berlin: Springer-Verlag, 1973); Elazar Uchupi, S.A. Swift, and D.A. Ross, "Gas Venting and Late Quaternary Sedimentation in the Persian (Arabian) Gulf," *Marine Geology* 129 (1996): 237–69; Michael Brookfield, "The Desertification of the Egyptian Sahara during the Holocene (the Last 10,000 Years) and Its Influence on the Rise of Egyptian Civilization," in *Landscapes and Societies: Selected Cases*, ed. I. Peter Martini and Ward Chesworth, 91–108 (Dordrecht, Netherlands: Springer, 2010); Arie S. Issar, *Water Shall Flow from the Rock: Hydrogeology and Climate in the Lands of the Bible* (Berlin: Springer-Verlag, 1990); Thorkild Jacobsen and Robert M. Adams, "Salt and Silt in Ancient Mesopotamian Agriculture," *Science* 128 (1958): 1251–58; Thorkild Jacobsen, *Salinity and Irrigation Agriculture in Antiquity: Diyala Basin Archaeological Report on Essential Results, 1957–58*, Bibliotheca Mesopotamica, vol. 14 (Malibu, CA: Undena Publications, 1982).

16. On the environmental history of China from antiquity to the present, see Mark Elvin, *The Retreat of the Elephants: An Environmental History of China* (New Haven, CT: Yale University Press, 2004); Mark Elvin and Liu Ts'ui-Jung, eds., *Sediments of Time: Environment and Society in Chinese History* (Cambridge: Cambridge University Press, 1998). On South Asia, see, for example, Richard H. Grove, Vinita Damodaran, and Satpal Sangwan, eds., *Nature and the Orient: The Environmental History of South and Southeast Asia* (Delhi: Oxford University Press, 1998).

17. For some of this history, see Robert O. Collins, *The Nile* (New Haven, CT: Yale University Press, 2002).

18. For works that use Ottoman archives for environmental history, see, for example, S. White, *Climate of Rebellion*; Alan Mikhail, *Nature and Empire in Ottoman Egypt: An Environmental History* (Cambridge: Cambridge University Press, 2011); Selçuk Dursun, "Forest and the State: History of Forestry and Forest Administration in the Ottoman Empire" (PhD diss., Sabancı University, 2007).

19. For studies that use these archives for environmental history, see, for example, Diana K. Davis, *Resurrecting the Granary of Rome: Environmental History and French Colonial Expansion in North Africa* (Athens: Ohio University Press, 2007); Khazeni, *Tribes and Empire*; Sandra M. Sufian, *Healing the Land and the Nation: Malaria and the Zionist Project in Palestine, 1920–1947* (Chicago: University of Chicago Press, 2007).

20. For a representative sampling of some of this work, see William C. Brice, ed., *The Environmental History of the Near and Middle East since the Last Ice Age* (London: Academic Press, 1978).

21. Karl W. Butzer, *Early Hydraulic Civilization in Egypt: A Study in Cultural Ecology* (Chicago: University of Chicago Press, 1976); Peter Christensen, *The Decline of Iranshahr: Irrigation and Environments in the History of the Middle East, 500 B.C. to A.D. 1500* (Copenhagen: Museum Tusculanum Press, 1993).

22. J.M. Wagstaff, *The Evolution of Middle Eastern Landscapes: An Outline to A.D. 1840* (London: Croon Helm, 1985). To be accurate, Wagstaff's book does address the post-1500 period, but the bulk of the work (over two-

thirds) concerns the earlier period. Carlos E. Cordova, *Millennial Landscape Change in Jordan: Geoarchaeology and Cultural Ecology* (Tucson: University of Arizona Press, 2007); Russell Meiggs, *Trees and Timber in the Ancient Mediterranean World* (Oxford: Clarendon Press, 1982); J.V. Thirgood, *Man and the Mediterranean Forest: A History of Resource Depletion* (London: Academic Press, 1981); Robert McC. Adams, *Land behind Baghdad: A History of Settlement on the Diyala Plains* (Chicago: University of Chicago Press, 1965). See also the very useful discussion of long-term environmental landscape change in Edmund Burke III, "The Transformation of the Middle Eastern Environment, 1500 B.C.E.–2000 C.E.," in *The Environment and World History*, ed. Edmund Burke III and Kenneth Pomeranz, 81–117 (Berkeley: University of California Press, 2009).

23. Christensen, *Decline of Iranshahr*; J.R. McNeill, *Mountains of the Mediterranean*.

24. See the relevant sections of J.R. McNeill, *Mountains of the Mediterranean*. This study focuses on five mountain ranges around the Mediterranean. In addition to the Rif and Taurus Mountains, in Morocco and Turkey, respectively, McNeill also examines mountain ranges in Greece, Spain, and Italy.

25. In addition, see Rhoads Murphey, "The Decline of North Africa since the Roman Occupation: Climatic or Human?," *Annals of the Association of American Geographers* 41 (1951): 116–32. This study uses empirical evidence from North Africa from the Roman period to the present to challenge some tenets of the historical and geographic literature on climate change.

26. Karl A. Wittfogel, *Oriental Despotism: A Comparative Study of Total Power* (New Haven, CT: Yale University Press, 1957). For a more abbreviated form of his main arguments, see Karl A. Wittfogel, "The Hydraulic Civilizations," in *Man's Role in Changing the Face of the Earth*, ed. William L. Thomas Jr., 152–64 (Chicago: University of Chicago Press, 1956).

27. This argument is advanced in the case of Egypt more thoroughly in Mikhail, *Nature and Empire*.

28. Mohamed Reda Sbeinati, Ryad Darawcheh, and Mikhail Mouty, "The Historical Earthquakes of Syria: An Analysis of Large and Moderate Earthquakes from 1365 B.C. to 1900 A.D.," *Annals of Geophysics* 48 (2005): 347–435; Elizabeth Zachariadou, ed., *Natural Disasters in the Ottoman Empire* (Rethymnon, Greece: Crete University Press, 1999); Yaron Ayalon, *Natural Disasters in the Ottoman Empire: Plague, Famine, and Other Misfortunes* (Cambridge: Cambridge University Press, 2015); N.N. Ambraseys and C.F. Finkel, *The Seismicity of Turkey and Adjacent Areas: A Historical Review, 1500–1800* (Istanbul: Eren, 1995); N.N. Ambraseys, C.P. Melville, and R.D. Adams, *The Seismicity of Egypt, Arabia and the Red Sea: A Historical Review* (Cambridge: Cambridge University Press, 1994); Nicholas Ambraseys, *Earthquakes in the Eastern Mediterranean and Middle East: A Multidisciplinary Study of Seismicity up to 1900* (Cambridge: Cambridge University Press, 2009); N.N. Ambraseys and C.P. Melville, *A History of Persian Earthquakes* (Cambridge: Cambridge University Press, 1982).

29. On Aswan, see Yusuf A. Shibl, *The Aswan High Dam* (Beirut: Arab Institute for Research and Publishing, 1971); Hussein M. Fahim, *Dams, People*

and Development: The Aswan High Dam Case (New York: Pergamon Press, 1981). See also the relevant sections of John Waterbury, *Hydropolitics of the Nile Valley* (Syracuse, NY: Syracuse University Press, 1979); Collins, *Nile*. On the construction of dams during the period that came to be known as Iran's First Seven Year Plan, from 1948 to 1955, see Peter Beaumont, "Water Resource Development in Iran," *Geographical Journal* 140 (1974): 418–31; Gordon R. Clapp, "Iran: A TVA for the Khuzestan Region," *Middle East Journal* 11 (1957): 1–11. On dams in Turkey, see J.R. McNeill, *Something New under the Sun: An Environmental History of the Twentieth-Century World* (New York: Norton, 2000), 123.

30. On the Suez Canal, see D.A. Farnie, *East and West of Suez: The Suez Canal in History, 1854–1956* (Oxford: Clarendon Press, 1969); John Marlowe, *World Ditch: The Making of the Suez Canal* (New York: Macmillan, 1964). On the Don-Volga Canal, see Halil İnalcık, "The Origins of the Ottoman-Russian Rivalry and the Don-Volga Canal, 1569," *Les annales de l'Université d'Ankara* 1 (1946–47): 47–106; A.N. Kurat, "The Turkish Expedition to Astrakhan and the Problem of the Don-Volga Canal," *Slavonic and East European Review* 40 (1961): 7–23. On the Karun River project, see Khazeni, *Tribes and Empire*, 23–25. On the construction of a pipeline between the Red and Dead Seas, see Basel N. Asmar, "The Science and Politics of the Dead Sea: Red Sea Canal or Pipeline," *Journal of Environment and Development* 12 (2003): 325–39. On Turkey's Southeast Anatolia Development Project (GAP), see Ali Çarkoğlu and Mine Eder, "Development *alla Turca*: The Southeastern Anatolia Development Project (GAP)," in *Environmentalism in Turkey: Between Democracy and Development?*, ed. Fikret Adaman and Murat Arsel, 167–84 (Aldershot, UK: Ashgate, 2005); Ali Çarkoğlu and Mine Eder, "Domestic Concerns and the Water Conflict over the Euphrates-Tigris River Basin," *Middle Eastern Studies* 37 (2001): 41–71; Leila Harris, "Postcolonialism, Postdevelopment, and Ambivalent Spaces of Difference in Southeastern Turkey," *Geoforum* 39 (2008): 1698–1708.

31. I have in mind here works like S. White, *Climate of Rebellion*; Khazeni, *Tribes and Empire*; Sufian, *Healing the Land and the Nation*; Wolf-Dieter Hütteroth, "Ecology of the Ottoman Lands," in *The Later Ottoman Empire, 1603–1839*, vol. 3 of *The Cambridge History of Turkey*, ed. Suraiya N. Faroqhi, 18–43 (Cambridge: Cambridge University Press, 2006); Dursun, "Forest and the State"; Mikhail, *Nature and Empire*. See also several of the essays in Suraiya Faroqhi, ed., *Animals and People in the Ottoman Empire* (Istanbul: Eren, 2010).

32. On some of the dangers of environmental determinism in the study of the Middle Eastern environment, see Diana K. Davis, "Power, Knowledge, and Environmental History in the Middle East and North Africa," *International Journal of Middle East Studies* 42 (2010): 657–59.

33. William Cronon, *Changes in the Land: Indians, Colonists, and the Ecology of New England*, 1st rev. ed. (New York: Hill and Wang, 2003), 13.

34. Ibid., 15.

35. In addition to the works I discuss in detail here, see also Arie S. Issar and Mattanyah Zohar, *Climate Change—Environment and Civilization in the Middle East* (Berlin: Springer, 2004).

36. One of the classic studies of the impact of climate change on European history is Emmanuel Le Roy Ladurie, *Times of Feast, Times of Famine: A History of Climate since the Year 1000*, trans. Barbara Bray (Garden City, NY: Doubleday, 1971). The study of the historical consequences of climate change is now an enormous and fast-moving field. For recent and accessible takes on the subject, see Brian Fagan, *The Long Summer: How Climate Changed Civilization* (New York: Basic Books, 2004); Eugene Linden, *The Winds of Change: Climate, Weather, and the Destruction of Civilizations* (New York: Simon and Schuster, 2006). Some of the earliest attempts to integrate climate change into a history of the Middle East include Murphey, "Decline of North Africa"; William Griswold, "Climatic Change: A Possible Factor in the Social Unrest of Seventeenth Century Anatolia," in *Humanist and Scholar: Essays in Honor of Andreas Tietze*, ed. Heath W. Lowry and Donald Quataert, 37–57 (Istanbul: Isis Press, 1993). The best recent studies of the impacts of climate change in the Middle East are Bulliet, *Cotton, Climate, and Camels*; S. White, *Climate of Rebellion*.

37. Bulliet, *Cotton, Climate, and Camels*, 69–95. For further discussion of climate change in Iran and other parts of the Middle East in this period, see Ronnie Ellenblum, *The Collapse of the Eastern Mediterranean: Climate Change and the Decline of the East, 950–1072* (Cambridge: Cambridge University Press, 2012).

38. S. White, *Climate of Rebellion*. Literature on the general crisis of the seventeenth century began in the 1950s with E.J. Hobsbawm, "The General Crisis of the European Economy in the 17th Century," *Past and Present* 5 (1954): 33–53; E.J. Hobsbawm, "The Crisis of the 17th Century—II," *Past and Present* 6 (1954): 44–64. For a recent series of studies on the topic, see Jonathan Dewald, Geoffrey Parker, Michael Marmé, and J.B. Shank, "*AHR* Forum: The General Crisis of the Seventeenth Century Revisited," *American Historical Review* 113 (2008): 1029–99. For a recent global analysis of climate change and the general crisis of the seventeenth century, see Geoffrey Parker, *Global Crisis: War, Climate Change and Catastrophe in the Seventeenth Century* (New Haven, CT: Yale University Press, 2013).

39. Karen Barkey, *Bandits and Bureaucrats: The Ottoman Route to State Centralization* (Ithaca, NY: Cornell University Press, 1994); William Griswold, *The Great Anatolian Rebellion, 1000–1020/1591–1611* (Berlin: Klaus Schwarz Verlag, 1983).

40. S. White, *Climate of Rebellion*.

41. Oman, Robock, Stenchikov, and Thordarson, "High-Latitude Eruptions Cast Shadow," L18711.

42. For a general introduction to the Tanzimat, see M. Şükrü Hanioğlu, *A Brief History of the Late Ottoman Empire* (Princeton, NJ: Princeton University Press, 2008).

43. On these famines, see Erler, *Kuraklık ve Kıtlık*; Pehlivan, "Beyond 'The Desert and the Sown'"; Donald Quataert, *The Ottoman Empire, 1700–1922*, 2nd ed. (Cambridge: Cambridge University Press, 2005), 114–15.

44. For discussions of environmental determinism in the study of the Middle East and North Africa, see D. Davis, "Power, Knowledge, and Envi-

ronmental History"; Bulliet, *Cotton, Climate, and Camels*, vii–x; Murphey, "Decline of North Africa."

45. Edmund Burke III, "The Big Story: Human History, Energy Regimes, and the Environment," in *The Environment and World History*, ed. Edmund Burke III and Kenneth Pomeranz (Berkeley: University of California Press, 2009), 35.

46. On this point, see Vaclav Smil, *Energy in World History* (Boulder, CO: Westview Press, 1994); Vaclav Smil, *Energy in Nature and Society: General Energetics of Complex Systems* (Cambridge: Massachusetts Institute of Technology Press, 2008); Stephen J. Pyne, *World Fire: The Culture of Fire on Earth* (Seattle: University of Washington Press, 1997); Stephen J. Pyne, *Vestal Fire: An Environmental History, Told through Fire, of Europe and Europe's Encounter with the World* (Seattle: University of Washington Press, 1997). For recent work in this regard on the Middle East, see Toby Craig Jones, *Desert Kingdom: How Oil and Water Forged Modern Saudi Arabia* (Cambridge, MA: Harvard University Press, 2010); Mitchell, "Carbon Democracy."

47. Analyses of the various impacts of animals on Middle Eastern history include Faroqhi, *Animals and People*; Annemarie Schimmel, *Islam and the Wonders of Creation: The Animal Kingdom* (London: al-Furqān Islamic Heritage Foundation, 2003); Mohamed Hocine Benkheira, Catherine Mayeur-Jaouen, and Jacqueline Sublet, *L'animal en islam* (Paris: Indes savantes, 2005); Basheer Ahmad Masri, *Animal Welfare in Islam* (Markfield, Leicestershire, UK: Islamic Foundation, 2007); Richard C. Foltz, *Animals in Islamic Tradition and Muslim Cultures* (Oxford: Oneworld, 2006); Thomas T. Allsen, *The Royal Hunt in Eurasian History* (Philadelphia: University of Pennsylvania Press, 2006); Richard W. Bulliet, *The Camel and the Wheel* (New York: Columbia University Press, 1990); Bulliet, *Cotton, Climate, and Camels*; Alan Mikhail, *The Animal in Ottoman Egypt* (New York: Oxford University Press, 2014); Suraiya Faroqhi, "Camels, Wagons, and the Ottoman State in the Sixteenth and Seventeenth Centuries," *International Journal of Middle East Studies* 14 (1982): 523–39.

48. Richard Bulliet, "The Camel and the Watermill," *International Journal of Middle East Studies* 42 (2010): 666–68.

49. Burke, "Big Story," 33–53. For another study that gives energy utilization primacy of place in a long story of human history, see J.R. McNeill, "The First Hundred Thousand Years," in *The Turning Points of Environmental History*, ed. Frank Uekoetter, 13–28 (Pittsburgh: University of Pittsburgh Press, 2010). For a general discussion of energy regimes, see J.R. McNeill, *Something New under the Sun*, 296–324.

50. Bulliet, "Camel and the Watermill."

51. The classic studies of plague in the Middle East are Dols, *Black Death in the Middle East*; Daniel Panzac, *La peste dans l'Empire Ottoman, 1700–1850* (Leuven, Belgium: Association pour le Développement des Études Turques, 1985). See also the following important recent study: Nükhet Varlık, *Plague and Empire in the Early Modern Mediterranean World: The Ottoman Experience, 1347–1600* (Cambridge: Cambridge University Press, 2015).

52. For a recent review of this literature, see Sam White, "Rethinking

Disease in Ottoman History," *International Journal of Middle East Studies* 42 (2010): 549–67.

53. The sources for the history of disease in the Middle East are rather nonspecific in their definition of outbreaks and epidemics. It is therefore difficult to know the exact character of most instances of diseases described as "plague" (*ṭāʿūn* in Arabic, *taun* in Turkish) or as an "epidemic" (*wabāʾ* in Arabic, *veba* in Turkish). Nevertheless, other circumstantial evidence can often help in determining whether or not a particular epidemic was, for example, *Yersinia pestis* (bubonic plague) or something else entirely. For a very useful treatment of some of the difficulties involved in correctly identifying diseases in Arabic and Ottoman source materials, see S. White, "Rethinking Disease in Ottoman History," 555–58.

54. Dols, "Second Plague," 169, 176. For the period from 1416 to 1514, David Neustadt (Ayalon), reports that an outbreak of plague struck Egypt an average of once every seven years. David Neustadt (Ayalon), "The Plague and Its Effects upon the Mamlûk Army," *Journal of the Royal Asiatic Society of Great Britain and Ireland* (1946): 68. See also: Dols, *Black Death in the Middle East*, 223–24; Panzac, *La peste*, 197–207.

55. Dols, "Second Plague," 168–69, 175–76; André Raymond, "Les Grandes Épidémies de peste au Caire aux XVIIe and XVIIIe siècles," *Bulletin d'Études Orientales* 25 (1973): 203–10. The plague epidemics included in Raymond's study represent a subset of those cited by Dols.

56. Michael W. Dols, "Plague in Early Islamic History," *Journal of the American Oriental Society* 94 (1974): 381; Raymond, "Les Grandes Épidémies de peste au Caire," 208–9. My discussion of the relationships between the movements of peoples and goods and the spread of plague is informed by W. McNeill, *Plagues and Peoples*; Abu-Lughod, *Before European Hegemony*.

57. Raymond, "Les Grandes Épidémies de peste au Caire," 208–9; Dols, "Second Plague," 179–80. Dols compiles a list of plagues that came to Egypt and North Africa from the Sudan and Central Africa on the basis of Georg Sticker, *Abhandlungen aus der Seuchengeschichte und Seuchenlehre* (Giessen, Germany: A. Töpelmann, 1908–12). For more on plague in the Sudan, see Terence Walz, *Trade between Egypt and Bilād as-Sūdān, 1700–1820* (Cairo: Institut français d'archéologie orientale, 1978), 200–201.

58. J.R. McNeill, *Something New under the Sun*, 195–96. On cholera in India, see David Arnold, *Colonizing the Body: State Medicine and Epidemic Disease in Nineteenth-Century India* (Berkeley: University of California Press, 1993), 159–99. On the disease in the United States, see Charles E. Rosenberg, *The Cholera Years: The United States in 1832, 1849, and 1866* (Chicago: University of Chicago Press, 1987). On cholera in Europe, see Richard J. Evans, *Death in Hamburg: Society and Politics in the Cholera Years, 1830–1910* (New York: Oxford University Press, 1987); François Delaporte, *Disease and Civilization: The Cholera in Paris, 1832*, trans. Arthur Goldhammer (Cambridge: Massachusetts Institute of Technology Press, 1986); Catherine J. Kudlick, *Cholera in Post-Revolutionary Paris: A Cultural History* (Berkeley: University of California Press, 1996).

59. Valeska Huber, *Channelling Mobilities: Migration and Globalisation*

in the Suez Canal Region and Beyond, 1869–1914 (Cambridge: Cambridge University Press, 2013), 241–71; Michael Christopher Low, "Empire and the Hajj: Pilgrims, Plagues, and Pan-Islam under British Surveillance, 1865–1908," *International Journal of Middle East Studies* 40 (2008): 269–90.

60. LaVerne Kuhnke, *Lives at Risk: Public Health in Nineteenth-Century Egypt* (Berkeley: University of California Press, 1990), 95.

61. On the history of quarantine in the Ottoman Empire, see Birsen Bulmuş, *Plague, Quarantines and Geopolitics in the Ottoman Empire* (Edinburgh: Edinburgh University Press, 2012).

62. The full details of this episode are given in Kuhnke, *Lives at Risk*, 107–8.

63. Huber, *Channelling Mobilities*, 241–71; J.R. McNeill, *Something New under the Sun*, 196.

64. For some of the existing literature on these topics, see Jeff Albert, Magnus Bernhardsson, and Roger Kenna, eds., *Transformations of Middle Eastern Natural Environments: Legacies and Lessons*, Bulletin Series, no. 103 (New Haven, CT: Yale School of Forestry and Environmental Sciences, 1998); Sharif S. Elmusa, ed., *Culture and the Natural Environment: Ancient and Modern Middle Eastern Texts*, vol. 26, no. 1, *Cairo Papers in Social Science* (Cairo: American University in Cairo Press, 2003); D. Davis, *Resurrecting the Granary of Rome*, 131–76; Diana K. Davis, "Potential Forests: Degradation Narratives, Science, and Environmental Policy in Protectorate Morocco, 1912–1956," *Environmental History* 10 (2005): 211–38; Stuart Schoenfeld, ed., *Palestinian and Israeli Environmental Narratives: Proceedings of a Conference Held in Association with the Middle East Environmental Futures Project* (Toronto: York University, 2005); Sufian, *Healing the Land and the Nation*; Richard C. Foltz, "Is There an Islamic Environmentalism?," *Environmental Ethics* 22 (2000): 63–72.

65. For some of the literature on Islamic environmental ethics, see Mawil Izzi Dien, *The Environmental Dimensions of Islam* (Cambridge: Lutterworth Press, 2000); Mawil Izzi Dien, "Islam and the Environment: Theory and Practice," *Journal of Beliefs and Values* 18 (1997): 47–57; Yūsuf al-Qaraḍāwī, *Riʿāyat al-Bīʾah fī Sharīʿat al-Islām* (Cairo: Dār al-Shurūq, 2001); Harfiyah Abdel Haleem, ed., *Islam and the Environment* (London: Ta-Ha Publishers, 1998); Ziauddin Sardar, ed., *An Early Crescent: The Future of Knowledge and the Environment in Islam* (London: Mansell, 1989); Richard C. Foltz, Frederick M. Denny, and Azizan Baharuddin, eds., *Islam and Ecology: A Bestowed Trust* (Cambridge, MA: Harvard University Press, 2003); Fazlun M. Khalid with Joanne O'Brien, eds., *Islam and Ecology* (New York: Cassell, 1992); Richard C. Foltz, ed., *Environmentalism in the Muslim World* (New York: Nova Science Publishers, 2005); Foltz, "Is There an Islamic Environmentalism?"

66. From Clarence Glacken's *Traces on the Rhodian Shore* to Ramachandra Guha's and others' writings on environmentalism in India, much work exists on Jewish, Buddhist, Christian, Hindu, and Shamanist views of nature. Islam and its various traditions deserve no less. Clarence J. Glacken, *Traces on the Rhodian Shore: Nature and Culture in Western Thought from Ancient*

Times to the End of the Eighteenth Century (Berkeley: University of California Press, 1967); Madhav Gadgil and Ramachandra Guha, *This Fissured Land: An Ecological History of India* (Berkeley: University of California Press, 1993); Ramachandra Guha and J. Martinez-Alier, *Varieties of Environmentalism: Essays North and South* (London: Earthscan Publications, 1997). On Buddhism, see David E. Cooper and Simon P. James, *Buddhism, Virtue and Environment* (Aldershot, UK: Ashgate, 2005). On Judaism, see Martin D. Yaffe, ed., *Judaism and Environmental Ethics: A Reader* (Lanham, MD: Lexington Books, 2001). See also *The Encyclopedia of Religion and Nature* for a useful beginning on the relationships between various religious traditions and the environment. Bron R. Taylor, ed., *The Encyclopedia of Religion and Nature*, 2 vols. (London: Thoemmes Continuum, 2005).

67. Leila Harris, "Water and Conflict Geographies of the Southeastern Anatolia Project," *Society and Natural Resources* 15 (2002): 743–59; Harris, "Postcolonialism, Postdevelopment, and Ambivalent Spaces of Difference"; Ali Ihsan Bagis, "Turkey's Hydropolitics of the Euphrates-Tigris Basin," *International Journal of Water Resources Development* 13 (1997): 567–82.

CHAPTER 1

1. For general histories of Ottoman Egypt, see Michael Winter, *Egyptian Society under Ottoman Rule, 1517–1798* (London: Routledge, 1992); Stanford J. Shaw, *The Financial and Administrative Organization and Development of Ottoman Egypt, 1517–1798* (Princeton, NJ: Princeton University Press, 1962); ʿAbd al-Raḥīm ʿAbd al-Raḥman ʿAbd al-Raḥīm, *al-Rīf al-Miṣrī fī al-Qarn al-Thāmin ʿAshar* (Cairo: Maktabat Madbūlī, 1986); Laylā ʿAbd al-Laṭīf Aḥmad, *al-Idāra fī Miṣr fī al-ʿAṣr al-ʿUthmānī* (Cairo: Maṭbaʿat Jāmiʿat ʿAyn Shams, 1978); Laylā ʿAbd al-Laṭīf Aḥmad, *al-Mujtamaʿ al-Miṣrī fī al-ʿAṣr al-ʿUthmānī* (Cairo: Dār al-Kitāb al-Jāmiʿī, 1987); Laylā ʿAbd al-Laṭīf Aḥmad, *Tārīkh wa Muʾarrikhī Miṣr wa al-Shām ibbāna al-ʿAṣr al-ʿUthmānī* (Cairo: Maktabat al-Khānjī, 1980); ʿIrāqī Yūsuf Muḥammad, *al-Wujūd al-ʿUthmānī fī Miṣr fī al-Qarnayn al-Sādis ʿAshar wa al-Sābiʿ ʿAshar (Dirāsa Wathāʾiqiyya)* (Cairo: Markaz Kliyūbātrā lil-Kumbiyūtar, 1996); ʿIrāqī Yūsuf Muḥammad, *al-Wujūd al-ʿUthmānī al-Mamlūkī fī Miṣr fī al-Qarn al-Thāmin ʿAshar wa Awāʾil al-Qarn al-Tāsiʿ ʿAshar* (Cairo: Dār al-Maʿārif, 1985); André Raymond, *Artisans et commerçants au Caire au XVIIIe siècle*, 2 vols. (Damascus: Institut français de Damas, 1973–74).

For a comparative perspective on water management, see the following works on the subject in medieval and early modern Europe: André E. Guillerme, *The Age of Water: The Urban Environment in the North of France, A.D. 300–1800* (College Station: Texas A&M University Press, 1988); Paolo Squatriti, *Water and Society in Early Medieval Italy, AD 400–1000* (Cambridge: Cambridge University Press, 1998); Paolo Squatriti, ed., *Working with Water in Medieval Europe: Technology and Resource-Use* (Leiden, Netherlands: Brill, 2000); Roberta J. Magnusson, *Water Technology in the Middle Ages: Cities, Monasteries, and Waterworks after the Roman Empire* (Baltimore: Johns Hopkins University Press, 2001); Glick, *Irrigation and Hydraulic Technology*; Glick, *Irrigation and Society*; Jean-Pierre Goubert, *The Conquest *of Water: The Advent of Health in the Industrial Age*, trans. Andrew Wilson (Princeton, NJ: Princeton University Press, 1989); Patrick Fournier and Sandrine Lavaud, eds., *Eaux et conflits dans l'Europe médiévale et moderne* (Toulouse: Presses universitaires du Mirail, 2012); Michele Campopiano, "Rural Communities, Land Clearance and Water Management in the Po Valley in the Central and Late Middle Ages," *Journal of Medieval History* 39 (2013): 377–93.

2. Before the Aswan Dams, the annual cycle of agricultural cultivation in Egypt was, of course, timed to the Nile's flood. Summer rains in the Ethiopian highlands swelled the river, causing it to rise at Aswan in the south of Egypt by June and in Cairo by early July. Water continued to rise through the summer until its peak in Cairo in late August or early September. It then began to fall steadily, reaching half of its flood height by the middle of November and its minimum by May before the cycle began anew. The onset of the flood in late summer was designated as the start of the agricultural year in Egypt.

3. Lands watered at the beginning of the annual agricultural cycle in September or October produced the major harvest of the year, consisting of wheat, barley, lentils, clover, flax, chickpeas, onions, and garlic. This was known as the winter crop. Lands were also planted and harvested from January through May using water stored in basins and canals, producing a second major yield for the agricultural year known as the summer crop and consisting mainly of wheat, barley, cotton, melons, sugarcane, and sesame. There was, of course, wide regional variation in the kinds and amounts of crops grown. Rice cultivation, for example, was concentrated in the north of Egypt, tobacco and sugarcane in the south, cotton in middle and northern Egypt, and flax in the interior of the delta and in the Fayyum oasis. Wheat was grown most everywhere. On food cultivation and the movement of agricultural resources from Egypt to other parts of the empire, see Mikhail, *Nature and Empire*, 82–123.

4. For discussions of irrigation in other regions of the early modern Ottoman Empire, see Rhoads Murphey, "The Ottoman Centuries in Iraq: Legacy or Aftermath? A Survey Study of Mesopotamian Hydrology and Ottoman Irrigation Projects," *Journal of Turkish Studies* 11 (1987): 17–29; *Encyclopaedia of Islam*, 2nd ed. (Leiden, Netherlands: Brill, 2006), s.v. "Māʾ. 8. Irrigation in the Ottoman Empire" (Halil İnalcık). For particularly useful treatments of the status of land, agriculture, and landholding in the Ottoman Empire, see Çağlar Keyder and Faruk Tabak, eds., *Landholding and Commercial Agriculture in the Middle East* (Albany: State University of New York Press, 1991); Huri İslamoğlu-İnan, *State and Peasant in the Ottoman Empire: Agrarian Power Relations and Regional Economic Development in Ottoman Anatolia during the Sixteenth Century* (Leiden, Netherlands: Brill, 1994); Huri İslamoğlu-İnan, ed., *The Ottoman Empire and the World-Economy* (Cambridge: Cambridge University Press, 1987).

5. For discussions of this source, see Nicolas Michel, "Les Dafātir al-ǧusūr, source pour l'histoire du réseau hydraulique de l'Égypte ottomane," *Annales Islamologiques* 29 (1995): 151–68; Mikhail, *Nature and Empire*, 40–46. The five registers under the archival heading of al-Jusūr al-Sulṭāniyya are numbered 784, 785, 786, 787, and 788 and are classified under the larger archival head-

ing of al-Rūznāma. The last two of these registers (787 and 788) are accounting books related to matters of irrigation, while the first three consist of records of individual canals organized by subprovince and then by village. What we know about the date of the compilation of these registers comes mainly from the section dealing with the subprovince of al-Minūfiyya northwest of Cairo. One of the cases in this section references the year 1539 or 1540 (946), but there is no date given for the actual recording of this case in the register. From other internal evidence in the register and from the scribal hand, it is clear, however, that this case was written into the registers of al-Jusūr al-Sulṭāniyya at some point in the seventeenth or eighteenth century. Thus, the descriptions of canals and the procedures for managing them presented in these registers were still of enough relevance in the seventeenth and eighteenth centuries to be copied from sources compiled hundreds of years earlier.

6. This was also the case in various other geographies. By way of comparison, see the Andean example in Paul B. Trawick, "Successfully Governing the Commons: Principles of Social Organization in an Andean Irrigation System," *Human Ecology* 29 (2001): 1–25; the Western Indian example in Arjun Appadurai, "Wells in Western India: Irrigation and Cooperation in an Agricultural Society," *Expedition* 26 (1984): 3–14; and the Balinese example in Clifford Geertz, "The Wet and the Dry: Traditional Irrigation in Bali and Morocco," *Human Ecology* 1 (1972): 26–31. Cases involving the sharing of a canal between various communities provide good examples of how the use of water also entailed the responsibility for the upkeep of irrigation works. These cases, moreover, also document the disputes this sharing of responsibility often produced. See, for example, DWQ, Maḥkamat al-Manṣūra, 12, p. 446, no case no. (15 Ca 1104/22 Jan. 1693); DWQ, Maḥkamat al-Baḥayra, 10, p. 220, case 523 (3 C 1191/8 July 1777); DWQ, Maḥkamat al-Manṣūra, 19, p. 374, no case no. (9 M 1124/16 Feb. 1712); DWQ, Maḥkamat al-Manṣūra, 19, p. 368, no case no. (A) (13 M 1125/9 Feb. 1713); DWQ, Maḥkamat al-Manṣūra, 19, p. 379, no case no. (23 M 1123/13 Mar. 1711); DWQ, Maḥkamat al-Manṣūra, 12, p. 425, no case no. (2 Za 1102/28 July 1691); DWQ, Maḥkamat al-Manṣūra, 24, p. 288, case 628 (28 N 1136/20 June 1724).

7. DWQ, al-Jusūr al-Sulṭāniyya, 784, pp. 182–83, no case no. (13 Ṣ).

8. Answers to these and other similar questions informed one of the most basic defining characteristics of canals in rural Egypt—the distinction between *sulṭānī* and *baladī* canals and the differences in the sorts of communities these waterways created. If a canal served a large group of peasants rather than the interests of a small privileged few, contributed to the common good, or aided in the achievement of equality among peasants, then it was considered a *sulṭānī* canal, the responsibilities of which fell on the Ottoman state in Egypt. *Baladī* canals, by contrast, were those that served the irrigation needs of one particular community and no one else. These were to be maintained by local leaders. Though both *sulṭānī* and *baladī* canals ultimately remained the property of the state, their upkeep often fell on the shoulders of Egyptian peasants who lived near them since they were the ones who most directly relied on these waterways. On the difference between *sulṭānī* and *baladī* canals and for useful schematic drawings of irrigation communities in the Egyptian

countryside, see Stuart J. Borsch, "Environment and Population: The Collapse of Large Irrigation Systems Reconsidered," *Comparative Studies in Society and History* 46 (2004): 458–60. For citations to relevant medieval texts on the differences between *sulṭānī* and *baladī* canals and dikes, see Sato Tsugitaka, *State and Rural Society in Medieval Islam: Sultans, Muqta's and Fallahun* (Leiden, Netherlands: Brill, 1997), 225–27.

9. DWQ, al-Jusūr al-Sulṭāniyya, 784, pp. 182–83, no case no. (13 Ş).

10. Ibid.

11. For more on these local "experts," see Mikhail, *Nature and Empire*, 66, 176–78. For other cases involving them, see BOA, Cevdet Nafia, 120 (Evasıt Ca 1125/5–14 June 1713); BOA, MM, 8:469 (Evasıt L 1180/12–21 Mar. 1767); DWQ, Maḥkamat al-Manṣūra, 18, p. 266, no case no. (B) (6 B 1121/10 Sept. 1709); BOA, MM, 9:424 (Evail C 1194/4–13 June 1780); DWQ, Maḥkamat Asyūṭ, 4, p. 206, case 645 (11 C 1156/2 Aug. 1743); DWQ, Maḥkamat Asyūṭ, 2, p. 238, case 566 (13 M 1108/11 Aug. 1696); DWQ, Maḥkamat Asyūṭ, 5, p. 179, case 343 (20 C 1189/17 Aug. 1775); DWQ, Maḥkamat Asyūṭ, 8, p. 260, case 563 (14 S 1211/18 Aug. 1796); DWQ, Maḍābiṭ al-Daqahliyya, 19, p. 299, case 878 (1185/1771 and 1772); DWQ, Maḍābiṭ al-Daqahliyya, 19, p. 299, case 875 (1185/1771 and 1772); DWQ, Maḍābiṭ al-Daqahliyya, 19, p. 299, case 874 (1185/1771 and 1772); DWQ, Maḍābiṭ al-Daqahliyya, 19, p. 299, case 876 (1185/1771 and 1772); DWQ, Maḍābiṭ al-Daqahliyya, 19, p. 299, case 877 (1185/1771 and 1772); DWQ, Maḍābiṭ al-Daqahliyya, 19, p. 299, case 872 (1185/1771 and 1772); DWQ, Maḍābiṭ al-Daqahliyya, 19, p. 299, case 873 (1185/1771 and 1772); DWQ, Maḍābiṭ al-Daqahliyya, 19, p. 300, case 880 (1185/1771 and 1772).

12. Irrigation surveys were of course organized in all of Egypt's subprovinces after the Ottoman conquest in 1517 and in later decades as well. For the subprovince of al-Qalyūbiyya, see DWQ, al-Jusūr al-Sulṭāniyya, 786, p. 158v, no case no. (n.d.). On this register, see Michel, "Les Dafātir al-ğusūr," 152–56. For al-Fayyūm, see DWQ, al-Jusūr al-Sulṭāniyya, 785, p. 2v, no case no. (17 Z 956/5 Jan. 1550). On register number 785, see Michel, "Les Dafātir al-ğusūr," 160–63. For Beni Suef (al-Bahnasāwiyya), see DWQ, al-Jusūr al-Sulṭāniyya, 785, p. 4v, no case no. (Evahir Za 996/12–21 Oct. 1588). The Ottoman administration of Egypt used its already existing network of rural Islamic courts to distribute instructions for these surveys to its functionaries throughout the Egyptian countryside.

13. DWQ, Maḥkamat al-Manṣūra, 22, p. 228, case 497 (25 Za 1184/12 Mar. 1771).

14. For a similar case involving a silted-over canal, see DWQ, Rūznāma 4557, Daftar Irtifāʿ al-Miyāh bi-Baḥr Sayyidnā Yūsuf lihi al-Ṣalāh wa al-Salām ʿan al-Qabḍa al-Yūsufiyya Tābiʿ Wilāyat al-Fayyūm (Raqam al-Ḥifẓ al-Nauʿī 1, ʿAyn 59, Makhzin Turkī 1, Musalsal 4557), p. 14, no case no. (3 C 1200/2 Apr. 1786).

15. Opening too many ditches or auxiliary canals on a central artery, thus draining the main channel to a point where it was rendered unusable for irrigation, was a rather common problem in the Egyptian countryside. See, for example, DWQ, Maḥkamat al-Manṣūra, 18, p. 249 or p. 250, no case no. (B)

(23 Ş 1121/27 Oct. 1709); DWQ, Maḥkamat al-Manṣūra, 19, p. 386, no case no. (1 Ş 1122/24 Sept. 1710); DWQ, Maḥkamat al-Manṣūra, 19, p. 371, no case no. (9 Ş 1124/11 Sept. 1712).

16. For an example of a case involving Egyptian peasants petitioning the Ottoman state to open canals, see DWQ, Maḥkamat al-Manṣūra, 12, p. 426, no case no. (15 Za 1102/10 Aug. 1691).

17. The text of this case from the registers of the court of al-Manṣūra does not give any indication that the Divan of Egypt sent this request all the way to Istanbul for direction. In many cases, as we will see, petitions of this sort and other requests and disputes were sent from Cairo to Istanbul for resolution. Since no major disagreement or point of contention existed in this case, however, there was really no reason to send it to Istanbul.

18. The firman sent back to the court of al-Manṣūra sanctioning the peasants to do as they pleased was purposefully written in Arabic. The Ottoman imperial bureaucracy of Egypt functioned predominantly in Ottoman Turkish, although Arabic was also widely used. The fact that the Divan of Egypt purposefully chose to write this firman in Arabic, and to state explicitly in the text that it did so, underscores that this imperial body sought to communicate directly with the countryside's rural cultivators, not the institution of the court. This direct communication between the imperial bureaucracy and Egyptian peasants shows that the empire understood that peasants were the most important actors in the project of maintaining this canal.

19. For additional examples of such peasant initiatives from the court of al-Manṣūra, see DWQ, Maḥkamat al-Manṣūra, 4, p. 108, case 281 (1 M 1075/24 July 1664); DWQ, Maḥkamat al-Manṣūra, 12, p. 448, no case no. (11 C 1104/17 Feb. 1693); DWQ, Maḥkamat al-Manṣūra, 16, p. 402, no case no. (B) (30 Za 1115/5 Apr. 1704). For other cases specifically about the repair of al-Baḥr al-Ṣaghīr, see DWQ, Maḥkamat al-Manṣūra, 14, p. 100, case 219 (12 L 1110/13 Apr. 1699); DWQ, Maḥkamat al-Manṣūra, 1, p. 232, no case no. (Z 1058/Dec. 1648 and Jan. 1649); DWQ, Maḥkamat al-Manṣūra, 18, p. 244 or p. 245, no case no. (14 S 1122/14 Apr. 1710); DWQ, Maḥkamat al-Manṣūra, 19, p. 373, no case no. (8 R 1124/14 May 1712).

20. For additional cases illustrating the effective authority of local peasant communities over irrigation works in their villages, see DWQ, Maḥkamat al-Manṣūra, 16, p. 47, case 86 (13 B 1115/23 Nov. 1703); DWQ, Maḥkamat al-Manṣūra, 19, p. 33, case 88 (11 C 1122/7 Aug. 1710); DWQ, Maḥkamat al-Manṣūra, 18, p. 245 or p. 246, no case no. (A) (15 Z 1122/3 Feb. 1711).

21. DWQ, Maḥkamat Asyūṭ, 2, p. 235, case 558 (22 Z 1107/23 July 1696).

22. The seven Ottoman military regiments of Egypt were the ʿAzeban, Çavuşan, Çerakise, Gönüllüyan, Mustahfızan (Janissaries), Müteferrika, and Tüfenkciyan. On these military ranks, see Shaw, *Financial and Administrative Organization*, 189–210; Winter, *Egyptian Society*, 37–43; ʿAbd al-Raḥīm, *al-Rīf al-Miṣrī*, 71–81; Jane Hathaway, *The Politics of Households in Ottoman Egypt: The Rise of the Qazdağlıs* (Cambridge: Cambridge University Press, 1997), 5–16.

23. BOA, Cevdet Nafia, 120 (Evasıt Ca 1125/5–14 June 1713).

24. Notions of life and death abound in descriptions of land and water in

the archival and literary record of Ottoman Egypt. Unirrigated—and hence barren—land was often described as dead or lifeless (*mawāt* in Arabic, *mevat* in Ottoman Turkish). See, for example, DWQ, al-Jusūr al-Sulṭāniyya, 786, pp. 113v–114r, no case no. (18 Z 1117/2 Apr. 1706). Moreover, one of the words most often used to describe the irrigation of land was *iḥyāʾ*, meaning "enlivening," "revitalization," "revival," and the like. For examples of the use of this word in this context, see DWQ, Maḥkamat al-Manṣūra, 19, p. 374, no case no. (9 M 1124/16 Feb. 1712); BOA, MM, 6:238 (Evasıt Ca 1158/12–21 June 1745). Thus, images of life and death were intimately connected to the ability of water to make possible or to preclude the productivity of land.

25. One engineering *zirāʿ* (*zirāʿ al-handasa*) equaled 0.656 meters. Walther Hinz, *Islamische Masse und Gewichte umgerechnet ins metrische System* (Leiden, Netherlands: Brill, 1955), 58. The Egyptian purse equaled 25,000 *para*, the official Ottoman name given to the *niṣf fiḍḍa*. In Stanford J. Shaw's words, "The silver coin in common use during Mamlûk and Ottoman times in Egypt was called *nıṣf fiḍḍe* colloquially and *para* officially." Shaw, *Financial and Administrative Organization*, 65n169.

26. Tax farmers were tasked with delivering to the state a set amount of tax revenue from a plot of land every year and for ensuring general order on the land. In exchange, they had the right to raise additional profits for themselves. Their balancing act was thus to collect as much personal surplus profit as they could while not spurring disorder or sparking resistance from those peasants working underneath them.

27. DWQ, Mahkamat al-Manṣūra, 16, p. 47, case 86 (13 B 1115/23 Nov. 1703).

28. For other examples of court-ordered inspections of irrigation works, see DWQ, Maḥkamat Asyūṭ, 1, p. 201, case 583 (12 Za 1067/22 Aug. 1657); DWQ, Maḥkamat Asyūṭ, 1, p. 287, case 844 (18 Za 1068/18 Aug. 1658); DWQ, Maḥkamat al-Manṣūra, 22, p. 236, no case no. (13 S 1152/21 May 1739).

29. Generally on village abandonment by Egyptian peasants during the Ottoman period, see Nicolas Michel, "Migrations de paysans dans le Delta du Nil au début de l'époque ottoman," *Annales Islamologiques* 35 (2001): 241–90; Nicolas Michel, "Villages désertés, terres en friche et reconstruction rurale en Égypte au début de l'époque ottoman," *Annales Islamologiques* 36 (2002): 197–251; Zayn al-ʿĀbidīn Shams al-Dīn Najm, "Tasaḥḥub al-Fallāḥīn fī ʿAṣr Muḥammad ʿAlī, Asbābuhu wa Natāʾijuhu," *Egyptian Historical Review* 36 (1989): 259–316; Naṣra ʿAbd al-Mutajallī, "al-Muqāwama bil-Tasaḥḥub fī Rīf Miṣr al-ʿUthmāniyya," in *al-Rafḍ wa al-Iḥtijāj fī al-Mujtamaʿal-Miṣrī fī al-ʿAṣr al-ʿUthmānī*, ed. Nāṣir Ibrāhīm and Raʾūf ʿAbbās, 127–36 (Cairo: Markaz al-Buḥūth wa al-Dirāsāt al-Ijtimāʿiyya, 2004). For a comparative example of village abandonment by peasants in Ottoman Palestine, see Amy Singer, "Peasant Migration: Law and Practice in Early Ottoman Palestine," *New Perspectives on Turkey* 8 (1992): 49–65.

30. DWQ, Maḥkamat al-Manṣūra, 9, p. 36, case 83 (20 B 1099/20 May 1688). *Jalabī* is the Arabized form of the Ottoman Turkish title *Çelebi*.

31. As the name of this village suggests, Famm Ẓāfir was strategically situated at the mouth of the canal of Jisr al-Ẓafar and thus very much controlled the supply of water to villages farther downstream.

32. This tax farmer was Riḍwān ibn al-Marḥūm al-Amīr Jaʿfar.

33. Tellingly, the phrase "*ḥifẓ wa ḥarāsa*" (*hıfz ü hiraset* in Ottoman Turkish) was used to refer not only to the upkeep and protection of Egyptian irrigation works, but also to the Ottoman state's finances and well-being. For examples of these latter uses of the phrase, see TSMA, E. 664/52 (n.d.); TSMA, E. 664/63 (Evail Ra 1159/24 Mar.–2 Apr. 1746).

34. Michael Adas collectively refers to peasant flight, village abandonment, and other similar actions as "avoidance protest." For his useful analysis of these forms of peasant resistance, see Michael Adas, "From Avoidance to Confrontation: Peasant Protest in Precolonial and Colonial Southeast Asia," *Comparative Studies in Society and History* 23 (1981): 217–47.

35. For an example of the Ottoman bureaucracy's flexibility in responding effectively to environmental changes that impacted the agricultural productivity of land in Egypt, see DWQ, Maḥkamat al-Baḥayra, 14, p. 390, case 612 (23 L 1200/18 Aug. 1786).

36. For examples of peasant petitions for greater allotments of water from Egypt's rural irrigation network, see DWQ, Rūznāma 4557, Daftar Irtifāʿ al-Miyāh bi-Baḥr Sayyidnā Yūsuf lihi al-Ṣalāh wa al-Salām ʿan al-Qabḍa al-Yūsufiyya Tābiʿ Wilāyat al-Fayyūm (Raqam al-Ḥifẓ al-Nauʿī 1, ʿAyn 59, Makhzin Turkī 1, Musalsal 4557), p. 11, no case no. (21 Ra 1192/19 Apr. 1778); DWQ, Rūznāma 4557, Daftar Irtifāʿ al-Miyāh bi-Baḥr Sayyidnā Yūsuf lihi al-Ṣalāh wa al-Salām ʿan al-Qabḍa al-Yūsufiyya Tābiʿ Wilāyat al-Fayyūm (Raqam al-Ḥifẓ al-Nauʿī 1, ʿAyn 59, Makhzin Turkī 1, Musalsal 4557), p. 12, no case no. (21 Ra 1195/17 Mar. 1781); DWQ, Rūznāma 4557, Daftar Irtifāʿ al-Miyāh bi-Baḥr Sayyidnā Yūsuf lihi al-Ṣalāh wa al-Salām ʿan al-Qabḍa al-Yūsufiyya Tābiʿ Wilāyat al-Fayyūm (Raqam al-Ḥifẓ al-Nauʿī 1, ʿAyn 59, Makhzin Turkī 1, Musalsal 4557), p. 13, no case no. (9 S 1197/13 Jan. 1783); DWQ, Maḥkamat Asyūṭ, 2, p. 235, case 558 (22 Z 1107/22 July 1696).

37. In the following cases, the Ottoman administration of Egypt sought to move quickly enough to repair a broken canal embankment before all of the canal's waters wastefully spilled out of the waterway: DWQ, Maḥkamat al-Manṣūra, 18, p. 266, no case no. (A) (21 C 1121/27 Aug. 1709); DWQ, Maḥkamat al-Manṣūra, 18, p. 266, no case no. (D) (21 C 1121/27 Aug. 1709); DWQ, Maḥkamat al-Manṣūra, 18, p. 266, no case no. (C) (6 B 1121/10 Sept. 1709).

38. For examples of the various means through which the Ottoman administration of Egypt financed irrigation repairs, see DWQ, Maḥkamat al-Manṣūra, 12, p. 422, no case no. (29 Ş 1102/28 May 1691); DWQ, Maḥkamat al-Manṣūra, 12, p. 424, no case no. (5 L 1102/1 July 1691); DWQ, Maḥkamat al-Manṣūra, 12, p. 423, no case no. (25 L 1102/21 July 1691); BOA, MM, 1:310 (1 M 1126/18 Jan. 1724); BOA, MM, 5:46 (Evail Ca 1146/9–18 Oct. 1733); DWQ, Maḥkamat al-Manṣūra, 12, p. 445, no case no. (B) (4 Ca 1104/11 Jan. 1693); DWQ, Maḥkamat al-Manṣūra, 18, p. 249 or p. 250,

no case no. (A) (1 Ş 1121/5 Oct. 1709); DWQ, Maḥkamat al-Manṣūra, 19, no page no., no case no. (9 Ra 1122/8 May 1710); DWQ, Maḥkamat al-Manṣūra, 15, p. 205, no case no. (20 L 1113/19 Mar. 1702); BOA, MM, 8:337 (Evahir Z 1178/10–19 June 1765); BOA, MM, 6:558 (Evahir N 1162/4–13 Sept. 1749); BOA, MM, 5:189 (Evasıt Ş 1147/6–15 Jan. 1735); DWQ, Rūznāma 4557, Daftar Irtifāʿ al-Miyāh bi-Baḥr Sayyidnā Yūsuf lihi al-Ṣalāh wa al-Salām ʿan al-Qabḍa al-Yūsufiyya Tābiʿ Wilāyat al-Fayyūm (Raqam al-Ḥifẓ al-Nauʿī 1, ʿAyn 59, Makhzin Turkī 1, Musalsal 4557), p. 22, no case no. (28 R 1127/2 May 1715). For examples of the theft of funds made available for irrigation repairs, see DWQ, Maḥkamat al-Manṣūra, 51, p. 159, case 297 (Ca 1203/Jan. and Feb. 1789); DWQ, Maḥkamat al-Manṣūra, 19, p. 368, no case no. (B) (2 M 1125/29 Jan. 1713).

39. For studies that stress the oppressive nature of Ottoman rule in rural Egypt, see ʿAbd al-Raḥīm, *al-Rīf al-Miṣrī*; ʿAbd al-Raḥman al-Rāfʿī, *ʿAṣr Muḥammad ʿAlī* (Cairo: Dār al-Maʿārif, 1989).

CHAPTER 2

1. The sedimentary processes that formed the Nile delta began during the Upper Miocene. Scot E. Smith and Adel Abdel-Kader, "Coastal Erosion along the Egyptian Delta," *Journal of Coastal Research* 4 (1988): 245–55; Rushdi Said, *The Geological Evolution of the River Nile* (New York: Springer-Verlag, 1981).

2. Herodotus, *The History*, trans. David Grene (Chicago: University of Chicago Press, 1987), 2.5.

3. Approximately half of the total sediment load of 125 million tons carried by the river each year into Egypt was discharged to the Mediterranean. Smith and Abdel-Kader, "Coastal Erosion," 249. For further analysis of these delta sediments, see Janusz Dominik and Daniel Jean Stanley, "Boron, Beryllium and Sulfur in Holocene Sediments and Peats of the Nile Delta, Egypt: Their Use as Indicators of Salinity and Climate," *Chemical Geology* 104 (1993): 203–16.

4. Herodotus, *History*, 2.15.

5. John Antes, *Observations on the Manners and Customs of the Egyptians, the Overflowing of the Nile and Its Effects; with Remarks on the Plague and Other Subjects. Written during a Residence of Twelve Years in Cairo and Its Vicinity* (London: printed for J. Stockdale, 1800), 64–65. For a study of the delta's fossilized mollusks, see Maria Pia Bernasconi, Daniel Jean Stanley, and Italo Di Geronimo, "Molluscan Faunas and Paleobathymetry of Holocene Sequences in the Northeastern Nile Delta, Egypt," *Marine Geology* 99 (1991): 29–43.

6. Antes, *Observations*, 74–75.

7. Ibid., 75.

8. Smith and Abdel-Kader, "Coastal Erosion"; Abdel-Aziz I. Kashef, "Salt-Water Intrusion in the Nile Delta," *Ground Water* 21 (1983): 160–67; Omran E. Frihy, Alfy M. Fanos, Ahmed A. Khafagy, and Paul D. Komar, "Patterns of Nearshore Sediment Transport along the Nile Delta, Egypt," *Coastal*

Engineering 15 (1991): 409–29. See also J.R. McNeill, *Something New under the Sun*, 166–73.

9. Mohamed A.K. Elsayed, Nazeih A. Younan, Alfy M. Fanos, and Khalid H. Baghdady, "Accretion and Erosion Patterns along Rosetta Promontory, Nile Delta Coast," *Journal of Coastal Research* 21 (2005): 413. From 1900 to 1964 (when construction of the Aswan High Dam began), the average rate of erosion was much lower: about fifty meters per year. As with most aspects of environmental or historical change, this phenomenon was neither completely linear nor universal. Even into the nineteenth century, some parts of the delta continued to expand into the sea. For example, between 1500 and 1900, sections of the Rosetta promontory advanced into the sea at an average rate of twenty-five meters per year.

10. Smith and Abdel-Kader, "Coastal Erosion," 249.

11. Much of the writing of environmental history has privileged the stories of very big changes to the natural world—coastal erosion, climate change, deforestation, species extinction, and so forth. As this chapter attempts to illustrate, however, the minutiae of these enormous ecological processes as well as other independent small-scale environmental changes are of no less historical and environmental significance and are thus equally deserving of our attention.

12. For the Arabic critical edition of this very important text and an English translation, see Yūsuf ibn Muḥammad al-Shirbīnī, *Kitāb Hazz al-Quḥūf bi-Sharḥ Qaṣīd Abī Shādūf*, ed. and trans. Humphrey Davies, 2 vols. (Leuven, Belgium: Peeters, 2005–7).

13. For a comparative example of how shifting riverine ecologies affected Ottoman relations with local populations, see the following analysis of the Euphrates in Iraq: Faisal Husain, "In the Bellies of the Marshes: Water and Power in the Countryside of Ottoman Baghdad," *Environmental History* 19 (2014): 638–64.

14. In addition to the cases about dredging discussed here, see also Mikhail, *Nature and Empire*, 38–81.

15. On al-Manzala, see Muḥammad Ramzī, *al-Qāmūs al-Jughrāfī lil-Bilād al-Miṣriyya min ʿAhd Qudamāʾ al-Miṣriyyīn ilā Sanat 1945*, 6 vols. in 2 pts. (Cairo: al-Hayʾa al-Miṣriyya al-ʿĀmma lil-Kitāb, 1994), pt. 2, 1:203–4.

16. DWQ, Maḥkamat al-Manṣūra, 16, p. 397, no case no. (A) (6 Za 1115/12 Mar. 1704).

17. The *para* was the basic unit of currency in Ottoman Egypt. On its history, conversion, and usage, see Shaw, *Financial and Administrative Organization*, xxii.

18. For the breakdown of these payments by village, see DWQ, Maḥkamat al-Manṣūra, 16, p. 397, no case no. (B) (6 Za 1115/12 Mar. 1704).

19. Another case from the court of al-Manṣūra from late June 1692 concerned a community of water formed around the shared usage of a canal known as Ḥammām. One section of this canal was badly in need of dredging, and people from villages along its length came to the court to push the Ottoman administration to dredge this part of the waterway because the earth near it was dry, hard, and stony. If immediate action was not taken to dredge this shared canal, water would not reach these villages' fields, and they would

remain bare, unfit for cultivation, and destitute. The Ottoman imperial administration heeded the peasants' warnings in this case and appointed one of its local functionaries to oversee the dredging and cleaning of the canal. DWQ, Maḥkamat al-Manṣūra, 12, p. 447, no case no. (A) (10 L 1103/25 June 1692).

20. For an example of the ease with which the canal could be drained, see ʿAbd al-Raḥman ibn Ḥasan al-Jabartī, *ʿAjāʾib al-Āthār fī al-Tarājim wa al-Akhbār*, ed. ʿAbd al-Raḥīm ʿAbd al-Raḥman ʿAbd al-Raḥīm, 4 vols. (Cairo: Maṭbaʿat Dār al-Kutub al-Miṣriyya, 1998), 4:31. In this case, a military regiment dammed the canal in two places and was then able to siphon off all of the trapped water to ground its enemy's ships.

21. See, for example, DWQ, Maḥkamat al-Baḥayra, 7, p. 56, case 112 (2 Ra 1171/14 Nov. 1757).

22. DWQ, Maḥkamat al-Baḥayra, 5, pp. 9–10, case 15 (12 Ş 1164/6 July 1751).

23. DWQ, Maḥkamat al-Baḥayra, 5, pp. 172–73, case 302 (10 Ş 1165/22 June 1752).

24. For references to this waterwheel, see DWQ, Maḥkamat al-Baḥayra, 5, pp. 9–10, case 15 (12 Ş 1164/6 July 1751); DWQ, Maḥkamat al-Baḥayra, 5, pp. 172–73, case 302 (10 Ş 1165/22 June 1752). For descriptions of other kinds of irrigation works built at the mouth of the canal to aid its current, see DWQ, Maḥkamat al-Baḥayra, 5, p. 6, case 11 (30 B 1164/24 June 1751); DWQ, Maḥkamat al-Baḥayra, 5, p. 314, case 389 (10 Ş 1165/22 June 1752). On the village of al-Raḥmāniyya, see Ramzī, *al-Qāmūs al-Jughrāfī*, pt. 2, 2:305.

25. BOA, MM, 5:393 (Evahir L 1150/10–20 Feb. 1738).

26. There is no indication as to how the sultan's imperial council reached this population figure, which seems rather exaggerated for the mid-eighteenth century. The consensus in the historical literature puts Alexandria's population in this period closer to ten thousand or twenty thousand. See, for example, Daniel Panzac, "Alexandrie: Peste et croissance urbaine (XVIIe–XIXe siècles)," in *Population et santé dans l'Empire ottoman (XVIIIe–XXe siècles)*, 45–55 (Istanbul: Isis, 1996); Michael J. Reimer, "Ottoman Alexandria: The Paradox of Decline and the Reconfiguration of Power in Eighteenth-Century Arab Provinces," *Journal of the Economic and Social History of the Orient* 37 (1994): 107–46.

27. BOA, MM, 8:139 (Evasıt L 1176/24 Apr.–4 Mar. 1763).

28. DWQ, Maḥkamat al-Manṣūra, 24, p. 288, case 628 (28 N 1136/20 June 1724). On the village of Kafr Ghannām, see Ramzī, *al-Qāmūs al-Jughrāfī*, pt. 2, 1:199. On al-Hajārsa, see ibid., pt. 2, 1:128.

29. Cases like this one about the dredging of a shared canal seem to indicate that there was an imaginary line assumed to run down the middle of canals dividing them into two equal parts. We are given no indication as to how this line was established or how the complicated work of dredging only *half* a canal was organized and actually accomplished.

30. The operative phrase is *in ḥaṣala fīhi khalal . . . kāna dhālika muqābilan bi-arwāḥihim*. For another case that cites a similar threat of execution, see DWQ, Maḥkamat al-Manṣūra, 9, p. 205, case 466 (Evail L

1100/19–28 July 1689). The notion that a village head could be killed for failing to properly maintain irrigation works is one that was established in the founding Ottoman law code (*kanunname*) of Egypt promulgated in 1525, less than a decade after the Ottoman conquest of Egypt in 1517. For the relevant sections of this law code, see Ömer Lûtfi Barkan, *Kanunlar*, vol. 1 of *XV ve XVIinci asırlarda Osmanlı İmparatorluğunda Ziraî Ekonominin Hukukî ve Malî Esasları*, İstanbul Üniversitesi Yayınlarından 256 (Istanbul: Bürhaneddin Matbaası, 1943), 360–61. For an Arabic translation, see Aḥmad Fuʾād Mutawallī, trans. and intro., *Qānūn Nāmah Miṣr, alladhī Aṣdarahu al-Sulṭān al-Qānūnī li-Ḥukm Miṣr* (Cairo: Maktabat al-Anjlū al-Miṣriyya, 1986), 30–31.

31. DWQ, Maḥkamat al-Manṣūra, 7, p. 310, case 767 (19 S 1093/27 Feb. 1682).

32. On Nūb Ṭarīf see Ramzī, *al-Qāmūs al-Jughrāfī*, pt. 2, 1:196. On Ṭummāy see ibid., pt. 2, 1:192.

33. Its name meaning "garden" or "meadow," the island of al-Rauḍa is a long, slender piece of land (approximately three kilometers long and half a kilometer wide at its thickest point) separated from the east bank of the Nile by a thin section of the river and from the west bank by a much wider stretch of water. The most detailed account of the island and its history is the following fifteenth-century work: Jalāl al-Dīn al-Suyūṭī, *Kawkab al-Rauḍa*, ed. Muḥammad al-Shashtāwī (Cairo: Dār al-Āfāq al-ʿArabiyya, 2002).

The Ottoman historian Muṣṭafā ʿĀlī, who visited Egypt at the end of the sixteenth century, describes the Nilometer as "indeed one of the rare creations of the world and of the curious works resembling magical devices." Andreas Tietze, *Muṣṭafā ʿĀlī's Description of Cairo of 1599: Text, Transliteration, Translation, Notes* (Vienna: Verlag der Österreichischen Akademie der Wissenschaften, 1975), 30. More generally on the Nilometer, see William Popper, *The Cairo Nilometer: Studies in Ibn Taghrî Birdî's Chronicles of Egypt, I* (Berkeley: University of California Press, 1951); Amīn Sāmī, *Taqwīm al-Nīl*, 5 vols. in 3 pts. (Cairo: Dār al-Kutub wa al-Wathāʾiq al-Qawmiyya, 2003), pt. 1, 65–95; Nicholas Warner, *The True Description of Cairo: A Sixteenth-Century Venetian View*, 3 vols. (Oxford: Arcadian Library in association with Oxford University Press, 2006), 2:123–25.

34. Al-Jabartī, *ʿAjāʾib al-Āthār* (1998), 2:363–64.

35. Ibid., 2:363.

36. Ibid., 2:363–64. A few decades later, in May 1812, the Nile in Cairo dried to such an extent that it was possible to walk on the exposed riverbed nearly the entire width of the river from Bulaq to Imbāba. Ibid., 4:246.

37. DWQ, Maḍābiṭ al-Daqahliyya, 34, pp. 93–94, case 198 (21 S 1211/25 Aug. 1796). On the village of Qūjindīma, see Ramzī, *al-Qāmūs al-Jughrāfī*, pt. 1, no vol. no., 354. On Ṭalkhā, see ibid., pt. 2, 2:88.

38. For cases in which the presence of irrigation works on an island was used as evidence of the continued cultivation of that island by a particular group of peasants, see DWQ, Maḥkamat al-Manṣūra, 7, p. 90, case 240 (3 Ca 1091/31 May 1680).

39. DWQ, Maḥkamat al-Baḥayra, 21, p. 152, case 298 (12 Ş 1206/4 Apr. 1792); DWQ, Maḥkamat al-Baḥayra, 21, p. 480, case 943 (25 B 1206/19 Mar.

1792). On Nitmā, see Ramzī, *al-Qāmūs al-Jughrāfī*, pt. 2, 2:340. On Kafr al-Gharīb, see ibid., pt. 2, 2:169.

40. Of note here is the use of the colloquial Egyptian Arabic verb declension in the following passage: "*ahālī al-nāḥyatayn al-madhkūratayn biyazra'ū aṭyānahā*" (the people of the two aforementioned villages cultivate its [the island's] lands). DWQ, Maḥkamat al-Baḥayra, 21, p. 480, case 943 (25 B 1206/19 Mar. 1792). This and other instances of the use of colloquial Egyptian Arabic in the normally formal Arabic texts of Islamic court cases are important examples of the ways in which Egyptian peasants were able literally to interject their voices into the official workings of the court and of other bureaucratic structures in rural Ottoman Egypt.

41. The operative phrase in the following case is "*ḥukm mā kānū 'alayhi min qadīm al-zamān*" (in accordance with what they practiced from times of old). DWQ, Maḥkamat al-Baḥayra, 21, p. 152, case 298 (12 Ş 1206/4 Apr. 1792). In the other case, the phrase is "*ḥukm mā kānū awwal 'alā qadīmihim*" (in accordance with the precedent of their customary practice). DWQ, Maḥkamat al-Baḥayra, 21, p. 480, case 943 (25 B 1206/19 Mar. 1792).

42. As an example of these efforts on the part of the Ottoman state in Egypt, consider the speed with which the two court cases related to the previous island dispute were adjudicated and implemented. Only seventeen days elapsed between the date of the original petition and the date of the second case confirming that the orders of the firman issued in response to this petition were implemented. In other words, in two and a half weeks, a petition was written, delivered to Cairo, and heard by the high divan of Egypt; this petition was discussed and a decision was made; the piece of paper on which the divan's firman was written was transported from Cairo to Damanhūr, the seat of the court of al-Baḥayra (a distance of roughly one hundred miles); it was then delivered to the appropriate representatives of the subprovince; these men discussed the best way to address the situation; their plan was implemented; the results of this implementation were heard by the court; and the case reporting these results was recorded in the registers of this legal body. On Damanhūr and its hinterland, see Ramzī, *al-Qāmūs al-Jughrāfī*, pt. 2, 2:282–97. Not only does the speed with which all of this occurred suggest a very efficient process of Ottoman imperial rule that relied on networks of communication, roads, necessary material objects like paper, and the timely actions of administrators to undertake all the steps in these bureaucratic processes, but it also serves as further evidence of the imperative to quickly establish and uphold precedent.

43. On Ottoman Iraq, see Husain, "In the Bellies of the Marshes."

CHAPTER 3

1. Understandably, more has been written about Cairo than any other city or region in Ottoman Egypt. For scholarship on other parts of Ottoman Egypt, see the following. On al-Daqahliyya and al-Manṣūra in the Ottoman period, see Kenneth M. Cuno, *The Pasha's Peasants: Land, Society, and Economy in Lower Egypt, 1740–1858* (Cambridge: Cambridge University Press, 1992); Nāṣira 'Abd al-Mutajallī Ibrāhīm 'Alī, "al-Daqahliyya fī al-'Aṣr al-'Uthmānī" (MA thesis, 'Ayn Shams University, 2005). On Alexandria, see Reimer,

"Ottoman Alexandria"; Nāṣir ʿUthmān, "Maḥkamaṭ Rashīd ka-Maṣdar li-Dirāsat Tijārat al-Nasīj fī Madīnat al-Iskandariyya fī al-ʿAṣr al-ʿUthmānī," *al-Rūznāma: al-Ḥauliyya al-Miṣriyya lil-Wathāʾiq* 3 (2005): 355–85. On Ottoman Rosetta, see Ṣalāḥ Aḥmad Harīdī ʿAlī, "al-Ḥayāh al-Iqtiṣādiyya wa al-Ijtimāʿiyya fī Madīnat Rashīd fī al-ʿAṣr al-ʿUthmānī, Dirāsa Wathāʾiqiyya," *Egyptian Historical Review* 30–31 (1983–84): 327–78. On Ottoman Upper Egypt (al-*Ṣaʿīd*), see Laylā ʿAbd al-Laṭīf Aḥmad, *al-Ṣaʿīd fī ʿAhd Shaykh al-ʿArab Hammām* (Cairo: al-Hayʾa al-Miṣriyya al-ʿAmma lil-Kitāb, 1987); Muḥammad ibn Muḥammad Ḥāmid al-Marāghī al-Jirjāwī, *Tārīkh Wilāyat al-Ṣaʿīd fī al-ʿAṣrayn al-Mamlūkī wa al-ʿUthmānī: al-Musammā bi-"Nūr al-ʿUyūn fī Dhikr Jirjā min ʿAhd Thalāthat Qurūn,"* ed. Aḥmad Ḥusayn al-Namakī (Cairo: Maktabat al-Nahḍa al-Miṣriyya, 1998). On al-Minūfiyya, see Yāsir ʿAbd al-Minʿam Maḥārīq, *al-Minūfiyya fī al-Qarn al-Thāmin ʿAshar* (Cairo: al-Hayʾa al-Miṣriyya al-ʿAmma lil-Kitāb, 2000). On rural Ottoman Egypt in general, see ʿAbd al-Raḥīm, *al-Rīf al-Miṣrī.*

2. Aḥmad ibn ʿAlī al-Maqrīzī, *al-Mawāʿiẓ wa al-Iʿtibār bi-Dhikr al-Khiṭaṭ wa al-Āthār*, 2 vols. (Būlāq, Egypt: Dār al-Ṭibāʿa al-Miṣriyya, 1853), 1:245. For al-Maqrīzī's full description of Fayyum, see ibid., 1:241–50.

3. On the fact that Fayyum and Beni Suef were often treated as one administrative unit, see Aḥmad al-Damurdāshī Katkhudā ʿAzabān, *Kitāb al-Durra al-Muṣāna fī Akhbār al-Kināna*, ed. ʿAbd al-Raḥīm ʿAbd al-Raḥman ʿAbd al-Raḥīm (Cairo: Institut français d'archéologie orientale, 1989), 42. There is one register of the court of Beni Suef available to researchers in the Egyptian National Archives. It is incorrectly cataloged as register 120 of the court of al-Bāb al-ʿAlī in Cairo. It covers part of the year 1639. This register is referenced in Galal H. El-Nahal, *The Judicial Administration of Ottoman Egypt in the Seventeenth Century* (Minneapolis: Bibliotheca Islamica, 1979), 77.

4. In 2008, register 120 of the court of al-Bāb al-ʿAlī remained the only (albeit miscataloged) register from the court of either Fayyum or Beni Suef available in the Egyptian National Archives. Given that between 1979 (the publication date of El-Nahal's work) and 2008 this register remained the only one from either of these two courts, it seems likely that these records are no longer extant.

5. Jane Hathaway describes the period from roughly 1650 to 1750 as "a relatively unexplored backwater of the Ottoman Egyptian subfield." Hathaway, *Politics of Households*, 15.

6. For some histories of Ottoman Egypt based on these archival collections, see Hathaway, *Politics of Households*; Jane Hathaway, *A Tale of Two Factions: Myth, Memory, and Identity in Ottoman Egypt and Yemen* (Albany: State University of New York Press, 2003); Shaw, *Financial and Administrative Organization*. On the potential of using these archives for the history of Ottoman Egypt, see Stanford J. Shaw, "The Ottoman Archives as a Source for Egyptian History," *Journal of the American Oriental Society* 83 (1962): 447–52.

7. In these cases, the names of individual *mültezimīn*, Ottoman officials, *kāşifler*, and engineers are often given. In contrast, the *ahālī* and *ehl-i vukūf* are cited only as collective social groups without any further differentiation.

8. The so-called restoration of the beylicate (the name given to the power

regime of the Mamluk beys) at the end of the eighteenth century looms large in much of this scholarship. Perhaps the most seminal work in establishing this historiographical perspective as the dominant mode of evaluating the eighteenth century, and even earlier periods, is the chapter entitled "The Ascendancy of the Beylicate in Eighteenth-Century Egypt," in P.M. Holt, *Egypt and the Fertile Crescent, 1516–1922: A Political History*, 85–101 (Ithaca, NY: Cornell University Press, 1966). Reading earlier periods of Ottoman rule through this late eighteenth-century lens, many scholars have too often and too quickly emphasized the Mamluk character of various political, military, and social phenomena, creating a teleology that can only lead to the Mamluks at the end of the century. On this point, Jane Hathaway writes of "an urge to link Ottoman to Mamluk Egypt via the beylicate. This is accompanied by a tendency to refer to the late eighteenth century as the culmination of the beylicate's evolution." Hathaway, *Politics of Households*, 15.

9. Both of these conflicts were between rival military factions and were carried out largely in the streets of Cairo. On 1711, see André Raymond, "Une 'révolution' au Caire sous les Mamelouks: La crise de 1123/1711," *Annales Islamologiques* 6 (1966): 95–120. On the 1740s, see the relevant sections of al-Damurdāshī Katkhudā ʿAzabān, *Kitāb al-Durra al-Muṣāna*. For secondary studies, see Holt, *Egypt and the Fertile Crescent*, 85–101; Daniel Crecelius, "Egypt in the Eighteenth Century," in *Modern Egypt, from 1517 to the End of the Twentieth Century*, vol. 2 of *The Cambridge History of Egypt*, ed. M.W. Daly, 59–86 (Cambridge: Cambridge University Press, 1998).

10. For a discussion of the politics of irrigation in Ottoman Iraq, see Husain, "In the Bellies of the Marshes."

11. For classic examples of this model, see Halil İnalcık, "Centralization and Decentralization in Ottoman Administration," in *Studies in Eighteenth Century Islamic History*, ed. Thomas Naff and Roger Owen, 27–52 (Carbondale: Southern Illinois University Press, 1977); Şerif Mardin, "Center-Periphery Relations: A Key to Turkish Politics," *Daedalus* 102 (1973): 169–91. For some recent critiques of the model, see Dina Rizk Khoury, *State and Provincial Society in the Ottoman Empire: Mosul, 1540–1834* (Cambridge: Cambridge University Press, 1997); Leslie Peirce, *Morality Tales: Law and Gender in the Ottoman Court of Aintab* (Berkeley: University of California Press, 2003). See also Dina Rizk Khoury, "The Ottoman Centre versus Provincial Power-Holders: An Analysis of the Historiography," in *The Later Ottoman Empire, 1603–1839*, vol. 3 of *The Cambridge History of Turkey*, ed. Suraiya N. Faroqhi, 135–56 (Cambridge: Cambridge University Press, 2006).

12. For a very useful recent work reconceptualizing peripheries in the early modern Spanish Empire, see Bartolomé Yun Casalilla, *Las redes del imperio: Élites sociales en la articulación de la Monarquía Hispánica, 1492–1714* (Madrid: Marcial Pons; Seville: Universidad Pablo de Olavide, 2009). My thanks to Yuen-Gen Liang for bringing this study to my attention.

13. Karen Barkey, *Empire of Difference: The Ottomans in Comparative Perspective* (Cambridge: Cambridge University Press, 2008), 9.

14. For Ottoman orders to send Anatolian lumber to Egypt, see BOA, Cevdet Nafia, 644 (28 R 1190/15 June 1776); BOA, Cevdet Nafia, 302 (23 Za

1216/28 Mar. 1802); DWQ, Maḥkamat Rashīd, 132, p. 199, case 308 (16 Ş 1137/29 Apr. 1725); DWQ, Maḥkamat Rashīd, 132, p. 88, case 140 (17 Ş 1137/30 Apr. 1725). For other examples of the importation of wood from Anatolia to Egypt, see al-Jabartī, *ʿAjāʾib al-Āthār* (1998), 4:245–46, 4:254, 4:255, 4:400. On the movement of wood from the Black Sea coast to Egypt, see BOA, Cevdet Bahriye, 5701 (n.d.). The shipbuilding concerns of the Ottoman navy were often a primary factor in the movement of lumber between Anatolia (and other regions of the empire) and Egypt. On this point, see Palmira Brummett, *Ottoman Seapower and Levantine Diplomacy in the Age of Discovery* (Albany: State University of New York Press, 1994), 96, 115–16, 144, 174. For a discussion and analysis of Ottoman Egypt's dual system of grain export and timber import, see Mikhail, *Nature and Empire*, 82–169.

15. This movement of food was a two-way street. During instances of famine in Egypt, food was transferred from elsewhere around the empire to the province. In the following case, for example, shortages of rice in Egypt were met by shipments from Crete. See TSMA, E. 2444/107 (n.d.).

16. In addition to the discussion below of grains moving from Fayyum to the Hijaz, see also some of the many cases about grain shipments from Upper Egypt to the region. For example, DWQ, Maḥkamat Manfalūṭ, 2, p. 189, case 631 (24 Ca 1179/8 Nov. 1765); DWQ, Maḥkamat Manfalūṭ, 2, p. 190, case 632 (20 C 1179/4 Dec. 1765); DWQ, Maḥkamat Manfalūṭ, 2, p. 190, case 633 (3 Z 1180/2 May 1767); DWQ, Maḥkamat Asyūṭ, 2, p. 235, no case no. (23 Z 1107/23 July 1696).

17. DWQ, Maḥkamat Rashīd, 125, p. 328, case 540 (8 Za 1132/11 Sept. 1720); DWQ, Maḥkamat Rashīd, 132, p. 196, case 298 (25 R 1137/10 Jan. 1725).

18. DWQ, Maḥkamat Rashīd, 125, p. 333, case 548 (23 L 1132/28 Aug. 1720).

19. DWQ, Maḥkamat Rashīd, 125, p. 287, case 452 (13 Ca 1132/22 Mar. 1720).

20. DWQ, Maḥkamat Rashīd, 125, pp. 323–24, case 530 (12 M 1133/13 Nov. 1720). Unlike these other locales, Morocco was not a part of the Ottoman Empire in the early eighteenth century.

21. Some of these many cases include DWQ, Maḥkamat Rashīd, 125, p. 319, case 517 (28 Ra 1133/27 Jan. 1721); DWQ, Maḥkamat Rashīd, 154, p. 3, case 5 (6 C 1159/25 June 1746); DWQ, Maḥkamat Rashīd, 154, p. 10, no case no. (A) (29 S 1161/29 Feb. 1748).

22. DWQ, Maḥkamat Rashīd, 125, p. 147, case 257 (27 Z 1132/29 Oct. 1720).

23. DWQ, Maḥkamat Rashīd, 125, p. 287, case 452 (13 Ca 1132/22 Mar. 1720).

24. DWQ, Maḥkamat Rashīd, 154, p. 341, no case no. (22 M 1163/1 Jan. 1750). For cases involving the sending of provisions and military supplies from Egypt to Crete, see TSMA, E. 5207/62 (Evail M 1057/6–15 Feb. 1647); TSMA, E. 664/55 (n.d.).

25. See, for example, DWQ, Maḥkamat Rashīd, 125, p. 319, case 517 (28 Ra 1133/27 Jan. 1721); DWQ, Maḥkamat Rashīd, 154, p. 3, case 6 (8

R 1162/27 Mar. 1749); DWQ, Maḥkamat Rashīd, 154, p. 2, no case no. (15 R 1162/3 Apr. 1749); DWQ, Maḥkamat Rashīd, 154, p. 3, case 4 (12 M 1162/2 Jan. 1749); DWQ, Maḥkamat Rashīd, 154, p. 341, no case no. (4 Ra 1163/11 Feb. 1750).

26. DWQ, Maḥkamat Rashīd, 154, p. 10, no case no. (21 S 1161/21 Feb. 1748).

27. For a very small sampling of the thousands of cases involving shipments of grain from Egypt to Istanbul, see DWQ, Maḥkamat Rashīd, 122, p. 67, case 113 (21 Ca 1131/11 Apr. 1719); DWQ, Maḥkamat Rashīd, 123, p. 142, case 241 (25 B 1131/14 June 1719); DWQ, Maḥkamat Rashīd, 124, p. 253, case 352 (1 Ca 1132/10 Mar. 1720); DWQ, Maḥkamat Rashīd, 125, p. 318, case 516 (26 Ra 1133/25 Jan. 1721); DWQ, Maḥkamat Rashīd, 146, p. 139, case 116 (1 C 1153/24 Aug. 1740); DWQ, Maḥkamat Rashīd, 148, p. 176, case 219 (21 Z 1154/27 Feb. 1742); DWQ, Maḥkamat Rashīd, 154, p. 182, case 203 (25 Z 1162/6 Dec. 1749); DWQ, Maḥkamat Rashīd, 157, p. 324, case 319 (15 R 1166/19 Feb. 1753).

28. The best treatment of Ottoman Egypt's many trading links remains Raymond, *Artisans et commerçants*.

29. Much of this is recounted in Abu-Lughod, *Before European Hegemony*.

30. Daniel Panzac, *La caravane maritime: Marins européens et marchands ottomans en Méditerranée (1680–1830)* (Paris: CNRS éditions, 2004); Daniel Panzac, "International and Domestic Maritime Trade in the Ottoman Empire during the 18th Century," *International Journal of Middle East Studies* 24 (1992): 189–206; Daniel Crecelius and Hamza ʿAbd al-ʿAziz Badr, "French Ships and Their Cargoes Sailing between Damiette and Ottoman Ports, 1777–1781," *Journal of the Economic and Social History of the Orient* 37 (1994): 251–86; Raymond, *Artisans et commerçants*.

31. Nelly Hanna, *Making Big Money in 1600: The Life and Times of Ismaʿil Abu Taqiyya, Egyptian Merchant* (Syracuse, NY: Syracuse University Press, 1998).

32. See, for example, Haggai Erlich and Israel Gershoni, eds., *The Nile: Histories, Cultures, Myths* (Boulder, CO: Lynne Rienner, 2000); Haggai Erlich, *The Cross and the River: Ethiopia, Egypt, and the Nile* (Boulder, CO: Lynne Rienner, 2002). The focus of these works is more the economic and cultural relations forged along the Nile than the environmental history of the river or the ecological communities created through the shared usage of water, which are my interests here. More generally on commercial relations between the Sudan and Egypt in the eighteenth century, see Walz, *Trade between Egypt and Bilād as-Sūdān*.

33. See, for example, Tuchscherer, *Le commerce du café*.

34. In this vein, see Panzac, *La peste*; Daniel Panzac, *Quarantaines et lazarets: l'Europe et la peste d'Orient (XVIIe–XXe siècles)* (Aix-en-Provence, France: Édisud, 1986); Dols, "Second Plague"; Kuhnke, *Lives at Risk*.

35. Peter Gran, *Islamic Roots of Capitalism: Egypt, 1760–1840* (Austin: University of Texas Press, 1979); Khaled el-Rouayheb, "Was There a Revival of Logical Studies in Eighteenth-Century Egypt?," *Die Welt des Islams* 45 (2005):

1–19. In this regard, see also Khaled el-Rouayheb, "Sunni Muslim Scholars on the Status of Logic, 1500–1800," *Islamic Law and Society* 11 (2004): 213–32; Khaled el-Rouayheb, "Opening the Gate of Verification: The Forgotten Arab-Islamic Florescence of the 17th Century," *International Journal of Middle East Studies* 38 (2006): 263–81. Generally on the various impacts of North Africans in Ottoman Egypt, see ʿAbd al-Raḥīm ʿAbd al-Raḥman ʿAbd al-Raḥīm, *Wathāʾiq al-Maḥākim al-Sharʿiyya al-Miṣriyya ʿan al-Jāliya al-Maghāribiyya ibbāna al-ʿAṣr al-ʿUthmānī*, ed. and intro. ʿAbd al-Jalīl al-Tamīmī (Zaghwān, Tunisia: Markaz al-Dirāsāt wa al-Buḥūth al-ʿUthmāniyya wa al-Mūrīskiyya wa al-Tawthīq wa al-Maʿlūmāt, 1992); Ḥusām Muḥammad ʿAbd al-Muʿṭī, "al-Buyūt al-Tijāriyya al-Maghribiyya fī Miṣr fī al-ʿAṣr al-ʿUthmānī" (PhD diss., Mansura University, 2002); Ḥusām Muḥammad ʿAbd al-Muʿṭī, "Riwāq al-Maghāriba fī al-Jāmiʿ al-Azhar fī al-ʿAṣr al-ʿUthmānī," *al-Rūznāma: al-Ḥauliyya al-Miṣriyya lil-Wathāʾiq* 3 (2005): 165–204; ʿAbd Allah Muḥammad ʿAzabāwī, "al-ʿAlāqāt al-ʿUthmāniyya-al-Maghribiyya fī ʿAhd Kullin min Maulāya Muḥammad (1757–1790) wa Ibnihi Yazīd (1790–1792)," *Egyptian Historical Review* 30–31 (1983–84): 379–413.

36. Panzac, "International and Domestic Maritime Trade," 194–95.

37. For a detailed account of the food provisioning of the Hijaz from Egypt in the eighteenth century, see Ḥusām Muḥammad ʿAbd al-Muʿṭī, *al-ʿAlāqāt al-Miṣriyya al-Ḥijāziyya fī al-Qarn al-Thāmin ʿAshar* (Cairo: al-Hayʾa al-Miṣriyya al-ʿĀmma lil-Kitāb, 1999), 131–41. ʿAbd al-Muʿṭī estimates the following total yearly averages for shipments of food from Egypt to the Hijaz in the eighteenth century: 30,000 *ardabb*s of wheat, 15,000 *ardabb*s of *fūl*, 5,000 *ardabb*s of lentils, and 500 *ardabb*s of rice. The value of an *ardabb* varied greatly over the course of the long eighteenth century from a minimum of 75 liters in 1665 to 184 liters in 1798. Raymond, *Artisans et commerçants*, 1:LVII; Hinz, *Islamische Masse und Gewichte*, 39–40.

38. On the administrative unity between Lower Egypt (the Nile delta) and Fayyum in the realm of land tenure during the Ottoman period, see Cuno, *Pasha's Peasants*, 66. On shifts in trading patterns in the fifteenth century, see Nelly Hanna, *An Urban History of Būlāq in the Mamluk and Ottoman Periods* (Cairo: Institut français d'archéologie orientale, 1983), 7–32. Daniel Crecelius argues that Qusayr saw a resurgence in importance as a Red Sea trading port at the end of the eighteenth century. Daniel Crecelius, "The Importance of Qusayr in the Late Eighteenth Century," *Journal of the American Research Center in Egypt* 24 (1987): 53–60.

39. The maintenance of these *awqāf* was one of the key realms of grandee influence in the countryside. This influence, however, does not seem to have extended to water management or the repair of irrigation works in Fayyum.

40. Shaw, *Financial and Administrative Organization*, 269–70. One of the most famous and lucrative of these *awqāf* in the late seventeenth century was very near Fayyum and possibly administratively connected to it. It consisted of a group of nine villages in al-Bahnasa (Beni Suef) controlled by Ḥasan Aghā Bilīfyā, a Faqari grandee and commander of the Gönüllüyan military bloc. Jane Hathaway, "The Role of the Kızlar Ağası in 17th–18th Century Ottoman Egypt," *Studia Islamica* 75 (1992): 153–58; Hathaway, *Politics of Households*,

157–60; Jane Hathaway, "Egypt in the Seventeenth Century," in *Modern Egypt, from 1517 to the End of the Twentieth Century*, vol. 2 of *The Cambridge History of Egypt*, ed. M.W. Daly (Cambridge: Cambridge University Press, 1998), 50.

41. For descriptions of the topography, geography, and history of Fayyum in various periods, see A.E.R. Boak, "Irrigation and Population in Faiyum, the Garden of Egypt," *Geographical Review* 16 (1926): 353–64; R. Neil Hewison, *The Fayoum: A Practical Guide* (Cairo: American University in Cairo Press, 1984), 1–17; W. Willcocks and J.I. Craig, *Egyptian Irrigation*, 2 vols. (London: E. & F.N. Spon, 1913), 1:441–47; al-Maqrīzī, *al-Khiṭaṭ*, 1:245–48; *Encyclopaedia of Islam*, 2nd ed., s.v. "al-Fayyūm" (P.M. Holt).

42. The Fayyum was often termed "the lowest of land" (*asfal al-arḍ*). Abū ʿUthmān al-Nābulusī al-Ṣafadī, *Tārīkh al-Fayyūm wa Bilādihi* (Beirut: Dār al-Jīl, 1974), 9. This text was written in 1243 during an inspection of Fayyum commissioned for administrative purposes by the Ayyubid sultan Ṣāliḥ Najm al-Dīn al-Ayyūb. Al-Nābulusī (as the author is most commonly known) writes that he spent a little over two months in Fayyum in order to compose this text on Fayyum's geography, people, industries, and built environment. Ibid., 8. Though written nearly five hundred years before the eighteenth century, this is one of the few narrative sources devoted solely to Fayyum and represents the most complete cadastral survey to have survived from medieval Egypt. I will therefore make reference to it where appropriate for comparative purposes. I do not mean to imply, of course, that what al-Nābulusī writes in his account obtained until the first half of the eighteenth century. For another edition of this account and for a useful collection of essays about the text, see Fuat Sezgin, Mazen Amawi, Carl Ehrig-Eggert, and Eckhard Neubauer, eds., *Studies of the Faiyūm Together with* Tārīḫ al-Faiyūm wa-Bilādihī *by Abū ʿUṯmān an-Nābulusī (d. 1261)*, Islamic Geography, vol. 54 (Frankfurt am Main: Institute for the History of Arabic-Islamic Science at the Johann Wolfgang Goethe University, 1992). See also Yossef Rapoport, "Invisible Peasants, Marauding Nomads: Taxation, Tribalism, and Rebellion in Mamluk Egypt," *Mamlūk Studies Review* 8 (2004): 1–22; G. Keenan, "Fayyum Agriculture at the End of the Ayyubid Era: Nabulsi's *Survey*," in *Agriculture in Egypt: From Pharaonic to Modern Times*, ed. Alan K. Bowman and Eugene Rogan, 287–99 (Oxford: Oxford University Press for the British Academy, 1999).

43. Hewison, *Fayoum*, 2.

44. In the thirteenth century, its land was said to be made of the perfect combination of unadulterated alluvial deposits (*al-iblīz al-maḥḍ*) and composite soils (*al-ṭīn al-mukhtaliṭ*). Al-Nābulusī, *Tārīkh al-Fayyūm* (1974), 5.

45. The total percentage of salts in the lake is 1.34. Sodium chloride represents 0.92 percent of the total. Boak, "Irrigation and Population in Faiyum," 356.

46. On Birkat Qārūn, see Butzer, *Early Hydraulic Civilization*, 36–38, 92–93, 108; Ali Shafei Bey, "Fayoum Irrigation as Described by Nabulsi in 1245 A.D. with a Description of the Present System of Irrigation and a Note on Lake Moeris," in *Studies of the Faiyūm Together with* Tārīḫ al-Faiyūm wa-Bilādihī *by Abū ʿUṯmān an-Nābulusī (d. 1261)*, ed. Fuat Sezgin, Mazen Amawi, Carl

Ehrig-Eggert, and Eckhard Neubauer, Islamic Geography, vol. 54 (Frankfurt am Main: Institute for the History of Arabic-Islamic Science at the Johann Wolfgang Goethe University, 1992), 308–9. The presence, size, shape, and location of the ancient ancestor of Birkat Qārūn, Lake Moeris, have been subjects of much scholarly debate. See, for example, Gertrude Caton-Thompson and E.W. Gardner, "Recent Work on the Problem of Lake Moeris," *Geographical Journal* 73 (1929): 20–58; J.A.S. Evans, "Herodotus and the Problem of the 'Lake of Moeris,'" *Classical World* 56 (1963): 275–77. As suggested by this last citation, much of the literature is concerned with explaining Herodotus's description of the lake. The relevant section of his *History* is 2.148–50.

47. For technical details and drawings of drainage and discharge in Fayyum, see Willcocks and Craig, *Egyptian Irrigation*, 1:442–47; Shafei Bey, "Fayoum Irrigation," 286–309. On the problem of drainage in modern Egypt, see M.H. Amer and N.A. de Ridder, eds., *Land Drainage in Egypt* (Cairo: Drainage Research Institute, 1989); H.J. Nijland, ed., *Drainage along the River Nile* (Egypt: Egyptian Public Authority for Drainage Projects; Netherlands: Directorate-General of Public Works and Water Management, 2000).

48. For a thirteenth-century description of this ridge, see al-Nābulusī, *Tārīkh al-Fayyūm* (1974), 5, 7.

49. On Baḥr Yūsuf, see Shafei Bey, "Fayoum Irrigation," 298–99; Helen Anne B. Rivlin, *The Agricultural Policy of Muḥammad ʿAlī in Egypt* (Cambridge, MA: Harvard University Press, 1961), 238–39; Willcocks and Craig, *Egyptian Irrigation*, 1:441–44; Butzer, *Early Hydraulic Civilization*, 16, 36–38, 53.

50. We are fortunate to have an example of the financial and administrative organization of the distribution of this water in Ottoman Fayyum. See DWQ, al-Rūznāma, Daftar Irtifāʿ al-Miyāh bi-Baḥr Sayyidnā Yūsuf lihi al-Ṣalāh wa al-Salām ʿan al-Qabḍa al-Yūsufiyya Tābiʿ Wilāyat al-Fayyūm (Raqam al-Ḥifẓ al-Nauʿī 1, ʿAyn 59, Makhzin Turkī 1, Musalsal 4557). This register consists of copies of Ottoman firmans, Arabic administrative cases, and financial calculations related to how much water (these amounts were known as *qabaḍāt al-miyāh*) various villages and regions of Fayyum were to receive for a given year. It is best thought of as a detailed bureaucratic accounting book of how much water went where. It includes entries for the *hijrī* years 948, 1017, 1027, 1066, 1091, 1102, 1109, 1114, 1116, 1125, 1127, 1128, 1129, 1130, 1187, 1192, 1195, 1197, 1200, and 1207.

51. During the Ottoman period, both of these structures were part of a class of irrigation works in Egypt known as *al-jusūr al-sulṭāniyya*. Again, if a canal or irrigation feature served a large group of peasants rather than the interests of a small privileged few, contributed to the common good, or aided in the achievement of equality among peasants, then it was considered a *sulṭānī* irrigation work, the responsibilities of which fell on the Ottoman state in Egypt. *Baladī* works, by contrast, were those that served the irrigation needs of one particular community and no one else, and they were to be maintained by locals. For cases explicitly stating the imperial (*sulṭānī*) status of the dam of al-Gharaq and the dike of al-Lāhūn, see BOA, MM, 3:11 (Evasıt B 1131/30 May–8 June 1719); BOA, MM, 4:36 (Evail Za 1139/20–

29 June 1727). For cases involving repairs to Fayyumi irrigation works other than these two, see BOA, MM, 5:301 (Evahir L 1148/5–14 Mar. 1736); BOA, MM, 5:475 (Evahir S 1152/30 May–8 June 1739). On the medieval history of the difference between *sulṭānī* and *baladī* irrigation works, see Borsch, "Environment and Population"; Tsugitaka, *State and Rural Society*, 225–27.

52. For descriptions of the construction and maintenance of the dike of al-Lāhūn in the thirteenth century, see al-Nābulusī, *Tārīkh al-Fayyūm* (1974), 12, 15–17.

53. BOA, Cevdet Nafia, 458 (9 Ra 1158/11 Apr. 1745).

54. Not everyone was pleased with the quality of the canal's waters. In the mid-thirteenth century, al-Nābulusī complained that water in Baḥr Yūsuf was extremely vile (*radī'*). Passing through wide areas of gluey muddy earth (*arḍ ṭīniyya lazija*), this water collected all sorts of rotten human and animal waste, making it into a stagnant and boggy pool (*mā' baṭīḥa wa-naq'a*). Al-Nābulusī, *Tārīkh al-Fayyūm* (1974), 9–10. More generally on the waters of Baḥr Yūsuf, see ibid., 17.

55. In 1695, for example, the Nile's flood was much greater than expected. It overwhelmed Fayyum's irrigation features and was said to have killed many people in the region and to have destroyed large areas of agricultural land. Al-Damurdāshī Katkhudā ʿAzabān, *Kitāb al-Durra al-Muṣāna*, 30.

56. As it was in the thirteenth century as well. For examples of repairs to these dams and other irrigation works in the thirteenth century, see al-Nābulusī, *Tārīkh al-Fayyūm* (1974), 6, 12, 16.

57. BOA, MM, 5:111 (Evail M 1147/3–12 June 1734).

58. On some of the particularities of agricultural cultivation in Fayyum, see Watson, *Agricultural Innovation*, 17, 28, 40.

59. P.M. Holt, "The Beylicate in Ottoman Egypt during the Seventeenth Century," *Bulletin of the School of Oriental and African Studies* 24 (1961): 220–21.

60. For comparisons with other instances of repair work in Ottoman Cairo, see Nelly Hanna, *Construction Work in Ottoman Cairo (1517–1798)* (Cairo: Institut français d'archéologie orientale, 1984); Doris Behrens-Abouseif, *Egypt's Adjustment to Ottoman Rule: Institutions, Waqf, and Architecture in Cairo, 16th and 17th Centuries* (Leiden, Netherlands: Brill, 1994).

61. For another case that makes these points explicitly, see BOA, MM, 8:66 (Evail N 1175/26 Mar.–4 Apr. 1762).

62. BOA, İbnülemin Umur-i Nafia, 94 (Evasıt Ra 1121/21–30 May 1709). The Egyptian Arabic word *sharāqī* refers to land that is not reached by water and is hence parched and dry. In contrast to *būr* land, which is uncultivatable wasteland, *sharāqī* earth has the potential for cultivation given the proper amount of water.

63. No information is given about the specific identity of these engineers. By comparison, al-Nābulusī offers tantalizingly suggestive information that in similar situations of repair in the thirteenth century, the engineers consulted in these cases were indeed local Fayyumis. He writes that people of the village (*ahl al-qarya*) who were often consulted in irrigation repair jobs were known as engineers (*yuʿrifūn bi-l-muhandisīn*). Al-Nābulusī adds that this title had

nothing to do with any technical training in engineering or related sciences. It seems, rather, that it was given to those with expertise and experience in the local environment of whatever irrigation work was under repair. Abū ʿUthmān al-Nābulusī al-Ṣafadī al-Shāfiʿī, *Tārīkh al-Fayyūm wa Bilādihi* (Cairo: al-Maṭbaʿa al-Ahliyya, 1898), 16. My thanks to Yossef Rapoport for bringing this passage to my attention. Of course, this thirteenth-century account says nothing specifically about our case from 1709. Rural engineers are the focus of chapter 5.

64. The Egyptian purse equaled 25,000 *para*, the basic unit of currency in Ottoman Egypt. On the *para*, see Shaw, *Financial and Administrative Organization*, xxii. Repairs to irrigation works in Fayyum were usually funded from the Egyptian *irsaliye*. The *irsaliye* represented the overall revenue garnered from an Ottoman province in any given year, and it was the responsibility of the provincial governor to send these funds to Istanbul. The Egyptian *irsaliye* was historically the largest in the empire. Diverting a portion of these funds away from the imperial coffers to finance needed irrigation repairs in the province was clearly undesirable from the perspective of the imperial administration. For cases concerning various aspects of the organization of the yearly Egyptian *irsaliye*, see TSMA, E. 664/4 (n.d.); TSMA, E. 664/64 (1 C 1059/12 June 1649); TSMA, E. 5207/57 (Evail B 1056/12–21 Aug. 1646); TSMA, E. 5207/58 (Evasıt B 1056/22–31 Aug. 1646); TSMA, E. 7016/95 (n.d.); TSMA, E. 5207/49 (Evahir Ca 1056/5–14 July 1646); TSMA, E. 664/66 (n.d.); TSMA, E. 4675/2 (20 N 1061/6 Sept. 1651); TSMA, E. 3522 (24 Ş 1148/8 Jan. 1736). For a detailed discussion of the *irsaliye*, see Shaw, *Financial and Administrative Organization*, 283–312, 399–401. For a detailed accounting of each component of the *irsaliye* from 1596 to 1597, see Stanford J. Shaw, *The Budget of Ottoman Egypt, 1005–1006/1596–1597* (The Hague: Mouton, 1968). For a case in which the dam of al-Gharaq was repaired using funds from the *cizye*, see BOA, Cevdet Nafia, 2570 (Evahir Ş 1174/28 Mar.–6 Apr. 1761).

65. This again points to the necessity and utility of considering the Ottoman imperial record for the history of Fayyum and Egypt more generally.

66. Al-Jabartī, *ʿAjāʾib al-Āthār* (1998), 1:69.

67. Al-Jabartī was born in 1753 and wrote much of his chronicle on the basis of earlier materials. Thus, he almost certainly took his information about the repairs of 1709 from another source.

68. BOA, MM, 1:116 (Evail R 1122/30 May–8 June 1710).

69. Though this case specifically mentions imperial funds and others do not, financial losses were a consistent concern of the Ottoman bureaucracy in Egypt when dealing with repairs to the province's irrigation works.

70. The use of the title *kāşif* in this order provides a good case study for considering the nature of Mamluk influence in Ottoman Egypt by showing how the Ottoman bureaucracy both maintained and modified some elements of Mamluk rule (and of even earlier Egyptian polities) in its administration. The 1525 Ottoman law code (*Kanunname*) retained the Mamluk term *kāşif* to refer to subprovincial governors. Despite the continued use of this term, however, the Mamluk system of land management based on the *iqṭāʿ* was wholly replaced by a very different arrangement that by the early years of the

seventeenth century had developed into a tax farming (*iltizām*) system that differentiated land administration in Egypt from the *tımār*s found in the rest of the empire. Thus, though technically *mültezimīn* (those who hold a tax farm), these landholders, as well as other officials, are variously referred to in the sources of the period using both "Mamluk" and "Ottoman" titles—*sancak*, *sancak beyi*, *mültezim*, *kāşif*, amir, and bey. Rather than reading this sort of slippery mélange of Mamluk, Ottoman, and other titles as "proof" of the continuation of *only* Mamluk influence in Egypt, I see it more as evidence of a new and different configuration of social and political power in Egypt that characterized the conglomerated and accumulated nature of Ottoman imperial rule. This paragraph relies heavily on Hathaway, *Politics of Households*, 9; Holt, "Beylicate in Ottoman Egypt during the Seventeenth Century."

71. In an apparent mistake or change, the amount reported to be released at the end of this order is eleven Egyptian purses and 11,650 *para*.

72. BOA, MM, 1:167 (Evasıt S 1123/31 Mar.–9 Apr. 1711).

73. The length of the *zirā*ʿ varied greatly in Ottoman Egypt. Here I take one *zirā*ʿ to equal 63 centimeters. According to the standard work on Islamic weights and measures, the *zirā*ʿ varied from 58 to 68 centimeters. Hinz, *Islamische Masse und Gewichte*, 56. According to Gábor Ágoston, one *zirā*ʿ equaled 75.8 centimeters. Gábor Ágoston, *Guns for the Sultan: Military Power and the Weapons Industry in the Ottoman Empire* (Cambridge: Cambridge University Press, 2005), 247.

74. BOA, MM, 4:36 (Evail Za 1139/20–29 June 1727).

75. For critical readings of these kinds of petitions to the imperial centers of Cairo and Istanbul, see James Edward Baldwin, "Islamic Law in an Ottoman Context: Resolving Disputes in Late Seventeenth and Early Eighteenth-Century Cairo" (PhD diss., New York University, 2010), especially the chapter entitled "Petitioning the Sultan"; Majdī Jirjis, "Manhaj al-Dirāsāt al-Wathāʾiqiyya wa Wāqiʿ al-Baḥth fī Miṣr," *al-Rūznāma: al-Ḥauliyya al-Miṣriyya lil-Wathāʾiq* 2 (2004): 237–87. On petitioning in the Ottoman Empire more generally, see Halil İnalcık, "Şikâyet Hakkı: *ʿArż-i Ḥâl ve ʿArż-i Maḥzar*'lar," in *Osmanlı'da Devlet, Hukuk, Adâlet*, 49–71 (Istanbul: Eren Yayıncılık, 2000).

76. BOA, MM, 6:149 (Evasıt Ca 1157/22 June–1 July 1744).

77. Ibid.; BOA, MM, 5:696 (Evahir M 1155/28 Mar.–6 Apr. 1742); BOA, MM, 6:2 (Evahir S 1156/16–25 Apr. 1743).

78. Mehmed Paşa al-Yedekçi was appointed vali of Egypt in September 1743. He was described as a man of perception and sound judgment (*rüyet ve tefekkür*) who possessed the essence of trust and ability (*cevher-i sadākat ve kifayet*) and who had within him the germ of grace and faithfulness (*maya-ı himmet ve emānet*). BOA, MM, 6:37 (Evasıt Ra 1156/5–14 May 1743). For more on the period of his rule, see also al-Jabartī, *ʿAjāʾib al-Āthār* (1998), 1:260–61.

79. BOA, MM, 6:238 (Evasıt Ca 1158/11–20 June 1745).

80. Ibid. For more on the removal of Mehmed Paşa, see BOA, Cevdet Nafia, 458 (9 Ra 1158/11 Apr. 1745).

81. As was customary practice, the new vali Raghib Paşa requested from the imperial administration that these funds be taken from the Egyptian *irsaliye* of 1743–44. No doubt existed that this money was necessary to revive

the villages of Fayyum (*akālīm-i fayyum'un ihyasine peyda olmak*). However, given Mehmed Paşa's dishonesty in repairing the dam of al-Gharaq, the sultan shirked the custom of taking funds from the *irsaliye*. He ordered, instead, that the portion of these funds (over twenty-eight Egyptian purses) needed to repair those sections of the dam that should have been repaired by Mehmed Paşa in the first place were to be secured from the former vali as a form of punishment for his theft. This amount was, of course, in addition to the seven purses he had stolen from the state and promised to repay. The remainder of the necessary funds (over twelve Egyptian purses needed to repair newly damaged areas) was to be taken from the profits of villages "sold" as *ḥulvān* (a fee paid to the treasury for rights to a tax farm).

82. In the spring of 1746, the dam of al-Gharaq was repaired for the last time, after five consecutive years of intense repairs. BOA, MM, 6:295 (Evahir Ra 1159/13–22 Apr. 1746). Much of the dam was in near total ruin in 1746, and Raghib Paşa was thus ordered to fix as much of the structure as possible in as permanent a manner as possible to prevent the need for any future repairs. The sultan's government allocated the funds required for these repairs from the Egyptian *irsaliye* of 1744–45: 15 Egyptian purses and 5,250 *para* to repair a section measuring 22,350 square *zirā*ʿ (8,870.7 square meters). Here again each square *zirā*ʿ (0.3969 square meters) cost 15 *para* to repair. The dam continued to function without incident for a few years after 1746. BOA, MM, 6:557 (Evahir N 1162/4–13 Sept. 1749).

CHAPTER 4

1. For general histories of early nineteenth-century Egypt, see, for example, al-Rāfʿī, *ʿAṣr Muḥammad ʿAlī*; Afaf Lutfi al-Sayyid Marsot, *Egypt in the Reign of Muhammad Ali* (Cambridge: Cambridge University Press, 1984).

2. For a very useful examination of the function and maintenance of these elite political and economic alliances in the nineteenth and twentieth centuries, see Raouf Abbas and Assem El-Dessouky, *The Large Landowning Class and the Peasantry in Egypt, 1837–1952*, trans. Amer Mohsen with Mona Zikri, ed. Peter Gran (Syracuse, NY: Syracuse University Press, 2011). On the continuation of this brand of politics after the 1952 revolution, see Leonard Binder, *In a Moment of Enthusiasm: Political Power and the Second Stratum in Egypt* (Chicago: University of Chicago Press, 1978). While my analysis is essentially in line with the former study, I see the creation of elite crony politics as a result of processes occurring in the second half of the eighteenth century, not as the product of the age of Mehmet ʿAli in the first half of the nineteenth century. For analyses of aspects of peasant opposition to these political forces in the nineteenth and twentieth centuries, see Nathan J. Brown, *Peasant Politics in Modern Egypt: The Struggle against the State* (New Haven, CT: Yale University Press, 1990); Fred H. Lawson, "Rural Revolt and Provincial Society in Egypt, 1820–1824," *International Journal of Middle East Studies* 13 (1981): 131–53.

3. For discussions of this historiography, see Khaled Fahmy, *All the Pasha's Men: Mehmed Ali, His Army and the Making of Modern Egypt* (Cambridge: Cambridge University Press, 1997), 1–37; Khaled Fahmy, *Mehmed Ali: From*

Ottoman Governor to Ruler of Egypt (Oxford: Oneworld Publications, 2009), 112–27; Ehud R. Toledano, "Mehmet Ali Paşa or Muhammad Ali Basha? An Historiographic Appraisal in the Wake of a Recent Book," *Middle Eastern Studies* 21 (1985): 141–59; Alan Mikhail, "Unleashing the Beast: Animals, Energy, and the Economy of Labor in Ottoman Egypt," *American Historical Review* 118 (2013): 319–21.

4. For discussions of some of these various late eighteenth-century imperial processes and stresses, see Virginia H. Aksan, *An Ottoman Statesman in War and Peace: Ahmed Resmi Efendi, 1700–1783* (Leiden, Netherlands: Brill, 1995), 100–205; Virginia H. Aksan, *Ottoman Wars, 1700–1870: An Empire Besieged* (Harlow, UK: Routledge, 2007); Baki Tezcan, *The Second Ottoman Empire: Political and Social Transformation in the Early Modern World* (Cambridge: Cambridge University Press, 2010); Reşat Kasaba, *The Ottoman Empire and the World Economy: The Nineteenth Century* (Albany: State University of New York Press, 1988).

5. On the classical system of Ottoman administration, see Halil İnalcık, *The Ottoman Empire: The Classical Age, 1300–1600*, trans. Norman Itzkowitz and Colin Imber (New York: Praeger Publishers, 1973).

6. Daniel Crecelius, *The Roots of Modern Egypt: A Study of the Regimes of ʿAli Bey al-Kabir and Muhammad Bey Abu al-Dhahab, 1760–1775* (Minneapolis: Bibliotheca Islamica, 1981).

7. Ibid., 79–91, 159–68.

8. Hathaway, *Politics of Households*; Hathaway, *Tale of Two Factions*; Stanford J. Shaw, "Landholding and Land-Tax Revenues in Ottoman Egypt," in *Political and Social Change in Modern Egypt: Historical Studies from the Ottoman Conquest to the United Arab Republic*, ed. P.M. Holt, 91–103 (London: Oxford University Press, 1968); Holt, *Egypt and the Fertile Crescent*, 85–101.

9. For a fuller treatment of the intertwining phenomena discussed in this paragraph, see Mikhail, "Unleashing the Beast."

10. On some of these changes in rural agriculture, see Cuno, *Pasha's Peasants*; Kenneth M. Cuno, "Commercial Relations between Town and Village in Eighteenth and Early Nineteenth-Century Egypt," *Annales Islamologiques* 24 (1988): 111–35; Alan R. Richards, "Primitive Accumulation in Egypt, 1798–1882," in *The Ottoman Empire and the World-Economy*, ed. Huri İslamoğlu-İnan, 203–43 (Cambridge: Cambridge University Press, 1987).

11. My study therefore seeks to make a modest contribution to a venerable tradition of labor and working-class history in Egypt by offering both an earlier and an environmental perspective. For some of the major English-language studies of labor politics in Egypt in the nineteenth and twentieth centuries, see Joel Beinin and Zachary Lockman, *Workers on the Nile: Nationalism, Communism, Islam, and the Egyptian Working Class, 1882–1954* (Princeton, NJ: Princeton University Press, 1987); Ellis Goldberg, *Tinker, Tailor, and Textile Worker: Class and Politics in Egypt, 1930–1952* (Berkeley: University of California Press, 1986); Marsha Pripstein Posusney, *Labor and the State in Egypt: Workers, Unions, and Economic Restructuring* (New York: Columbia University Press, 1997). For a broader regional perspective, see Joel Beinin, *Workers*

and Peasants in the Modern Middle East (Cambridge: Cambridge University Press, 2001); Zachary Lockman, ed., *Workers and Working Classes in the Middle East: Struggles, Histories, Historiographies* (Albany: State University of New York Press, 1994); Ellis Jay Goldberg, ed., *The Social History of Labor in the Middle East* (Boulder, CO: Westview Press, 1996).

12. For an analysis of such cases, see Mikhail, *Nature and Empire*, 38–81.

13. On animal labor in Ottoman Egypt, see Alan Mikhail, "Animals as Property in Early Modern Ottoman Egypt," *Journal of the Economic and Social History of the Orient* 53 (2010): 621–52.

14. For a comparative perspective on this phenomenon, see Appadurai, "Wells in Western India."

15. On the involvement of these workers in irrigation projects in Ottoman Iraq, see Murphey, "Ottoman Centuries in Iraq," 23, 27.

16. One of these categories of worker was *al-mudamasīn*. For a canal repair project that employed both ditchdiggers (*al-hufrāʾ*) and *al-mudamasīn*, see DWQ, Maḥkamat Asyūṭ, 1, p. 201, case 583 (12 Za 1067/22 Aug. 1657).

17. For cases involving lifters and carriers, see DWQ, Maḥkamat Rashīd, 132, pp. 200–201, case 311 (3 N 1137/16 May 1725); DWQ, Maḥkamat al-Baḥayra, 5, p. 314, case 389 (10 Ş 1165/22 June 1752); DWQ, Maḥkamat Rashīd, 145, p. 126, case 101 (30 Z 1151/9 Apr. 1739).

18. The term in Arabic is *aṣḥāb al-idrāk*. For cases involving various of these early modern experts, see BOA, Cevdet Nafia, 120 (Evasıt Ca 1125/5–14 June 1713); BOA, MM, 8:469 (Evasıt L 1180/12–21 Mar. 1767); DWQ, Maḥkamat al-Manṣūra, 3, pp. 282 and 285, case 876 (18 N 1066/11 July 1656); DWQ, Maḥkamat Asyūṭ, 4, p. 206, case 645 (11 C 1156/2 Aug. 1743); DWQ, Maḥkamat Asyūṭ, 2, p. 238, case 566 (13 M 1108/11 Aug. 1696).

19. DWQ, Maḥkamat al-Manṣūra, 3, pp. 282 and 285, case 876 (18 N 1066/11 July 1656); DWQ, Maḥkamat al-Manṣūra, 3, p. 51, case 168 (5 B 1063/1 June 1653).

20. DWQ, Maḥkamat al-Manṣūra, 2, pp. 272 and 292, no case no. (1 Ca 1062/10 Apr. 1652).

21. DWQ, Maḥkamat al-Manṣūra, 17, p. 383, no case no. (11 M 1119/14 Apr. 1707). Animals were very often key actors in the digging and dredging of canals and in the reinforcement of canal embankments. For an example of the utilization of a group of buffalo cows in the repair of a canal's embankments in the subprovince of al-Daqahliyya, see DWQ, Maḥkamat al-Manṣūra, 3, p. 10, case 31 (19 S 1063/19 Jan. 1653). Renting these animals was the largest expense incurred in these repairs. Camels were likewise often used to clear debris and mud that hindered the proper function of wells and other water sources. Al-Damurdāshī Katkhudā ʿAzabān, *Kitāb al-Durra al-Muṣāna*, 131.

22. DWQ, al-Jusūr al-Sulṭāniyya, 784, p. 131, no case no. (n.d.).

23. DWQ, al-Jusūr al-Sulṭāniyya, 784, p. 134, no case no. (n.d.). One *qaṣaba* equals 3.99 meters. Hinz, *Islamische Masse und Gewichte*, 63. There is some discrepancy about this conversion. In her discussion of the repair of the Maḥmūdiyya Canal in the early nineteenth century, Helen Anne B. Rivlin takes one *qaṣaba* to equal 3.64 meters. Rivlin, *Agricultural Policy*, 218. Elsewhere she writes that the *qaṣaba* ranged between 3.75 meters and 3.99 meters. Ibid.,

125. Unless otherwise noted, throughout this book I take one *qaṣaba* to equal 3.99 meters.

24. ʿAbd al-Raḥman ibn Ḥasan al-Jabartī, *ʿAbd al-Raḥman al-Jabartī's History of Egypt: ʿAjāʾib al-Āthār fī al-Tarājim wa al-Akhbār*, ed. Thomas Philipp and Moshe Perlmann, 4 vols. (Stuttgart: Franz Steiner Verlag, 1994), 1:483.

25. For a general discussion of corvée in the seventeenth and eighteenth centuries, see ʿAbd al-Ḥamīd Sulaymān, "al-Sukhra fī Miṣr fī al-Qarnayn al-Sābiʿ ʿAshar wa al-Thāmin ʿAshar, Dirāsa fī al-Asbāb wa al-Natāʾij," in *al-Rafḍ wa al-Iḥtijāj fī al-Mujtamaʿ al-Miṣrī fī al-ʿAṣr al-ʿUthmānī*, ed. Nāṣir Ibrāhīm and Raʾūf ʿAbbās, 89–126 (Cairo: Markaz al-Buḥūth wa al-Dirāsāt al-Ijtimāʿiyya, 2004).

26. On *ūsya* land, see ʿAbd al-Raḥīm, *al-Rīf al-Miṣrī*, 96–100; Cuno, *Pasha's Peasants*, 36–37, 67–69; al-Shirbīnī, *Hazz al-Quḥūf*, 2:328n2.

27. Cuno, *Pasha's Peasants*, 36–37.

28. Al-Shirbīnī, *Hazz al-Quḥūf*, 2:328.

29. The phrase here is "*al nās fī al-balad*." Ibid., 2:327, 2:329.

30. *Lisān al-ʿArab*, 4 vols. (Beirut: Dār Lisān al-ʿArab, 1970), s.v. "*ʿawana*"; Edward William Lane, *An Arabic-English Lexicon*, 8 vols. (Beirut: Librairie du Liban, 1968), s.v. "*ʿawana*."

31. Al-Shirbīnī, *Hazz al-Quḥūf*, 2:330. Emphasis in original.

32. Ibid., 2:327.

33. Ibid., 2:329.

34. *Lisān al-ʿArab*, s.v. "*sakhara*"; Lane, *Arabic-English Lexicon*, s.v. "*sakhara*." On the differences between the use of the terms *al-ʿauna* and *al-sukhra* to refer to forced labor, see al-Jabartī, *ʿAjāʾib al-Āthār* (1994), 3:344.

35. DWQ, Maḥkamat Rashīd, 151, pp. 366–69, case 413 (25 Ra 1160/6 Apr. 1747).

36. Shaw, *Financial and Administrative Organization*.

37. For earlier examples of repairs to grain storage facilities (*wakāla*s) in Rosetta, see DWQ, Maḥkamat Rashīd, 123, pp. 97–98, case 170 (28 Z 1131/11 Nov. 1719); DWQ, Maḥkamat Rashīd, 125, pp. 92–93, case 159 (20 L 1132/25 Aug. 1720); DWQ, Maḥkamat Rashīd, 134, p. 167, case 204 (30 S 1140/16 Oct. 1727); DWQ, Maḥkamat Rashīd, 142, p. 64, case 58 (14 Za 1149/16 Mar. 1737); DWQ, Maḥkamat Rashīd, 151, pp. 38–39, case 49 (28 Z 1158/20 Jan. 1746). For more on the construction and repair of these facilities, see Hanna, *Making Big Money in 1600*, 125–33; Hanna, *Construction Work*, 46. Generally on the function of *wakāla*s in Ottoman Egypt, see Raymond, *Artisans et commerçants*, 1:254–60. For a study of one particularly prominent Mamluk and Ottoman *wakāla*, see Muḥammad Ḥusām al-Dīn Ismāʿīl ʿAbd al-Fattāḥ and Suhayr Ṣāliḥ, "A Wikāla of Sulṭān Mu'ayyid: Wikālat 'Ūda Pasha," *Annales Islamologiques* 28 (1994): 71–96.

38. On the movement of grain from Ottoman Egypt, see Mikhail, *Nature and Empire*, 82–123.

39. Another example of the use of these materials for construction is the following case about repairs to a home in Rosetta: DWQ, Maḥkamat Rashīd, 124, p. 254, case 354 (26 Ra 1132/5 Feb. 1720).

40. For other examples of the use of these kinds of workers in construction projects, see al-Jabartī, *ʿAjāʾib al-Āthār* (1994), 2:398–99, 3:11.

41. For earlier examples of the use of engineers in construction projects, see DWQ, Maḥkamat al-Manṣūra, 4, p. 108, case 281 (Evail M 1075/25 July–3 Aug. 1664); DWQ, Maḥkamat al-Manṣūra, 7, p. 134, case 340 (7 Za 1091/29 Nov. 1680); DWQ, Maḥkamat al-Manṣūra, 16, p. 257, case 527 (18 Z 1116/13 Apr. 1705); BOA, İbnülemin Umur-i Nafia, 94 (Evasıt Ra 1121/21–30 May 1709); BOA, Cevdet Nafia, 120 (Evasıt Ca 1125/5–14 June 1713).

42. For another example of giving coffee and water to workers as their only provisions, see DWQ, Maḥkamat Rashīd, 142, p. 64, case 58 (14 Za 1149/16 Mar. 1737).

43. This was also the case in the construction of the Süleymaniye mosque complex in Istanbul from 1550 to 1557. Ömer Lütfi Barkan, *Süleymaniye Cami ve İmareti İnşaatı (1550–1557)*, 2 vols. (Ankara: Türk Tarih Kurumu, 1972–79).

44. DWQ, Maḥkamat Manfalūṭ, 3, pp. 264–65, case 557 (24 Ş 1223/15 Oct. 1808). On the village of Banī Kalb, see Ramzī, *al-Qāmūs al-Jughrāfī*, pt. 2, 4:77.

45. During other projects, work was usually suspended on Fridays. See, for example, DWQ, Maḥkamat Rashīd, 125, pp. 92–93, case 159 (20 L 1132/25 Aug. 1720).

46. Rushdi Said, *The Geology of Egypt* (New York: Elsevier Publishing, 1962), 8.

47. Some repair cases from Manfalūṭ specified that canal work should be carried out in winter when temperatures were cooler and demands for water not as high as in summer. See, for example, DWQ, Maḥkamat Manfalūṭ, 2, p. 183, case 619 (16 Ca 1179/31 Oct. 1765).

48. The word *anfār* has a curious history. As seen in this case, in the early nineteenth century it was used to refer collectively to individuals brought to work as faceless units of labor on enormous construction projects. A few decades later it would be the name given to the lowest order of soldier in the Egyptian army. Because of the similarity of this military meaning to the use of the word *subaltern* to refer to comparably low classes in Western armies, Egyptian historians have used the word *anfār* in the Arabic translation of the historiographical movement known as Subaltern Studies. Thus, the valences of the word's meanings include a progression from worker to solider to subaltern. For a discussion of some of these meanings and of the choice to use *dirāsāt al-anfār* rather than *dirāsāt al-tābiʿ* as the Arabic translation of Subaltern Studies, see Khālid Fahmī, *al-Jasad wa al-Ḥadātha: al-Ṭibb wa al-Qānūn fī Miṣr al-Ḥadītha*, trans. Sharīf Yūnis (Cairo: Dār al-Kutub wa al-Wathāʾiq al-Qawmiyya, 2004), 33n10; Ālan Mīkhāʾīl, "Tārīkh Dirāsāt al-Tābiʿ wa Naẓariyyatayn ʿan al-Sulṭa," in *Thaqāfat al-Nukhba wa Thaqāfat al-ʿĀmma fī Miṣr fī al-ʿAṣr al-ʿUthmānī*, ed. Nāṣir Aḥmad Ibrāhīm, 349–60 (Cairo: Markaz al-Buḥūth wa al-Dirāsāt al-Ijtimāʿiyya, 2008).

49. I borrow this phrase from the following: "The economic plan, survey map, record of ownership, forest management plan, classification of ethnicity, passbook, arrest record, and map of political boundaries acquire their force

from the fact that these synoptic data are the points of departure for reality as state officials apprehend and shape it. . . . Where there is no effective way to assert another reality, fictitious facts-on-paper can often be made eventually to prevail on the ground, because it is on behalf of such pieces of paper that police and army are deployed." James C. Scott, *Seeing Like a State: How Certain Schemes to Improve the Human Condition Have Failed* (New Haven, CT: Yale University Press, 1998), 83. While the Egyptian case bears some family resemblances to the situation Scott describes, it is important to note that the precocious bureaucratizing Egyptian state's "pieces of paper" did not invent a wholly new "reality" of labor in this period, but rather usurped and recast long-held local practices.

50. Al-Jabartī, *'Ajā'ib al-Āthār* (1994), 4:289.

51. Mehmet 'Ali seized some tax farms before 1812 too. For a description of what happened when his government took over various tax farms in 1808, for example, see ibid., 4:115.

52. Ibid., 4:289. For a discussion of this important passage, see Cuno, *Pasha's Peasants*, 5–6, 37.

53. On Ottoman Alexandria, see İdris Bostan, "An Ottoman Base in Eastern Mediterranean: Alexandria of Egypt in the 18th Century," in *Proceedings of the International Conference on Egypt during the Ottoman Era: 26–30 November 2007, Cairo, Egypt*, ed. Research Centre for Islamic History, Art and Culture, 63–77 (Istanbul: IRCICA, 2010); Reimer, "Ottoman Alexandria."

54. On the challenges of the journey between Alexandria and Cairo, see Alan Mikhail, "Anatolian Timber and Egyptian Grain: Things That Made the Ottoman Empire," in *Early Modern Things: Objects and Their Histories, 1500–1800*, ed. Paula Findlen (New York: Routledge, 2013), 280–81.

55. 'Umar Ṭūsūn, *Tārīkh Khalīj al-Iskandariyya al-Qadīm wa Tur'at al-Maḥmūdiyya* (Alexandria: Maṭba'at al-'Adl, 1942); Isabelle Hairy and Oueded Sennoune, "Géographie historique du canal d'Alexandrie," *Annales Islamologiques* 40 (2006): 247–78.

56. For statements to this effect, see BOA, HAT, 130/5404 (29 Z 1232/9 Nov. 1817); BOA, HAT, 795/36893 (29 Z 1235/7 Oct. 1820).

57. On the official naming of the canal in July 1820, see Ṭūsūn, *Tārīkh Khalīj al-Iskandariyya*, 127; al-Jabartī, *'Ajā'ib al-Āthār* (1994), 4:438.

58. Rivlin, *Agricultural Policy*, 219–20, 353n15. The estimate of 360,000 comes from M.A. Linant de Bellefonds, *Mémoires sur les principaux travaux d'utilité publiqué éxécutés en Egypte depuis la plus haute antiquité jusqu'à nos jours: Accompagné d'un atlas renfermant neuf planches grand in-folio imprimées en couleur* (Paris: Arthus Bertrand, 1872–73), 351. Without citation, Marsot writes that 250,000 peasants were brought to work on the canal. Marsot, *Egypt in the Reign of Muhammad Ali*, 151. In the summer of 1817, for example, Mehmet 'Ali ordered that for every ten people in a village, one adult male had to be sent for work on the canal. Al-Jabartī, *'Ajā'ib al-Āthār* (1994), 4:389.

59. On the population of Cairo, see Daniel Panzac, "Alexandrie: Évolution d'une ville cosmopolite au XIXe siècle," in *Population et santé dans l'Empire ottoman (XVIIIe–XXe siècles)* (Istanbul: Isis, 1996), 147. On the population

of Egypt as a whole, see Panzac, *La peste*, 271; André Raymond, "La population du Caire et de l'Égypte à l'époque ottomane et sous Muḥammad ʿAlî," in *Mémorial Ömer Lûtfi Barkan*, 169–78 (Paris: Librairie d'Amérique et d'Orient Adrien Maisonneuve, 1980). Justin A. McCarthy offers slightly lower population figures for both Cairo and the whole of Egypt in Justin A. McCarthy, "Nineteenth-Century Egyptian Population," *Middle Eastern Studies* 12 (1976): 1–39.

60. For further discussion of this number of dead, see Mikhail, *Nature and Empire*, 281–82, 289–90.

61. Al-Jabartī, *ʿAjāʾib al-Āthār* (1994), 4:427. For an earlier example of water pouring over laborers during work on a building project, see ibid., 3:349.

62. It was also one that spread beyond just the state's infrastructural projects. The following is a description of how an elite urbanite took to rebuilding properties he seized during one of several land grabs in early nineteenth-century Cairo: "By dint of harshness and sternness toward workers and suppliers, he completed his constructions in the shortest possible time. He permitted his workers no rest but kept them locked up until early morning and woke them up at the end of the night with the whip. They worked from the time of the Shāfiʿī prayer until shortly before sunset, even during the severe heat of Ramaḍān!" Ibid., 4:444.

63. On the construction of these auxiliary irrigation features, see BOA, HAT, 656/32064 (27 Z 1232/7 Nov. 1817); BOA, HAT, 130/5404 (29 Z 1232/9 Nov. 1817); BOA, HAT, 131/5411 (29 Z 1232/9 Nov. 1817); BOA, HAT, 342/19546 (17 C 1233/23 Apr. 1818); al-Jabartī, *ʿAjāʾib al-Āthār* (1994), 4:362, 4:390, 4:423–24; Mikhail, *Nature and Empire*, 249–83.

64. On the founding and function of the school, see al-Jabartī, *ʿAjāʾib al-Āthār* (1994), 4:359; Aḥmad ʿIzzat ʿAbd al-Karīm, *Tārīkh al-Taʿlīm fī ʿAṣr Muḥammad ʿAlī* (Cairo: Maktabat al-Nahḍa al-Miṣriyya, 1938), 359–75.

65. For earlier examples of Mehmet ʿAli's government's monopolization of certain kinds of infrastructural labor, see al-Jabartī, *ʿAjāʾib al-Āthār* (1994), 4:220. For examples of the importation of construction labor, see ibid., 4:396. On the monopolization of building materials, see ibid., 4:356, 4:441.

66. On the canal's costs, see Mikhail, *Nature and Empire*, 271–72, 280.

67. Al-Jabartī, *ʿAjāʾib al-Āthār* (1994), 4:438.

68. For examples of the problems associated with wind on the canal, see BOA, HAT, 593/29055 (29 Z 1235/7 Oct. 1820); DWQ, Maḥkamat al-Baḥayra, 38, p. 336, case 789 (8 M 1236/16 Oct. 1820); DWQ, Maḥkamat al-Baḥayra, 38, p. 336, case 791 (28 M 1236/5 Nov. 1820); DWQ, Maḥkamat al-Baḥayra, 38, p. 329, case 772 (8 Za 1237/27 July 1822).

69. DWQ, Maḥkamat al-Baḥayra, 38, p. 336, case 789 (8 M 1236/16 Oct. 1820); DWQ, Maḥkamat al-Baḥayra, 38, p. 336, case 790 (n.d.); DWQ, Maḥkamat al-Baḥayra, 38, p. 336, case 791 (28 M 1236/5 Nov. 1820); DWQ, Maḥkamat al-Baḥayra, 38, p. 336, case 792 (22 S 1236/28 Nov. 1820); DWQ, Maḥkamat al-Baḥayra, 38, p. 329, case 772 (8 Za 1237/27 July 1822); DWQ, Maḥkamat al-Baḥayra, 38, p. 335, case 788 (16 Za 1237/4 Aug. 1822).

70. BOA, HAT, 131/5411 (29 Z 1232/9 Nov. 1817).

71. BOA, HAT, 130/5404 (29 Z 1232/9 Nov. 1817).

72. BOA, HAT, 131/5411 (29 Z 1232/9 Nov. 1817).

73. Marlowe, *World Ditch*; Zachary Karabell, *Parting the Desert: The Creation of the Suez Canal* (New York: Vintage, 2003); Farnie, *East and West*; Darcy Grimaldo Grigsby, *Colossal: Engineering the Suez Canal, Statue of Liberty, Eiffel Tower, and Panama Canal: Transcontinental Ambition in France and the United States during the Long Nineteenth Century* (Pittsburgh: Periscope, 2012).

74. Mitchell, "Can the Mosquito Speak?"; Jennifer Leslee Derr, "Cultivating the State: Cash Crop Agriculture, Irrigation, and the Geography of Authority in Colonial Southern Egypt, 1868–1931" (PhD diss., Stanford University, 2009), 118–72.

75. Shibl, *Aswan High Dam*; Tom Little, *High Dam at Aswan: The Subjugation of the Nile* (New York: Methuen, 1965); Fahim, *Dams, People and Development*; Nancy Y. Reynolds, "Building the Past: Rockscapes and the Aswan High Dam in Egypt," in *Water on Sand: Environmental Histories of the Middle East and North Africa*, ed. Alan Mikhail, 181–205 (New York: Oxford University Press, 2013); Gilbert F. White, "The Environmental Effects of the High Dam at Aswan," *Environment* 30 (1988): 5–11, 34–40; Elizabeth Bishop, "Talking Shop: Egyptian Engineers and Soviet Specialists at the Aswan High Dam" (PhD diss., University of Chicago, 1997); Ahmad Shokr, "Watering a Revolution: The Aswan High Dam and the Politics of Expertise in Mid-century Egypt" (MA thesis, New York University, May 2008).

76. Jeannie Sowers, "Remapping the Nation, Critiquing the State: Environmental Narratives and Desert Land Reclamation in Egypt," in *Environmental Imaginaries of the Middle East and North Africa*, ed. Diana K. Davis and Edmund Burke III, 158–91 (Athens: Ohio University Press, 2011); Timothy Mitchell, "Dreamland," in *Rule of Experts: Egypt, Techno-Politics, Modernity*, 272–303 (Berkeley: University of California Press, 2002).

77. Farnie, *East and West*; Marlowe, *World Ditch*; Karabell, *Parting the Desert*. The British formally colonized Egypt in 1882, beginning a period in which numerous British engineers, hydrologists, geologists, and others with interests in the Nile came to Egypt to study various aspects of the river. Exploring its sources, irrigating more land in Egypt and the Sudan, building various dams and other irrigation works, and regulating water distribution were all a part of British efforts to manage the entire Nile system—an integral aspect of imperial attempts to control Egypt and protect the Suez Canal and British trade routes to India. For a study of irrigation by a British imperial official in this period, see Willcocks and Craig, *Egyptian Irrigation*. On British imperial designs for the Nile, see Terje Tvedt, *The River Nile in the Age of the British: Political Ecology and the Quest for Economic Power* (London: I.B. Tauris, 2004); Collins, *Nile*, 141–56.

78. J.R. McNeill, *Something New under the Sun*, 166–73; Reynolds, "Building the Past"; Shibl, *Aswan High Dam*, 73–123; Fahim, *Dams, People and Development*; Waterbury, *Hydropolitics*, 154–73, 210–41.

79. Sowers, "Remapping the Nation, Critiquing the State."

80. On the concentration of power and the cronyism that characterized this period, see al-Jabartī, *ʿAjāʾib al-Āthār* (1994), 4:441.

81. On the nineteenth- and early twentieth-century aspects of this history, see Abbas and El-Dessouky, *Large Landowning Class and the Peasantry in Egypt.*

82. For an illuminating discussion of the last two decades of environmental politics in Egypt, see Jeannie L. Sowers, *Environmental Politics in Egypt: Activists, Experts, and the State* (London: Routledge, 2013).

CHAPTER 5

1. For examples of this communication, see Mikhail, *Nature and Empire*, 38–66.

2. DWQ, Maḥkamat al-Manṣūra, 4, p. 108, case 281 (Evail M 1075/25 July–3 Aug. 1664). On the village of Shārimsāḥ, see Ramzī, *al-Qāmūs al-Jughrāfī*, pt. 2, 1:243.

3. These downstream villages included Bisāṭ, Kafr Tiqay, and al-Zaʿātra. See the text of this case for a complete list. On these three villages see, respectively, Ramzī, *al-Qāmūs al-Jughrāfī*, pt. 2, 1:242, 1:245, 1:246.

4. DWQ, Maḥkamat al-Manṣūra, 1, p. 84, case 197 (20 Z 1055/6 Feb. 1646).

5. DWQ, Maḥkamat al-Manṣūra, 7, p. 134, case 340 (7 Za 1091/29 Nov. 1680). On Ṭunāmil, see Ramzī, *al-Qāmūs al-Jughrāfī*, pt. 2, 1:174, 1:179.

6. The ten other villages were Dammās, Kafr al-Rūla, Minyyat Gharb, Durra, Tanbūl, Tūḥ, Nūr Ṭīq, al-Sandalāwī, Barhamnus, and Shubrahūr.

7. Again, one engineering *zirāʿ* (*zirāʿ al-handasa*) equals 0.656 meters. Hinz, *Islamische Masse und Gewichte*, 58.

8. DWQ, Maḥkamat al-Manṣūra, 16, p. 257, case 527 (18 Z 1116/13 Apr. 1705).

9. The use of state funds for this repair is specifically mentioned in the following case: DWQ, Maḥkamat al-Manṣūra, 16, p. 289 or 290, case 599 (17 S 1117/10 June 1705).

10. DWQ, Maḥkamat al-Manṣūra, 16, p. 257, case 527 (18 Z 1116/13 Apr. 1705).

11. Generally on Ottoman guilds, see Eunjeong Yi, *Guild Dynamics in Seventeenth-Century Istanbul: Fluidity and Leverage* (Leiden, Netherlands: Brill, 2004); Ammon Cohen, *The Guilds of Ottoman Jerusalem* (Leiden, Netherlands: Brill, 2001); Suraiya Faroqhi, *Artisans of Empire: Crafts and Craftspeople under the Ottomans* (London: I.B. Tauris, 2009); Suraiya Faroqhi and Randi Deguilhem, eds., *Crafts and Craftsmen of the Middle East: Fashioning the Individual in the Muslim Mediterranean* (London: I.B. Tauris, 2005).

12. DWQ, Maḥkamat al-Manṣūra, 16, p. 289 or 290, case 599 (17 S 1117/10 June 1705).

13. For further discussion of this point, see Mikhail, *Nature and Empire*, 124–25.

14. Ibid., 82–169.

15. All of these measurements and cost estimates come from the following: DWQ, Maḥkamat al-Manṣūra, 16, p. 257, case 527 (18 Z 1116/13 Apr. 1705).

16. DWQ, Maḥkamat al-Manṣūra, 16, p. 289 or 290, case 599 (17 S 1117/10 June 1705).

17. On the distinction between these two types of canal, see Borsch, "Environment and Population"; Tsugitaka, *State and Rural Society*, 225–27; Mikhail, *Nature and Empire*, 40–46.

18. BOA, İbnülemin Umur-i Nafia, 94 (Evasıt Ra 1121/21–30 May 1709).

19. BOA, MM, 1:167 (Evasıt S 1123/31 Mar.–9 Apr. 1711).

20. BOA, Cevdet Nafia, 120 (Evasıt Ca 1125/5–14 June 1713).

21. For more on these local experts, see Mikhail, *Nature and Empire*, 66, 176–78.

22. For more on Rūḥ al-Dīn Efendi's work in Egypt, see ibid., 260–63, 284. On his career in the imperial translation office in Istanbul, see Christine M. Philliou, *Biography of an Empire: Governing Ottomans in an Age of Revolution* (Berkeley: University of California Press, 2011), 91–93.

23. Al-Jabartī, *ʿAjāʾib al-Āthār* (1994), 4:359.

24. Related to surveying, another motivation behind the founding of the school was the desire to create a class of technocrats who could advise Mehmet ʿAli on his massive irrigation schemes and other infrastructural manipulation projects. Mikhail, *Nature and Empire*, 260–61.

25. Al-Jabartī, *ʿAjāʾib al-Āthār* (1994), 4:448.

26. For more on this dynamic, see Mikhail, *Nature and Empire*, 279, 288.

27. William M. Denevan, "The Pristine Myth: The Landscape of the Americas in 1492," *Annals of the Association of American Geographers* 82 (1992): 369–85; William Cronon, "The Trouble with Wilderness; or, Getting Back to the Wrong Nature," *Environmental History* 1 (1996): 7–28. For further discussion, see also the several essays on this topic and William Cronon's response in the same issue of *Environmental History*.

28. For an analysis of some of the political and ecological uses of a declensionist environmental narrative in colonial North Africa, see D. Davis, *Resurrecting the Granary of Rome*, 131–76; D. Davis, "Potential Forests."

29. Cronon, *Changes in the Land*, 13.

CHAPTER 6

1. DWQ, Maḥkamat al-Baḥayra, 21, p. 177, case 343 (24 N 1206/16 May 1792). For a description of the village of Surunbāy, see Ramzī, *al-Qāmūs al-Jughrāfī*, pt. 2, 2:270.

2. Another possible explanation for al-Ḥājj Muṣṭafā's seemingly futile action in bringing this case to court is that it was some sort of midfeud snapshot of a tumultuous relationship that had been going on for some time. In other words, perhaps al-Ḥājj Muṣṭafā brought this case to court in an attempt to get revenge on al-Shābb ʿAlī; or perhaps al-Ḥājj Muṣṭafā sought to raise in the minds of others in their community the possibility that al-Shābb ʿAlī was a trickster who acted dishonestly; perhaps al-Ḥājj Muṣṭafā even meant to suggest that al-Shābb ʿAlī came in the middle of the night and killed the ox himself. There are several plausible scenarios. The important point to take away is that the decision to bring a case to court had many possible motivations with various consequences. It is not outside the realm of possibility—indeed, it

is quite likely—that plaintiffs brought cases to court knowing full well they would lose them. The goal of bringing a case to court, in other words, was not always only to win it. On this and other strategic uses of Ottoman courts, see Boğaç A. Ergene, *Local Court, Provincial Society and Justice in the Ottoman Empire: Legal Practice and Dispute Resolution in Çankırı and Kastamonu (1652–1744)* (Leiden, Netherlands: Brill, 2003); Peirce, *Morality Tales.*

3. On the paper trade, see, for example, Nelly Hanna, *In Praise of Books: A Cultural History of Cairo's Middle Class, Sixteenth to the Eighteenth Century* (Syracuse, NY: Syracuse University Press, 2003). On coffee, see, for example, André Raymond, "A Divided Sea: The Cairo Coffee Trade in the Red Sea Area during the Seventeenth and Eighteenth Centuries," in *Modernity and Culture: From the Mediterranean to the Indian Ocean*, ed. Leila Tarazi Fawaz and C.A. Bayly, 46–57 (New York: Columbia University Press, 2002); André Raymond, "Une famille de grands négociants en café au Caire dans la première moitié du XVIIIe siècle: Les Sharāybī," in *Le commerce du café avant l'ère des plantations coloniales: Espaces, réseaux, sociétés (XVe–XIXe siècle)*, ed. Michel Tuchscherer, 111–24 (Cairo: Institut français d'archéologie orientale, 2001). On the Ottoman silk trade, see Halil İnalcık, "Bursa and the Silk Trade," in *An Economic and Social History of the Ottoman Empire*, 2 vols., ed. Halil İnalcık with Donald Quataert, 1:218–55 (Cambridge: Cambridge University Press, 1994). Generally on the wealth accumulation of merchants and traders in Ottoman Egypt, see Raymond, *Artisans et commerçants*; Hanna, *Making Big Money in 1600.*

4. Most studies of agricultural labor in the Ottoman Empire have—perhaps not surprisingly—focused on the role of peasants in this rural economy. For general historiographical discussions of peasants in Ottoman studies, see Suraiya Faroqhi, "Agriculture and Rural Life in the Ottoman Empire (ca 1500–1878) (A Report on Scholarly Literature Published between 1970 and 1985)," *New Perspectives on Turkey* 1 (1987): 3–34; Suraiya Faroqhi, "Ottoman Peasants and Rural Life: The Historiography of the Twentieth Century," *Archivum Ottomanicum* 18 (2000): 153–82. For further studies of Ottoman peasants and labor during the early modern period, see Suraiya Faroqhi, "The Peasants of Saideli in the Late Sixteenth Century," *Archivum Ottomanicum* 8 (1983): 215–50; Suraiya Faroqhi, "Rural Society in Anatolia and the Balkans during the Sixteenth Century, I," *Turcica* 9 (1977): 161–95; Suraiya Faroqhi, "Rural Society in Anatolia and the Balkans during the Sixteenth Century, II," *Turcica* 11 (1979): 103–53; İslamoğlu-İnan, *State and Peasant in the Ottoman Empire.* For the specific case of Egypt, see Cuno, *Pasha's Peasants*; ʿAbd al-Raḥīm, *al-Rīf al-Miṣrī*; Maḥārīq, *al-Minūfiyya.*

5. Although it perhaps goes without saying, this chapter and the next accept the premise that animals, humans, and others share a common history that deserves the attention of historians of all times and places. Since the writings of Charles Darwin in the second half of the nineteenth century, historians, naturalists, philosophers, biologists, ecologists, anthropologists, and others have fleshed out the outlines, potentials, and difficulties involved in thinking about the shared histories of animals and humans.

For example, cultural historians interested in subjects like the history of pet-keeping have made important contributions to the historiography of human affect and the study of class and domesticity. Kathleen Kete, *The Beast in the Boudoir: Petkeeping in Nineteenth-Century Paris* (Berkeley: University of California Press, 1994); Erica Fudge, *Pets* (Stocksfield, UK: Acumen, 2008); Katherine C. Grier, *Pets in America: A History* (Chapel Hill: University of North Carolina Press, 2006); Yi-Fu Tuan, *Dominance and Affection: The Making of Pets* (New Haven, CT: Yale University Press, 1984); Donna J. Haraway, *When Species Meet* (Minneapolis: University of Minnesota Press, 2008).

For their part, political, intellectual, and environmental historians have considered the history of animals in the service of fleshing out genealogies of concepts like brutality, animality, wildness, the exotic, tameness, humanity, morality, and human and animal rights. Harriet Ritvo, *The Animal Estate: The English and Other Creatures in the Victorian Age* (Cambridge, MA: Harvard University Press, 1987); Harriet Ritvo, *The Platypus and the Mermaid and Other Figments of the Classifying Imagination* (Cambridge, MA: Harvard University Press, 1997); Dominick LaCapra, *History and Its Limits: Human, Animal, Violence* (Ithaca, NY: Cornell University Press, 2009); Erica Fudge, *Brutal Reasoning: Animals, Rationality, and Humanity in Early Modern England* (Ithaca, NY: Cornell University Press, 2006); Erica Fudge, *Perceiving Animals: Humans and Beasts in Early Modern English Culture* (New York: St. Martin's Press, 2000); Nancy J. Jacobs, "The Great Bophuthatswana Donkey Massacre: Discourse on the Ass and the Politics of Class and Grass," *American Historical Review* 106 (2001): 485–507; Londa Schiebinger, "Why Mammals Are Called Mammals: Gender Politics in Eighteenth-Century Natural History," *American Historical Review* 98 (1993): 382–411; Keith Tester, *Animals and Society: The Humanity of Animal Rights* (London: Routledge, 1991); H. Peter Steeves, ed., *Animal Others: On Ethics, Ontology, and Animal Life* (Albany: State University of New York Press, 1999); Cary Wolfe, *What Is Posthumanism?* (Minneapolis: University of Minnesota Press, 2010); Joyce E. Salisbury, *The Beast Within: Animals in the Middle Ages* (New York: Routledge, 1994); Richard W. Bulliet, *Hunters, Herders, and Hamburgers: The Past and Future of Human-Animal Relationships* (New York: Columbia University Press, 2005); Anita Guerrini, *Experimenting with Humans and Animals: From Galen to Animal Rights* (Baltimore: Johns Hopkins University Press, 2003); Louise E. Robbins, *Elephant Slaves and Pampered Parrots: Exotic Animals in Eighteenth-Century Paris* (Baltimore: Johns Hopkins University Press, 2002); Roel Sterchx, *The Animal and the Daemon in Early China* (Albany: State University of New York Press, 2002). See also the relevant chapters of the following important volumes: Gregory M. Pflugfelder and Brett L. Walker, eds., *JAPANimals: History and Culture in Japan's Animal Life* (Ann Arbor: Center for Japanese Studies at the University of Michigan, 2005); Nigel Rothfels, ed., *Representing Animals* (Bloomington: Indiana University Press, 2002); Erica Fudge, ed., *Renaissance Beasts: Of Animals, Humans, and Other Wonderful Creatures* (Urbana: University of Illinois Press, 2004); Angela N.H. Creager and William Chester Jordan, eds., *The Animal/Human Boundary: Historical Perspectives* (Rochester, NY: University of Rochester Press, 2002); Mary J.

Henninger-Voss, ed., *Animals in Human Histories: The Mirror of Nature and Culture* (Rochester, NY: University of Rochester Press, 2002).

Some environmental and agricultural historians have focused on the implications of animal husbandry, migration, hunting, and species depletion for landscapes and human communities alike. Brett L. Walker, *The Lost Wolves of Japan* (Seattle: University of Washington Press, 2005); Elinor G.K. Melville, *A Plague of Sheep: Environmental Consequences of the Conquest of Mexico* (Cambridge: Cambridge University Press, 1997); Alfred W. Crosby, *Ecological Imperialism: The Biological Expansion of Europe, 900–1900* (Cambridge: Cambridge University Press, 2004), 171–94; Joseph E. Taylor III, *Making Salmon: An Environmental History of the Northwest Fisheries Crisis* (Seattle: University of Washington Press, 1999); Virginia DeJohn Anderson, *Creatures of Empire: How Domestic Animals Transformed Early America* (New York: Oxford University Press, 2004); Andrew C. Isenberg, *The Destruction of the Bison: An Environmental History, 1750–1920* (Cambridge: Cambridge University Press, 2000).

And institutions like the zoo, its antecedent the zoological park, and schools of veterinary medicine have all also received some attention from historians. On zoos and zoological parks, see Nigel Rothfels, *Savages and Beasts: The Birth of the Modern Zoo* (Baltimore: Johns Hopkins University Press, 2002); R.J. Hoage and William A. Deiss, eds., *New Worlds, New Animals: From Menagerie to Zoological Park in the Nineteenth Century* (Baltimore: Johns Hopkins University Press, 1996); Donna Haraway, "Teddy Bear Patriarchy: Taxidermy in the Garden of Eden, New York City, 1908–1936," *Social Text* 11 (1984): 20–64; Ian Jared Miller, *The Nature of the Beasts: Empire and Exhibition at the Tokyo Imperial Zoo* (Berkeley: University of California Press, 2013); Randy Malamud, *Reading Zoos: Representations of Animals and Captivity* (New York: New York University Press, 1998).

On veterinary medicine, see Joanna Swabe, *Animals, Disease, and Human Society: Human-Animal Relations and the Rise of Veterinary Medicine* (London: Routledge, 1999); Karen Brown and Daniel Gilfoyle, eds., *Healing the Herds: Disease, Livestock Economies, and the Globalization of Veterinary Medicine* (Athens: Ohio University Press, 2010); Susan D. Jones, *Valuing Animals: Veterinarians and Their Patients in Modern America* (Baltimore: Johns Hopkins University Press, 2002); Housni Alkhateeb Shehada, *Mamluks and Animals: Veterinary Medicine in Medieval Islam* (Leiden, Netherlands: Brill, 2013).

For works on the history of animals in various Islamic traditions of the Middle East, see Schimmel, *Islam and the Wonders of Creation*; Basheer Ahmad Masri, *Animals in Islam* (Petersfield, UK: Athene Trust, 1989); Masri, *Animal Welfare in Islam*; Foltz, *Animals in Islamic Tradition.*

6. On animal wealth in Ottoman Egypt in general, see ʿAbd al-Raḥīm, *al-Rīf al-Miṣrī*, 208–10. For a comparative example of the economics of animal husbandry in the early modern Netherlands, see Jan de Vries, *The Dutch Rural Economy in the Golden Age, 1500–1700* (New Haven, CT: Yale University Press, 1974), 137–44.

7. DWQ, Maḥkamat Manfalūṭ, 1, p. 31, case 59 (13 R 1212/4 Oct. 1797).

In the cases that follow, I assume the animals in question to be healthy and still of an age fit for work. No statement in any of these cases suggests otherwise. Furthermore, although the archival record is replete with references to animals as property in the countryside, it is difficult to determine precise prices for these animals and to trace how these prices fluctuated over time and from place to place. As such, I focus in this section primarily on the *relative* prices of different species.

8. See, for example, the following case that documents the relative numbers of animals in the possession of one Sulaymān Salām Dūjal before his death: DWQ, Maḥkamat Isnā, 3, p. 23, case 41 (8 L 1171/14 June 1758). The *jāmūsa* was clearly his most valuable possession, as he owned six of them, in comparison to four cows, a she-camel, and some sheep. For other cases that highlight the high value of the *jāmūsa*, see DWQ, Maḥkamat Manfalūṭ, 3, p. 58, case 108 (11 M 1266/26 Nov. 1849); DWQ, Maḥkamat Manfalūṭ, 3, p. 74, case 135 (n.d.).

9. DWQ, Maḥkamat Manfalūṭ, 3, pp. 5–7, case 5 (4 Ca 1228/5 May 1813).

10. DWQ, Maḥkamat al-Baḥayra, 38, p. 49, case 101 (11 Za 1233/12 Sept. 1818).

11. DWQ, Maḥkamat al-Baḥayra, 24, pp. 291–92, case 526 (20 B 1209/10 Feb. 1795). More than most animals, camels have received a good deal of attention from historians of the Middle East. See, for example, Bulliet, *Camel and the Wheel*; Bulliet, *Cotton, Climate, and Camels*; Halil İnalcık, "'Arab' Camel Drivers in Western Anatolia in the Fifteenth Century," *Revue d'Histoire Maghrebine* 10 (1983): 256–70; İnalcık, "Ottoman State," 1:38–39, 1:62–63; Faroqhi, "Camels, Wagons, and the Ottoman State"; Roger S. Bagnall, "The Camel, the Wagon, and the Donkey in Later Roman Egypt," *Bulletin of the American Society of Papyrologists* 22 (1985): 1–6.

12. DWQ, Maḥkamat al-Baḥayra, 11, pp. 45–56, case 97 (1 Z 1196/7 Nov. 1782).

13. DWQ, Maḥkamat al-Baḥayra, 23, p. 198, case 357 (Evahir L 1208/20–30 May 1794). On Bīwīṭ, see Ramzī, *al-Qāmūs al-Jughrāfī*, pt. 2, 2:268–69.

14. While the assets of this estate did indeed total 298 *niṣf faḍḍa*, 134.61 *niṣf faḍḍa* of this amount were taken from the estate to repay debts and cover other expenses. Thus, the total amount left for the inheritors of the estate was 163.39 *niṣf faḍḍa*.

15. As this case shows, the *jāmūsa* was not always the most expensive animal in an estate's inventory. Nevertheless, the general point remains that when it did appear in estates, it was the single most expensive domesticated animal *more often* than any other animal.

16. See, for example, DWQ, Maḥkamat Rashīd, 132, p. 278, case 419 (10 Ra 1138/15 Nov. 1725). The *jāmūsa* in this case was valued at 500 *niṣf faḍḍa* and the horse at 720 *niṣf faḍḍa*.

17. See, for instance, DWQ, Maḥkamat Rashīd, 139, pp. 107–8, case 172 (30 M 1146/12 July 1733). In this case, a *jāmūsa* fetched 770 *niṣf faḍḍa* while a horse only cost 210 *niṣf faḍḍa*.

18. For a case involving the transport of horses to stables in Manfalūṭ, see

DWQ, Maḥkamat Manfalūṭ, 1, pp. 207–8, case 513 (14 B 1098/26 May 1687). On the social status attached to horses in medieval Europe, see Salisbury, *Beast Within*, 28–31, 35.

19. Al-Damurdāshī Katkhudā ʿAzabān, *Kitāb al-Durra al-Muṣāna*, 16. For a comparative example of the gifting of birds in early modern Japan, see Martha Chaiklin, "Exotic-Bird Collecting in Early-Modern Japan," in *JAPANimals: History and Culture in Japan's Animal Life*, ed. Gregory M. Pflugfelder and Brett L. Walker (Ann Arbor: Center for Japanese Studies, University of Michigan, 2005), 132–39.

20. Stanford J. Shaw, ed. and trans., *Ottoman Egypt in the Eighteenth Century: The Niẓâmnâme-i Mıṣır of Cezzâr Aḥmed Pasha* (Cambridge, MA: Center for Middle Eastern Studies of Harvard University, 1964), 14–15.

21. On the gifting of animals other than horses, see al-Jabartī, *ʿAjāʾib al-Āthār* (1998), 1:407.

22. Shaw, *Niẓâmnâme-i Mıṣır*, 41. On Shaykh al-Humām more generally, see ibid., 44. On the village of Akhmīm, see Ramzī, *al-Qāmūs al-Jughrāfī*, pt. 2, 4:89–90.

23. DWQ, Maḥkamat Rashīd, 144, pp. 493–95, case 525 (25 Ra 1152/2 July 1739).

24. DWQ, Maḥkamat Manfalūṭ, 3, p. 111, case 216 (n.d.). For other examples of cases involving the sharing of a *jāmūsa*, see DWQ, Maḥkamat Manfalūṭ, 3, p. 130 or 156, case 318 (25 Ra 1264/1 Mar. 1848); DWQ, Maḥkamat Manfalūṭ, 3, p. 136 or 162, case 332 (4 Ra 1264/9 Feb. 1848); DWQ, Maḥkamat Manfalūṭ, 3, p. 166, case 340 (11 Ra 1264/16 Feb. 1848); DWQ, Maḥkamat Manfalūṭ, 3, p. 166, case 342 (13 Ra 1264/18 Feb. 1848). The adjective *red* here most likely referred to the reddish-brown hue of the animal's coat.

25. DWQ, Maḥkamat Manfalūṭ, 1, p. 223, case 549 (21 B 1098/2 June 1687).

26. For cases involving the sharing of a she-camel, see DWQ, Maḥkamat Manfalūṭ, 3, p. 46, case 93 (n.d.).

27. One could also similarly compile a hierarchy of the values of different animals in the Egyptian countryside based on the price of their skins. Indeed, a quite robust trade in different kinds of leather from Egypt developed during the Ottoman period. *Jāmūsa* leather was the most expensive, followed by ox and cow leather, with the skins of camels being the cheapest. For more on the trade in animal skins in Ottoman Egypt, see DWQ, Maḥkamat al-Baḥayra, 37, p. 203, case 423 (29 B 1232/15 June 1817); DWQ, Maḥkamat al-Baḥayra, 37, p. 204, case 426 (17 B 1232/3 June 1817).

28. According to Sufi tradition, it was wrong to use cows as beasts of burden. When riding a cow one day, a rider purportedly heard the animal exclaim, "I was not created for this!" Schimmel, *Islam and the Wonders of Creation*, 48. Schimmel does not provide a citation for this story or quotation.

29. For evidence of the presence of a donkey market in Cairo in the latter half of the seventeenth century, see Ibrāhīm ibn Abī Bakr al-Ṣawāliḥī al-ʿAwfī al-Ḥanbalī, *Tarājim al-Ṣawāʿiq fī Wāqiʿat al-Ṣanājiq*, ed. ʿAbd al-Raḥīm ʿAbd

al-Raḥman ʿAbd al-Raḥīm (Cairo: Institut français d'archéologie orientale, 1986), 54.

30. DWQ, Maḥkamat Manfalūṭ, 1, p. 69, case 144 (n.d.).

31. These tables were compiled on the basis of DWQ, Maḥkamat Isnā, 6, pp. 73–79, case 131 (12 L 1172/8 June 1759).

32. For additional probate cases involving shares in cows, other animals, and their relative costs, see DWQ, Maḥkamat Manfalūṭ, 1, pp. 72–73, case 152 (8 Ş 1212/25 Jan. 1798); DWQ, Maḥkamat Manfalūṭ, 3, pp. 92–93, case 174 (n.d.); DWQ, Maḥkamat Isnā, 6, p. 67, case 113 (11 L 1172/7 June 1759); DWQ, Maḥkamat Isnā, 8, pp. 111–12, case 177 (23 Ra 1173/13 Nov. 1759). For a case involving the sale of a camel, see DWQ, Maḥkamat Manfalūṭ, 3, p. 33, case 57 (n.d.).

33. DWQ, Maḥkamat Manfalūṭ, 3, pp. 4–5, case 3 (4 R 1228/6 Apr. 1813).

34. DWQ, Maḥkamat Rashīd, 154, p. 179, case 197 (15 Za 1162/27 Oct. 1749).

35. DWQ, Maḥkamat Rashīd, 148, pp. 67–68, case 86 (1 B 1154/11 Sept. 1741).

36. For another financial transaction involving slaves and animals and their relative prices, see al-Damurdāshī Katkhudā ʿAzabān, *Kitāb al-Durra al-Muṣāna*, 44.

37. There is no equivalent of *wergeld* in Islamic law. Historians of medieval Europe have usefully employed the *wergeld* values of humans and animals to assess the relative valuations of humans to animals. Thus, for example, the Visigoths considered a nobleman to be five hundred times the value of the most prized animal, and the Franks considered a stallion to be only half the value of a man's nose or ear. For the Burgundians, the tooth of an individual from the lowest classes was equal to the value of a noble hunting dog. Salisbury, *Beast Within*, 36–37.

38. On the various dairy products made from the milk of different farm animals and on their relative values, see al-Jabartī, *ʿAjāʾib al-Āthār* (1998), 1:184, 1:338–39, 2:205, 3:332, 3:469.

39. This was also the case in medieval Europe, where food animals were considered to be of less value than laboring animals. Salisbury, *Beast Within*, 34.

40. For a case illustrating the high cost of she-camels, see DWQ, Maḥkamat Rashīd, 130, pp. 172–76, case 245 (20 C 1136/16 Mar. 1724).

41. DWQ, Maḥkamat Isnā, 6, pp. 1–3, case 1 (23 Ca 1172/23 Jan. 1759). In addition to this she-camel, ʿAlī ʿAbd al-Qādir al-Ḥamdanī owned three camels, two cows (one of which is described as being yellow), and a donkey (described as black). These animals were a significant investment in capital and formed the bulk of his inheritable estate.

42. This is despite the fact that the Quran seems to privilege cow milk over camel milk: "In cattle too you have a worthy lesson. We give you to drink of that which is in their bellies, between the bowels and the blood-streams: pure milk, pleasant for those who drink it" (Quran, 16:66). See similarly Quran, 23:21, 36:73. On the differences between cow and camel milk in Islamic tradition, see Schimmel, *Islam and the Wonders of Creation*, 47–48.

43. For more on the importance of camels in Arab and Persian historical writings, see Schimmel, *Islam and the Wonders of Creation*, 46–47.

44. On the kinds and prices of meat products in eighteenth-century Egypt, see al-Jabartī, *'Ajā'ib al-Āthār* (1998), 1:184, 1:338–39, 2:198, 2:205, 2:274–75, 3:332, 3:469, 3:551. For a very useful analysis of the provisioning of meat to Istanbul, see Antony Greenwood, "Istanbul's Meat Provisioning: A Study of the *Celepkeşan* System" (PhD diss., University of Chicago, 1988).

45. Al-Jabartī, *'Ajā'ib al-Āthār* (1998), 1:545, 3:300.

46. These festivities also included fireworks displays, a grand procession to the Citadel, the firing of cannons, and a lavish banquet. Al-Damurdāshī Katkhudā 'Azabān, *Kitāb al-Durra al-Muṣāna*, 6, 9, 28, 40, 57, 69, 72, 75, 79–80, 84, 103, 113, 114, 122–23, 132–33, 142, 159, 162, 179, 199, 209, 212, 224, 237, 241, 248, 251, 252–54, 260.

47. Shaw, *Niẓâmnâme-i Mıṣır*, 47–48. The *okke* was an Ottoman dry measure equivalent to 400 *dirhem*, or 1.2828 kilograms (2.8281 pounds). Hinz, *Islamische Masse und Gewichte*, 24.

48. DWQ, Maḥkamat Isnā, 3, pp. 15–16, case 29 (8 L 1171/14 June 1758). On Aṣfūn, see Ramzī, *al-Qāmūs al-Jughrāfī*, pt. 2, 4:152.

49. For another example of a *waqf* endowment that included animals to be slaughtered for the poor and other guests, see al-Jabartī, *'Ajā'ib al-Āthār* (1998), 1:612.

50. On the use of islands as natural enclosures to confine animals, see Anderson, *Creatures of Empire*, 160.

51. DWQ, Maḥkamat Rashīd, 134, p. 332, case 443 (30 B 1140/11 Mar. 1728).

52. Al-Damurdāshī Katkhudā 'Azabān, *Kitāb al-Durra al-Muṣāna*, 123.

53. Ibid., 214; Shaw, *Financial and Administrative Organization*, 127; Shaw, *Niẓâmnâme-i Mıṣır*, 47. For an example of another tax farm of this sort, see Shaw, *Financial and Administrative Organization*, 135.

54. Shaw, *Financial and Administrative Organization*, 130, 301.

55. Ibid.; Shaw, *Niẓâmnâme-i Mıṣır*, 47n1.

56. Shaw, *Niẓâmnâme-i Mıṣır*, 19.

57. Shaw, *Financial and Administrative Organization*, 139.

58. The tax revenue raised from this corporation between 1742–43 and 1760–61 was 1,620 *para*. By way of comparison, the money raised during the same period from taxes on sellers of sardines in Cairo was 1,800 *para*. On the taxes levied on other corporations in Cairo, see ibid., 119.

59. DWQ, Maḥkamat al-Baḥayra, 8, p. 24, case 37 (19 S 1176/8 Sept. 1762).

60. Al-Jabartī, *'Ajā'ib al-Āthār* (1998), 1:538–39. On the cultivation of sugarcane in Egypt, see Tsugitaka, *State and Rural Society*, 211–20.

61. DWQ, Maḍābiṭ al-Daqahliyya, 20, pp. 110–11, no case no. (7 C 1188/15 Aug. 1774).

62. Al-Damurdāshī Katkhudā 'Azabān, *Kitāb al-Durra al-Muṣāna*, 131.

63. DWQ, Maḥkamat al-Manṣūra, 3, p. 10, case 31 (19 S 1063/19 Jan. 1653). On the village of Sandūb, see Ramzī, *al-Qāmūs al-Jughrāfī*, pt. 2, 1:220.

64. DWQ, Maḥkamat al-Baḥayra, 24, p. 56, case 102 (1 Ra 1209/26 Sept.

1794). One of these waterwheels was known as the waterwheel of Bāb al-Naṣr, and the other was known as al-Kitāniyya.

65. DWQ, al-Jusūr al-Sulṭāniyya, 784, p. 129, no case no. (n.d.). On the village of Shūbar, see Ramzī, *al-Qāmūs al-Jughrāfī*, pt. 2, 2:101.

66. See, for example, DWQ, Maḥkamat al-Baḥayra, 10, p. 101, case 229 (1 Ra 1190/20 Apr. 1776).

67. DWQ, Maḥkamat Rashīd, 132, pp. 200–201, case 311 (3 N 1137/ 16 May 1725).

68. See, for example, DWQ, Maḥkamat al-Baḥayra, 5, p. 314, case 389 (10 Ş 1165/22 June 1752); DWQ, Maḥkamat Rashīd, 145, p. 126, case 101 (30 Z 1151/9 Apr. 1739). In this latter case, several donkeys were hired to remove dirt from the area around a *ḥammām* under repair. These animal laborers cost more to hire for the project than the weigher (*qabbānī*) but less than the total price of the human workers brought to the construction site. Because the relative numbers of animals and humans are not given in this case, it is impossible to determine the cost of a single donkey as compared to that of an individual human.

69. Some of the many examples of such cases include the following: DWQ, Maḥkamat Isnā, 3, p. 22, case 40 (3 L 1171/9 June 1758); DWQ, Maḥkamat Isnā, 6, p. 31, case 42 (5 B 1172/5 Mar. 1759). The former case revolved around a dispute over the inheritance of a she-camel, while the latter was about a *jāmūsa*. In the following case, the inheritors of an estate fought over the ownership of a cow, a donkey, and four ewes: DWQ, Maḥkamat al-Baḥayra, 22, pp. 347–48, case 733 (15 Z 1207/24 July 1793). In a case from the same court of al-Baḥayra a few years later, three brothers fought over the ownership of a quarter of a *jāmūsa* calf that belonged to the deceased wife of one of the brothers. DWQ, Maḥkamat al-Baḥayra, 25, p. 17, case 30 (26 Z 1211/22 June 1797). Another inheritance dispute from the court of al-Baḥayra in the year 1793 between Sitīta bint Sīdī Aḥmad Turkī from the village of Maḥallat Marḥūm and her maternal uncle (*khāluhā*) al-Ḥājj Yūsuf ibn al-Ḥājj Muḥammad al-ʿUrf revolved around the ownership of twelve *jāmūsa*s, eight steers, one sheep, and a camel. DWQ, Maḥkamat al-Baḥayra, 23, p. 45, case 87 (1 Ra 1208/7 Oct. 1793). On the village of Maḥallat Marḥūm, see Ramzī, *al-Qāmūs al-Jughrāfī*, pt. 2, 2:107.

70. DWQ, Maḥkamat Manfalūṭ, 1, p. 296, case 738 (12 Z 1098/ 19 Oct. 1687).

71. DWQ, Maḥkamat al-Baḥayra, 26, p. 13, case 24 (Evahir M 1212/15–25 July 1797). The quarter of the deceased's calf sold for three *riyāl*s, while half of her *jāmūsa* sold for 38 *riyāl*s—further evidence of the relatively high value of the *jāmūsa* in comparison to other animals in rural Ottoman Egypt.

72. For example, in a case from the court of al-Baḥayra from the year 1793, a woman named Sālima bint Ḥasan ʿĀmir brought her former husband Muḥammad Khalīf to court to assert her right to the ownership of a fourth of a *jāmūsa* and an *ardabb* of wheat. In response to Sālima's claims, Muḥammad asserted that when he divorced her, she gave up her right to this property in exchange for his payment of three *riyāl*s to her. Continuing to press her claims, Sālima asked that Muḥammad produce evidence (*thubūt*) of this transaction.

He, in turn, showed the court a *ḥukm* witnessed and signed by one Sulaymān al-Laqqānī and one al-Sayyid Muṣṭafā ʿUlwān and also another unspecified written legal instrument (*wathīqa*). Satisfied with this evidence, the judge in the case ruled in favor of the defendant Muḥammad and ordered Sālima to drop any claims against Muḥammad for the property. DWQ, Maḥkamat al-Baḥayra, 22, p. 343, case 724 (13 Z 1207/22 July 1793).

73. DWQ, Maḥkamat al-Baḥayra, 5, p. 183, case 319 (26 Ş 1165/8 July 1752).

74. Bedouin raiders not only stole animals but also rode them effectively in their attacks. Indeed, this was one of the reasons for their skill and swiftness in battle. They were described as being those who ride horses and use spears (*ʿurbān sevārī ve sahib-i mezrāk olup*). Shaw, *Nizâmnâme-i Mısır*, 26–27.

75. For an example of this kind of case, see DWQ, Maḥkamat al-Manṣūra, 18, p. 267, no case no. (3 M 1120/26 Mar. 1708). This case is an Ottoman *buyuruldu* sent from Istanbul to the court of al-Manṣūra. As evidenced by this case and by numerous other Ottoman firmans and decrees sent to the courts of Egypt, the imperial administration was quite concerned about stopping the Bedouin theft of animals from Egyptian villages. For examples of the Bedouin plunder of animals during battle, see al-Jabartī, *ʿAjāʾib al-Āthār* (1998), 1:190, 1:543, 2:150.

76. For other examples of Bedouin raids involving the theft of animals, see al-Damurdāshī Katkhudā ʿAzabān, *Kitāb al-Durra al-Muṣāna*, 7–8, 41, 43–44, 129.

77. On the ability of medieval European animals to increase their owners' social status, see Salisbury, *Beast Within*, 27–32.

78. DWQ, Maḥkamat al-Baḥayra, 16, p. 232, case 403 (27 R 1201/16 Feb. 1787). On al-Raḥmāniyya, see Ramzī, *al-Qāmūs al-Jughrāfī*, pt. 2, 2:305.

79. Again, the description of the steer as red probably indicates that the animal had a reddish-brown coat.

80. For another example of the social status attached to the possession of animals, see al-Jabartī, *ʿAjāʾib al-Āthār* (1998), 3:272.

81. Many other questions are raised by this case. For instance, why did the case only come to court seventeen days after the wedding in which the defendant was seen with the animals? Was this simply a function of the timing of the general convening of the court to hear cases? Did the plaintiff perhaps wait to form a case? This seems unlikely given the simplicity of the case and the fact that the plaintiff's key piece of evidence, the testimony of his two witnesses, was probably immediately available after the night of the wedding. Perhaps the plaintiff attempted to approach the defendant before coming to court to solve the case between them without the need to make recourse to the legal institution. For a discussion of these and related questions of legal procedure, see Ergene, *Local Court, Provincial Society and Justice*; Peirce, *Morality Tales*.

82. DWQ, Maḥkamat Manfalūṭ, 3, p. 62, case 117 (n.d.). On the village of Jalda, see Ramzī, *al-Qāmūs al-Jughrāfī*, pt. 2, 4:46–47.

83. The description of the donkey as green most likely refers to a greenish-gray or grayish-brown coat color. The term *khaḍrāʾ* was often used to denote browns or grays with a tinge of green. For example, half-tanned leather, certain

varieties of sun-dried mud brick, and camel leather were often described with this same word. El-Said Badawi and Martin Hinds, *A Dictionary of Egyptian Arabic* (Beirut: Librairie du Liban, 1986), s.v. "*khāʾ*," "*ḍād*," "*rāʾ*."

84. The text of this case does not give any details as to how these two witnesses evidenced their knowledge of the hereditary link between these two donkeys. Perhaps they witnessed the young donkey's birth, or perhaps they had known Rushdān's donkey to be pregnant before it wandered away. Jenny pregnancies usually last eleven months—a time period that fits well with the time line in this case.

85. For similar court cases concerning the establishment of the legal ownership of disputed wandering donkeys, see DWQ, Maḥkamat Manfalūṭ, 3, p. 98, case 188 (8 R 1266/21 Feb. 1850); DWQ, Maḥkamat Manfalūṭ, 3, pp. 86–87, case 161 (n.d.).

86. DWQ, Maḥkamat al-Baḥayra, 16, p. 89, case 155 (16 L 1200/ 11 Aug. 1786).

87. DWQ, Maḥkamat al-Baḥayra, 23, p. 129, case 241 (21 C 1208/ 23 Jan. 1794). On Sunṭīs, see Ramzī, *al-Qāmūs al-Jughrāfī*, pt. 2, 2:287.

88. On the village of Zāwiyat Naʿīm, see Ramzī, *al-Qāmūs al-Jughrāfī*, pt. 2, 2:240–41.

89. On Ruzzāfa, see ibid., pt. 2, 2:261.

90. On the prohibitive costs of feeding domestic work animals, see Salisbury, *Beast Within*, 20.

91. For a comparative example of this phenomenon in colonial America, see Anderson, *Creatures of Empire*, 165–66.

92. On this kind of "communal oversight" of livestock, see ibid., 163.

93. In medieval Europe as well, a wandering animal nevertheless remained its owner's legal property. Salisbury, *Beast Within*, 32.

94. On the absence of animals in modern scholarship of the early modern period, see Fudge, *Brutal Reasoning*, 4–5.

95. For a study of estate inventories from Bursa, Istanbul, and Edirne that makes this point as well, see Halil İnalcık, "Capital Formation in the Ottoman Empire," *Journal of Economic History* 29 (1969): 97–140.

CHAPTER 7

1. John F. Richards, "Toward a Global System of Property Rights in Land," in *The Environment and World History*, ed. Edmund Burke III and Kenneth Pomeranz (Berkeley: University of California Press, 2009), 71. On these points, see also J. Richards, *Unending Frontier*, 617–22.

2. The Brenner debate is usefully summarized, with contributions by many of the scholars involved in these discussions, in T.H. Aston and C.H.E. Philpin, eds., *The Brenner Debate: Agrarian Class Structure and Economic Development in Pre-industrial Europe* (Cambridge: Cambridge University Press, 1985). The literature on divergence of course began with Kenneth Pomeranz, *The Great Divergence: China, Europe, and the Making of the Modern World Economy* (Princeton, NJ: Princeton University Press, 2000). For a recent consideration of the subject, see Jean-Laurent Rosenthal and R. Bin Wong, *Before and*

Beyond Divergence: The Politics of Economic Change in China and Europe (Cambridge, MA: Harvard University Press, 2011).

3. Kasaba, *Ottoman Empire and the World Economy*, 23–27.

4. Much of this is recounted in al-Rāfʿī, *ʿAṣr Muḥammad ʿAlī*; Marsot, *Egypt in the Reign of Muhammad Ali*. On the educational reforms of this period, see ʿAbd al-Karīm, *Tārīkh al-Taʿlīm*.

5. Raymond, "La population du Caire"; Panzac, "Alexandrie: Évolution d'une ville cosmopolite."

6. The histories of animals and of human-animal relations in Ottoman Egypt have not received much attention. For some of the current literature, see Michel Tuchscherer, "Some Reflections on the Place of the Camel in the Economy and Society of Ottoman Egypt," trans. Suraiya Faroqhi, in *Animals and People in the Ottoman Empire*, ed. Suraiya Faroqhi, 171–85 (Istanbul: Eren, 2010); Catherine Mayeur-Jaouen, "Badawi and His Camel: An Animal as the Attribute of a Muslim Saint in Mamluk and Ottoman Egypt," trans. Suraiya Faroqhi, in *Animals and People in the Ottoman Empire*, ed. Suraiya Faroqhi, 113–28 (Istanbul: Eren, 2010); Jean-Louis Bacqué-Grammont, Joséphine Lesur-Gebremariam, and Catherine Mayeur-Jaouen, "Quelques aspects de la faune nilotique dans la relation d'Evliyâ Çelebî, voyageur ottoman," *Journal Asiatique* 296 (2008): 331–74; Mikhail, *Animal*.

For works on the general cultural, symbolic, and religious history of animals in various Islamic traditions of the Middle East, see Schimmel, *Islam and the Wonders of Creation*; Benkheira, Mayeur-Jaouen, and Sublet, *L'animal en islam*; Sarra Tlili, *Animals in the Qurʾan* (Cambridge: Cambridge University Press, 2012); Masri, *Animals in Islam*; Masri, *Animal Welfare in Islam*; Foltz, *Animals in Islamic Tradition*; Allsen, *Royal Hunt*.

7. Kasaba, *Ottoman Empire and the World Economy*, 23–27.

8. J.R. McNeill offers the following definition of an "energy regime": "the collection of arrangements whereby energy is harvested from the sun (or uranium atoms), directed, stored, bought, sold, used for work or wasted, and ultimately dissipated." J.R. McNeill, *Something New under the Sun*, 297. More generally on the relationship between energy and economy, see ibid., 296–324.

9. Water power was clearly also important. But because of the vagaries of the Nile and rural cultivators' inability to efficiently, easily, and usefully harness its excessive force, water was never utilized to the same degree as animal power in Egypt. Wood and wind were other potential energy sources, but wood was in too short supply to be burned as fuel, and, except for river transport, wind does not seem to have powered much in the countryside. Thus, it was animals, both human and nonhuman but mostly nonhuman, that were the primary sources of energy in early modern Ottoman Egypt.

10. For studies that make this point in some detail, see Smil, *Energy in World History*; Smil, *Energy in Nature and Society*; Pyne, *World Fire*; Pyne, *Vestal Fire*; Paul Warde, *Ecology, Economy and State Formation in Early Modern Germany* (Cambridge: Cambridge University Press, 2006); Burke, "Big Story"; Mary C. Stiner and Gillian Feeley-Harnik, "Energy and Ecosystems," in *Deep History: The Architecture of Past and Present*, ed. Andrew Shryock

and Daniel Lord Smail, 78–102 (Berkeley: University of California Press, 2011).

11. In this regard, see Clay McShane, *Down the Asphalt Path: The Automobile and the American City* (New York: Columbia University Press, 1994).

12. This point is made convincingly in J. Richards, *Unending Frontier*; Fernand Braudel, *The Wheels of Commerce*, trans. Siân Reynolds, vol. 2 of *Civilization and Capitalism, 15th–18th Century* (London: Collins, 1982); Richard W. Bulliet, "History and Animal Energy in the Arid Zone," in *Water on Sand: Environmental Histories of the Middle East and North Africa*, ed. Alan Mikhail, 51–69 (New York: Oxford University Press, 2013).

13. On 1800 as the turning point between energy regimes, see, for example, Bruce Podobnik, "Toward a Sustainable Energy Regime: A Long-Wave Interpretation of Global Energy Shifts," *Technological Forecasting and Social Change* 62 (1999): 155–72; Burke, "Big Story," 35.

14. I use the terms *caloric motor*, *caloric energy*, and *caloric power* to refer to the work capacities of animals and humans fueled by the expenditure of food calories stored in their bodies.

15. Throughout the eighteenth century, certain areas of Europe also experienced successive waves of epizootics, suggesting that this was likely a Mediterranean-wide phenomenon in the early modern period. For a discussion of some of these European epizootics, see Karl Appuhn, "Ecologies of Beef: Eighteenth-Century Epizootics and the Environmental History of Early Modern Europe," *Environmental History* 15 (2010): 268–87.

16. BOA, HAT, 29/1361 (13 Ş 1198/1 July 1784). I use the term *plague* throughout as the translation for both *taun* (the Ottoman Turkish version of the Arabic *ṭāʿūn*) and *veba* (or *wabāʾ* in Arabic). The original sources use both words to refer to diseases of humans and animals alike. The human diseases termed "plague" no doubt included a wide variety of ailments beyond just *Yersinia pestis*—very likely anthrax, typhus, and any number of other parasitic diseases. As for the epizootics that are also termed *taun* and *veba*, they could have been anthrax, rinderpest, bovine pleuropneumonia, foot-and-mouth disease, or something else entirely. The larger and more important point to make—and probably part of the reason the original sources confusingly use disease terms so interchangeably—is that the same conditions made both humans vulnerable to epidemic diseases and animals vulnerable to epizootics. Following the sources, and with really no other logical choice, I have necessarily replicated this general and wide-ranging, if even nonspecific, use of the word *plague* to refer to human diseases. I also sometimes term epizootics "plague" when the sources direct this usage, but more generally I simply use the catchall descriptor *epizootic*. It would be several decades after this period before veterinary medicine would make possible clearer delineations of specific animal diseases. For a useful treatment of some of the difficulties involved in identifying human diseases in Ottoman Turkish and Arabic source materials, see S. White, "Rethinking Disease in Ottoman History," 555–58. Venetians as well were not always sure how to categorize their animals' diseases. On this point, see Appuhn, "Ecologies of Beef," 270, 279–81. In the case of the great bovine pestilence of the fourteenth century, the scholarly consensus is that

the disease was rinderpest. Timothy P. Newfield, "A Cattle Panzootic in Early Fourteenth-Century Europe," *Agricultural History Review* 57 (2009): 188; Philip Slavin, "The Great Bovine Pestilence and Its Economic and Environmental Consequences in England and Wales, 1318–50," *Economic History Review* 65 (2012): 1240. For a useful analysis of the history of anthrax that touches on both the disease's Ottoman past and its relationship to other animal maladies, see Susan D. Jones, *Death in a Small Package: A Short History of Anthrax* (Baltimore: Johns Hopkins University Press, 2010). There is shockingly little work on the history of veterinary medicine. For some of the current literature, see S. Jones, *Valuing Animals*; K. Brown and Gilfoyle, *Healing the Herds*; Shehada, *Mamluks and Animals*; Swabe, *Animals, Disease, and Human Society*; Louise Hill Curth, *The Care of Brute Beasts: A Social and Cultural Study of Veterinary Medicine in Early Modern England* (Leiden, Netherlands: Brill, 2010); Diana K. Davis, "Brutes, Beasts, and Empire: Veterinary Medicine and Environmental Policy in French North Africa and British India," *Journal of Historical Geography* 34 (2008): 242–67.

17. BOA, HAT, 28/1354 (7 Za 1198/22 Sept. 1784); Ismāʿīl ibn Saʿd al-Khashshāb, *Khulāṣat mā Yurād min Akhbār al-Amīr Murād*, ed. and trans. Hamza ʿAbd al-ʿAzīz Badr and Daniel Crecelius (Cairo: al-ʿArabī lil-Nashr wa al-Tawzīʿ, 1992), 24–25.

18. BOA, HAT, 28/1354 (7 Za 1198/22 Sept. 1784); al-Jabartī, *ʿAjāʾib al-Āthār* (1994), 2:139, 2:156–57. This situation of prolonged food shortages led to riots in Cairo in the spring of 1785. Ibid., 2:210, 3:100.

19. Al-Jabartī, *ʿAjāʾib al-Āthār* (1994), 2:139. For another example of the human consumption of diseased raw donkey and horse meat, see ibid., 2:397. These years' desperate circumstances of drought, famine, and disease resulted in part from the eruption of the Laki fissure in Iceland in 1783 and 1784, a climate event of truly global proportions that will be the subject of chapter 10. It contributed to a decade (1785–95) of cold winters, inadequate floods, depressed cultivation levels, and disease outbreaks throughout Egypt. On the eruption, see Oman, Robock, Stenchikov, and Thordarson, "High-Latitude Eruptions Cast Shadow." For a useful discussion of the connections between volcanic eruptions and climate change in world history, see William S. Atwell, "Volcanism and Short-Term Climatic Change in East Asian and World History, c. 1200–1699," *Journal of World History* 12 (2001): 29–98. The most sustained and rigorous study of the role of climate change in Ottoman history is S. White, *Climate of Rebellion*. For observations that make clear the relationships between weather, plague, and food supplies and their combined effects on populations in Egypt (and elsewhere) in 1791, see TNA, PC, 1/19/24; TNA, FO, 24/1, 191r–196v (12 July 1791), 197r–200v (7 Oct. 1791).

20. Al-Jabartī, *ʿAjāʾib al-Āthār* (1994), 2:228–29. And, al-Jabartī continues, "Other losses were on a similar scale."

21. Ibid., 2:241, 2:229.

22. Ibid., 2:232, 2:241.

23. For a recent treatment of the role of plague in Ottoman history, see S. White, "Rethinking Disease in Ottoman History," 549–67.

24. Al-Jabartī, *'Ajā'ib al-Āthār* (1994), 2:260, 2:263. For references to some of the human victims of this plague, see ibid., 2:275, 2:280, 2:282.

25. BOA, Cevdet Dahiliye, 1722 (Evasıt N 1205/15–24 May 1791); al-Jabartī, *'Ajā'ib al-Āthār* (1994), 2:315; al-Khashshāb, *Akhbār al-Amīr Murād*, 33–34.

26. BOA, HAT, 1399/56283 (29 Z 1205/29 Aug. 1791). There is no internal evidence for the date of this case. The date given is the one assigned by the BOA. Although the high number of deaths reported here is surely exaggerated, it indicates that this plague was considered of unmatched severity. For a discussion of how to interpret reported numbers of human plague casualties, see chapter 9. For a description of the "chaos" of 1791, see BOA, HAT, 1412/57500 (29 Z 1205/29 Aug. 1791). Again, there is no internal evidence for the date of this case. The date given is the one assigned by the BOA.

27. Ismā'īl ibn Sa'd al-Khashshāb, *Akhbār Ahl al-Qarn al-Thānī 'Ashar: Tārīkh al-Mamālīk fī al-Qāhira*, ed. 'Abd al-'Azīz Jamāl al-Dīn and 'Imād Abū Ghāzī (Cairo: al-'Arabī lil-Nashr wa al-Tawzī', 1990), 58. The extreme brutality and lethality of this plague epidemic is made clear by the fact that it is mentioned in almost all of the narrative sources of the period, whereas other instances of disease appear in only one or some of the accounts.

28. A group of thirty-seven letters sent by British consuls in several of these locales describe the desperate situation of plague throughout 1791. TNA, PC, 1/19/24. For more British correspondence about plague in Egypt in 1791 and 1792, see, for example, the following reports sent by the British consul in Egypt, George Baldwin, to London: TNA, FO, 24/1, 183r–185v (4 July 1791), 191r–196v (12 July 1791), 211r–212v (21 June 1792).

29. On the 1792 plague, see BOA, HAT, 209/11213 (29 Z 1206/18 Aug. 1792). As in previous cases, the date given here is that assigned by the BOA. There is no internal evidence for the date of this case.

30. For British concerns about food and water shortages in Egypt in this period, see Consul Baldwin to London, TNA, FO, 24/1, 197r–200v (7 Oct. 1791).

31. Al-Jabartī, *'Ajā'ib al-Āthār* (1994), 2:374. For more on the problem of worms in the countryside in this period, see ibid., 2:397.

32. BOA, HAT, 209/11213 (29 Z 1206/18 Aug. 1792). As in previous cases, the date given here is the one assigned by the BOA. There is no internal evidence for the date of this case.

33. Al-Jabartī, *'Ajā'ib al-Āthār* (1994), 2:374–75, 2:397.

34. BOA, HAT, 245/13801A (3 Za 1213/9 Apr. 1799); BOA, HAT, 240/13451 (29 N 1214/24 Feb. 1800).

35. Al-Jabartī, *'Ajā'ib al-Āthār* (1994), 3:149–50. On the role of fodder shortages in bovine malnutrition and death in early fourteenth-century England and Wales, see Slavin, "Great Bovine Pestilence," 1244–47.

36. For another instance of attempts to sell off starving animals before they died, see al-Jabartī, *'Ajā'ib al-Āthār* (1994), 3:402.

37. Ibid., 3:459. In the fall of 1791, for example, food and fodder supplies were so low that the price of grain tripled. TNA, FO, 24/1, 197r–200v (7 Oct. 1791).

38. Al-Jabartī, *'Ajā'ib al-Āthār* (1994), 3:158.

39. 'Abd Allāh Sharqāwī, *Tuḥfat al-Nāẓirīn fī man Waliya Miṣr min al-Mulūk wa al-Salāṭīn*, ed. Riḥāb 'Abd al-Ḥamīd al-Qārī (Cairo: Maktabat Madbūlī, 1996), 124–25. For other examples of how the monopolization of food supplies led to animal starvation and death, see al-Jabartī, *'Ajā'ib al-Āthār* (1994), 3:402.

40. BOA, HAT, 86/3520 (29 N 1216/3 Feb. 1802).

41. Al-Jabartī, *'Ajā'ib al-Āthār* (1994), 3:480.

42. This major storm hit when Egypt was already suffering from "various diseases" (*enva-ı emraz*) throughout the countryside and Cairo. BOA, HAT, 88/3601 (1 Za 1218/12 Feb. 1804).

43. Al-Jabartī, *'Ajā'ib al-Āthār* (1994), 4:117, 4:240. The description of hailstones as being the size of hens' eggs is used here as a standard trope to convey that they were exceptionally large. Even bigger hailstones were sometimes described as being the size of "grinding stones." Ibid., 4:126.

44. See, for example, ibid., 4:412.

45. Data from the great bovine pestilence in England and Wales in the early fourteenth century suggest that it took about twenty years to recover about 60 percent of the bovid population lost in that crisis, and that about 20 percent of the population never recovered. Slavin, "Great Bovine Pestilence," 1249–54.

46. Al-Jabartī, *'Ajā'ib al-Āthār* (1994), 2:232.

47. For a comparative example of the animal labor lost during bovine epizootics in early modern Venice, see Appuhn, "Ecologies of Beef," 279.

48. Al-Jabartī, *'Ajā'ib al-Āthār* (1994), 2:232.

49. Al-Khashshāb, *Akhbār al-Amīr Murād*, 24–26.

50. On various of these earlier instances of population reductions, see Nāṣir Aḥmad Ibrāhīm, *al-Azamāt al-Ijtimā'iyya fī Miṣr fī al-Qarn al-Sābi' 'Ashar* (Cairo: Dār al-Āfāq al-'Arabiyya, 1998); Stuart J. Borsch, *The Black Death in Egypt and England: A Comparative Study* (Austin: University of Texas Press, 2005); Borsch, "Environment and Population"; Butzer, *Early Hydraulic Civilization*; Dols, *Black Death in the Middle East*; Michael W. Dols, "The General Mortality of the Black Death in the Mamluk Empire," in *The Islamic Middle East, 700–1900: Studies in Social and Economic History*, ed. Abraham L. Udovitch, 397–428 (Princeton, NJ: Darwin Press, 1981); William F. Tucker, "Natural Disasters and the Peasantry in Mamlūk Egypt," *Journal of the Economic and Social History of the Orient* 24 (1981): 215–24.

51. Crecelius, *Roots of Modern Egypt*.

52. Ibid., 79–91, 159–68.

53. Hathaway, *Politics of Households*; Hathaway, *Tale of Two Factions*; Shaw, "Landholding and Land-Tax Revenues"; Holt, *Egypt and the Fertile Crescent*, 85–101. For a recent rethinking of the role of Ottoman provincial elites at the end of the eighteenth century, see Ali Yaycıoğlu, "Provincial Power-Holders and the Empire in the Late Ottoman World: Conflict or Partnership?," in *The Ottoman World*, ed. Christine Woodhead, 436–52 (New York: Routledge, 2012).

54. Cuno, "Commercial Relations between Town and Village"; A. Richards, "Primitive Accumulation in Egypt."

55. Richard W. Bulliet identifies four phases in the history of human-animal relations: the separation of humans from other animals, predomesticity, domesticity, and postdomesticity. For his full characterization of these periods and his history of the transitions between them, see Bulliet, *Hunters, Herders, and Hamburgers*. Although at first glance the economic transitions I am discussing here would seem to map onto Bulliet's study of changes in human-animal relations from the phase of domesticity to the first moments of post-domesticity, it should be noted that he specifically distances his analysis from any connection to modernization. Ibid., 36–37.

56. This is one of the primary points of contrast between the seemingly similar situations of Egypt at the end of the eighteenth century and northern Europe in the first half of the fourteenth century. As William Chester Jordan has shown, apart from potentially improving nutritional standards, famine and its attendant consequences in Europe between 1315 and 1322 did *not* alter social and economic relations in any lasting way as they did in late Ottoman Egypt. William Chester Jordan, *The Great Famine: Northern Europe in the Early Fourteenth Century* (Princeton, NJ: Princeton University Press, 1996). Philip Slavin has recently attempted to link the great famine and the Black Death through what he terms the great bovine pestilence. Slavin, "Great Bovine Pestilence." In this regard, see also Newfield, "Cattle Panzootic in Early Fourteenth-Century Europe."

57. Bruce McGowan, "Peasants and Pastoralists," in *An Economic and Social History of the Ottoman Empire*, 2 vols., ed. Halil İnalcık with Donald Quataert (Cambridge: Cambridge University Press, 1994), 2:692.

58. About this situation at the end of the eighteenth century, Reşat Kasaba writes that Ottoman peasants "were transformed from being peasant producers whose freedom and protection was institutionally guaranteed to the status of, at best, sharecroppers—but more commonly, indebted tenants or dispossessed wage laborers." Kasaba, *Ottoman Empire and the World Economy*, 26.

59. Al-Jabartī, *ʿAjāʾib al-Āthār* (1994), 2:241.

60. Cuno, *Pasha's Peasants*; A. Richards, "Primitive Accumulation in Egypt."

61. Kasaba, *Ottoman Empire and the World Economy*, 23–27.

62. Mikhail, *Nature and Empire*, 170–200.

63. On the beginnings of rail in Egypt, see Omar Abdel-Aziz Omar, "Anglo-Egyptian Relations and the Construction of the Alexandria-Cairo-Suez Railway (1833–1858)" (DPhil thesis, University of London, 1966). If the American example is any indication, it seems safe to assume that rail initially actually increased the use of horses in Egypt—both in the original construction of rail lines and in the movement of people and goods to, from, and between trains. Ann Norton Greene, *Horses at Work: Harnessing Power in Industrial America* (Cambridge, MA: Harvard University Press, 2008), 43–45, 75–82.

64. In certain parts of England in the late sixteenth century, land consolidation based on enclosure was similarly part of the commercial development of the countryside. In the English case, however, there is much evidence to

suggest that animal populations, especially those of sheep, actually increased markedly in lockstep with the concentration and commercialization of land. John Martin, "Sheep and Enclosure in Sixteenth-Century Northamptonshire," *Agricultural History Review* 36 (1988): 39–54.

65. Cuno, *Pasha's Peasants*, 27–32, 44–47.

66. In addition to Cuno's work, see Kasaba, *Ottoman Empire and the World Economy*; A. Richards, "Primitive Accumulation in Egypt"; Shaw, "Landholding and Land-Tax Revenues"; Shaw, *Financial and Administrative Organization*, 12–97; ʿAbd al-Raḥīm, *al-Rīf al-Miṣrī*, 83–143; Gabriel Baer, *A History of Landownership in Modern Egypt, 1800–1950* (London: Oxford University Press, 1962).

67. Crecelius, *Roots of Modern Egypt*.

68. Marsot, *Egypt in the Reign of Muhammad Ali*, 14–19; Fahmy, *Mehmed Ali*, 23–24.

69. The early nineteenth-century estates in which animals did not play a role are many. Of course, absence of evidence in thousands of cases is not evidence of complete absence, but animals appear far less frequently in the estate inventories of members of all classes of rural society between 1780 and 1820 than they did earlier. For a discussion of the increasing role of land as property in the first few decades of the nineteenth century, see Cuno, *Pasha's Peasants*, 103–17.

70. On usufruct as effective, though of course not legal, ownership, see ibid., 74–84; Gabriel Baer, "The Dissolution of the Egyptian Village Community," *Die Welt des Islams* 6 (1959): 59.

71. This kind of confiscation was known as *müsadere* in the Ottoman Empire. For a discussion of this phenomenon in the early modern period, see Rifaʿat ʿAli Abou-El-Haj, *Formation of the Modern State: The Ottoman Empire, Sixteenth to Eighteenth Centuries*, 2nd ed. (Syracuse, NY: Syracuse University Press, 2005), 48–49.

72. Cuno, *Pasha's Peasants*, 33–47.

73. The creation of large rural estates in Egypt further exacerbated animal shortages in the countryside by reducing the amount of available pastureland. As competing amirs, beys, and other Egyptian elites continued to seize land and other properties, less and less pastureland was left for animals to graze, making it much more difficult to maintain sizable herds in and around major towns and cities. Al-Jabartī, *ʿAjāʾib al-Āthār* (1994), 3:477–78, 4:380.

74. On the emergence of regularity and reproducibility as paradigms of labor and natural resource management in Ottoman Egypt, see Mikhail, *Nature and Empire*. For a study of similar developments in the United States and Europe, see Jennifer Karns Alexander, *The Mantra of Efficiency: From Waterwheel to Social Control* (Baltimore: Johns Hopkins University Press, 2008). See also Scott, *Seeing Like a State*.

75. For a general discussion of corvée in the seventeenth and eighteenth centuries, see Sulaymān, "al-Sukhra fī Miṣr fī al-Qarnayn al-Sābiʿ ʿAshar wa al-Thāmin ʿAshar."

76. As we saw in chapter 4, Yūsuf al-Shirbīnī, author of a seventeenth-century satire of country living, offers a useful description of early modern

forced labor, highlighting as well the integral role of animals in village communities. He writes: "The corvée is found only in those multazims' villages that include ūsya land [the only land over which multazims held direct administrative control]. . . . The multazim sends oxen, timber, plows, and whatever else is needed and appoints an agent to take charge of it and prepares a place for the timber and animals belonging to it. . . . He also delegates someone to spend money on the upkeep of the animals, etc., and to keep careful accounts. . . . In some villages the corvée applies to a number of men, fixed by household, for example. Thus they say, 'From such and such a household one man is to go, and from such and such two' according to the quota set for them in the distant or more recent past." Al-Shirbīnī, *Hazz al-Quḥūf*, 2:328–29. In cases such as this one, the animals used for labor were usually owned by the multazim overseeing the work. For other descriptions of corvée in al-Shirbīnī's text, see ibid., 2:13, 2:49, 2:79, 2:333.

77. This land was again known as *ūsya* land and was generally found only in the delta and Middle Egypt. The figure of 10 percent comes from the end of the eighteenth century. Cuno, *Pasha's Peasants*, 36–37.

78. On this point, al-Shirbīnī again offers the following verse: "And on the day when the corvée descends on the people in the village / Umm Waṭīf hides me in the oven." Al-Shirbīnī, *Hazz al-Quḥūf*, 2:327. For an extended commentary on this verse, see ibid., 2:327–31.

79. For these definitions, see *Lisān al-ʿArab*, s.v. "*ʿawana*," "*sakhara*"; Lane, *Arabic-English Lexicon*, s.v. "*ʿawana*," "*sakhara*."

80. Mikhail, *Nature and Empire*, 170–200.

81. Al-Jabartī, *ʿAjāʾib al-Āthār* (1994), 4:289. For a discussion of this important passage, see Cuno, *Pasha's Peasants*, 5–6, 37.

82. Rivlin, *Agricultural Policy*, 248.

83. Cuno, *Pasha's Peasants*, 115.

84. Rivlin, *Agricultural Policy*, 247.

85. The following are only some of the largest canal projects undertaken by Mehmet ʿAli's government: al-Faraʿūniyya in the central delta; Shibīn in the delta; al-Zaʿfarāniyya and al-Sharqāwiyya in al-Qalyūbiyya; al-Būhiyya in al-Daqahliyya; al-Wādī in al-Sharqiyya; al-Fashn in al-Minyā; and al-Shunhūriyya in Qinā. Details of these and other canal repairs are given in Rivlin, *Agricultural Policy*, 213–41; al-Rāfʿī, *ʿAṣr Muḥammad ʿAlī*, 487–95.

86. For instance, an important outcome of the changing economy of animals in this period that I do not address is what corvée meant for the laboring agricultural family as a whole. As men were taken away from their home villages to work on infrastructural projects, women and other familial relations often replaced men in the laboring economy of their home communities. For a discussion of this point, see Judith E. Tucker, *Women in Nineteenth-Century Egypt* (Cambridge: Cambridge University Press, 1985), 29. For a useful recent analysis of the important role of women in rural labor in early modern France, see Chandra Mukerji, *Impossible Engineering: Technology and Territoriality on the Canal du Midi* (Princeton, NJ: Princeton University Press, 2009).

87. Rivlin, *Agricultural Policy*, 243–45.

88. For later examples of the principle that the organization of corvée

labor was the responsibility of village heads, see Baer, "Dissolution of the Egyptian Village Community," 63–64, 66–68.

89. According to M.A. Linant de Bellefonds's estimates, in each year between 1820 and 1831, about sixty-seven thousand men were used to build canals, and another four hundred thousand were employed to clean them. Cuno, *Pasha's Peasants*, 122.

90. On the population of Cairo, see Panzac, "Alexandrie: Évolution d'une ville cosmopolite," 147. On the population of Egypt as a whole, see Panzac, *La peste*, 271; Raymond, "La population du Caire."

91. This was true, for example, during certain periods of repair work on the Maḥmūdiyya Canal. Marsot, *Egypt in the Reign of Muhammad Ali*, 151.

92. Rivlin, *Agricultural Policy*, 227–28.

93. Ibid., 231.

94. Marsot, *Egypt in the Reign of Muhammad Ali*, 152.

95. Rivlin, *Agricultural Policy*, 232. On the intimate connection between corvée and Mehmet ʿAli's military conscription efforts, see Khaled Fahmy, "The Era of Muhammad ʿAli Pasha, 1805–1848," in *Modern Egypt, from 1517 to the End of the Twentieth Century*, vol. 2 of *The Cambridge History of Egypt*, ed. M.W. Daly (Cambridge: Cambridge University Press, 1998), 163, 166.

96. Marsot, *Egypt in the Reign of Muhammad Ali*, 150.

97. 2 al-Maʿiyya al-Saniyya (1 Ramaḍān 1251/20 Dec. 1835), cited in Marsot, *Egypt in the Reign of Muhammad Ali*, 150–51.

98. 1 al-Maʿiyya al-Saniyya (1 Shaʿbān 1253/30 Oct. 1837) and 2 al-Maʿiyya al-Saniyya (23 Rabīʿ al-Awwal 1251/18 July 1835), both cited in Marsot, *Egypt in the Reign of Muhammad Ali*, 151.

99. For a discussion and analysis of this construction project, see Mikhail, *Nature and Empire*, 242–96.

100. Rivlin, *Agricultural Policy*, 219–20, 353n15. The estimate of 360,000 comes from Linant de Bellefonds, *Mémoires sur les principaux travaux d'utilité publiqué*, 351. Without citation, Marsot writes that 250,000 peasants were brought to work on the canal. Marsot, *Egypt in the Reign of Muhammad Ali*, 151.

101. Al-Jabartī, *ʿAjāʾib al-Āthār* (1994), 4:408.

102. Most of what they cleared from the city's streets was dumped into Cairo's main urban canal, known as al-Khalīj. Since this waterway also functioned as one of the city's main trash receptacles, the massive amounts of dirt, sand, and garbage deposited in it served to clog up the canal even more than it usually was. Ibid.

103. For a discussion of this number of dead, see Mikhail, *Nature and Empire*, 281–82, 289–90.

104. Al-Jabartī, *ʿAjāʾib al-Āthār* (1994), 4:427. Al-Jabartī's full description of that month's work on the Maḥmūdiyya is given on p. 88 above.

105. Numerous examples of this process from around the globe are given in J. Richards, "Toward a Global System of Property Rights in Land," 54–78.

106. For an illuminating analysis in this regard, see Bulliet, "History and Animal Energy in the Arid Zone"; Bulliet, "Camel and the Watermill."

CHAPTER 8

1. On the Ottoman-Mamluk confrontation and the conquest of Egypt, see Andrew C. Hess, "The Ottoman Conquest of Egypt (1517) and the Beginning of the Sixteenth-Century World War," *International Journal of Middle East Studies* 4 (1973): 55–76; Emire Cihan Muslu, "Ottoman-Mamluk Relations: Diplomacy and Perceptions" (PhD diss., Harvard University, 2007); Jean-Louis Bacqué-Grammont and Anne Kroell, *Mamlouks, ottomans et portugais en Mer Rouge: L'affaire de Djedda en 1517* (Cairo: Institut français d'archéologie orientale, 1988); Michel M. Mazzaoui, "Global Policies of Sultan Selim, 1512–1520," in *Essays on Islamic Civilization: Presented to Niyazi Berkes*, ed. Donald P. Little, 224–43 (Leiden, Netherlands: Brill, 1976); Winter, *Egyptian Society*, 1–17.

2. On the specific case of Ottoman rule in Egypt, see Winter, *Egyptian Society*; Shaw, *Financial and Administrative Organization*; Aḥmad, *al-Idāra fī Miṣr*; Aḥmad, *al-Mujtamaʿal-Miṣrī*; Aḥmad, *Tārīkh wa Muʾarrikhī Miṣr*; Muḥammad, *al-Wujūd al-ʿUthmānī fī Miṣr*; Muḥammad, *al-Wujūd al-ʿUthmānī al-Mamlūkī*; Raymond, *Artisans et commerçants*.

3. This chapter necessarily builds upon a rich literature on material culture and the role of commodities in Ottoman history. Coffee, tulips, textiles, food, soap, and clothing are some of the material goods Ottoman historians have usefully studied. As even this partial list shows, the main focus of work on Ottoman material culture has been on luxury items used and consumed mostly by urban elites. This chapter takes a different tack, seeking to expand our knowledge of Ottoman objects by shifting attention to things whose histories are mostly rural and whose consumption and use were more about supporting the logistical function of the empire than they were about leisure or projecting status or wealth. For illuminating studies of Ottoman material culture, see, for example, Donald Quataert, ed., *Consumption Studies and the History of the Ottoman Empire, 1550–1922: An Introduction* (Albany: State University of New York Press, 2000); Suraiya Faroqhi, *Towns and Townsmen in Ottoman Anatolia: Trade, Crafts, and Food Production in an Urban Setting, 1520–1650* (Cambridge: Cambridge University Press, 1984); Dana Sajdi, ed., *Ottoman Tulips, Ottoman Coffee: Leisure and Lifestyle in the Eighteenth Century* (London: I.B. Tauris, 2007); Amy Singer, ed., *Starting with Food: Culinary Approaches to Ottoman History* (Princeton, NJ: Markus Wiener Publishers, 2011); Suraiya Faroqhi and Christoph K. Neumann, eds., *Ottoman Costumes: From Textile to Identity* (Istanbul: Eren, 2004); James Grehan, *Everyday Life & Consumer Culture in 18th-Century Damascus* (Seattle: University of Washington Press, 2007).

4. Generally on the Ottomans in the Red Sea and Indian Ocean, see Salih Özbaran, *Ottoman Expansion towards the Indian Ocean in the 16th Century* (Istanbul: Bilgi University Press, 2009); Salih Özbaran, *The Ottoman Response to European Expansion: Studies on Ottoman-Portuguese Relations in the Indian Ocean and Ottoman Administration in the Arab Lands during the Sixteenth Century* (Istanbul: Isis Press, 1994); Salih Özbaran, "A Turkish Report on the Red Sea and the Portuguese in the Indian Ocean (1525)," *Arabian Studies* 4 (1978): 81–88; Salih Özbaran, "Ottoman Naval Power in the Indian

Ocean in the 16th Century," in *The Kapudan Pasha, His Office and His Domain: Halcyon Days in Crete IV*, ed. Elizabeth Zachariadou, 109–17 (Rethymnon: Crete University Press, 2002); Giancarlo Casale, *The Ottoman Age of Exploration* (New York: Oxford University Press, 2010); Giancarlo Casale, "The Ottoman Administration of the Spice Trade in the Sixteenth-Century Red Sea and Persian Gulf," *Journal of the Economic and Social History of the Orient* 49 (2006): 170–98; Bacqué-Grammont and Kroell, *Mamlouks, ottomans et portugais*; Anthony Reid, "Sixteenth-Century Turkish Influence in Western Indonesia," *Journal of South East Asian History* 10 (1969): 395–414; Michel Tuchscherer, "La flotte impériale de Suez de 1694 à 1719," *Turcica* 29 (1997): 47–69.

5. On the problem of wood for Ottoman naval construction, see Brummett, *Ottoman Seapower and Levantine Diplomacy*, 96, 115–16, 144, 174; İdris Bostan, *Osmanlı Bahriye Teşkilâtı: XVII. Yüzyılda Tersâne-i Âmire* (Ankara: Türk Tarih Kurumu Basımevi, 1992), 102–18; Casale, *Ottoman Age of Exploration*, 201–2; Colin H. Imber, "The Navy of Süleiman the Magnificent," *Archivum Ottomanicum* 6 (1980): 211–82; Murat Çizakça, "Ottomans and the Mediterranean: An Analysis of the Ottoman Shipbuilding Industry as Reflected by the Arsenal Registers of Istanbul, 1529–1650," in *Le genti del Mare Mediterraneo*, vol. 2, ed. Rosalba Ragosta, 773–89 (Naples: Lucio Pironti, 1981); Svat Soucek, "Certain Types of Ships in Ottoman-Turkish Terminology," *Turcica* 7 (1975): 233–49. For a useful comparative study of this problematic in early modern Venice, see Karl Appuhn, *A Forest on the Sea: Environmental Expertise in Renaissance Venice* (Baltimore: Johns Hopkins University Press, 2009).

6. On the tension between Egypt's relative agricultural wealth and its dearth of domestic wood supplies, see Mikhail, *Nature and Empire*, 82–169.

7. Egypt's lack of wood, though a new problem for the Ottomans, was one that had been faced by all political powers that ruled it since antiquity. Roger S. Bagnall, *Egypt in Late Antiquity* (Princeton, NJ: Princeton University Press, 1993), 41; Meiggs, *Trees and Timber*, 57–68; John Perlin, *A Forest Journey: The Story of Wood and Civilization* (Woodstock, VT: Countryman Press, 2005), 131–34; Thirgood, *Man and the Mediterranean Forest*, 87–94.

8. On the connections between Egypt and the Hijaz in the Ottoman period, see Suraiya Faroqhi, "Trade Controls, Provisioning Policies, and Donations: The Egypt-Hijaz Connection during the Second Half of the Sixteenth Century," in *Süleymân the Second and His Time*, ed. Halil İnalcık and Cemal Kafadar, 131–43 (Istanbul: Isis Press, 1993); Suraiya Faroqhi, "Red Sea Trade and Communications as Observed by Evliya Çelebi (1671–72)," *New Perspectives on Turkey* 5–6 (1991): 87–105; Suraiya Faroqhi, "Coffee and Spices: Official Ottoman Reactions to Egyptian Trade in the Later Sixteenth Century," *Wiener Zeitschrift für die Kunde des Morgenlandes* 76 (1986): 87–93; Michel Tuchscherer, "Commerce et production du café en Mer Rouge au XVIe siècle," in *Le commerce du café avant l'ère des plantations coloniales: Espaces, réseaux, sociétés (XVe–XIXe siècle)*, ed. Michel Tuchscherer, 69–90 (Cairo: Institut français d'archéologie orientale, 2001); ʿAbd al-Muʿṭī, *al-ʿAlāqāt al-Miṣriyya al-Ḥijāziyya*; Colin Heywood, "A Red Sea Shipping Register of the 1670s for

the Supply of Foodstuffs from Egyptian *Wakf* Sources to Mecca and Medina (Turkish Documents from the Archive of ʿAbdurrahman 'ʿAbdi' Pasha of Buda, I)," *Anatolia Moderna* 6 (1996): 111–74.

9. Özbaran, *Ottoman Expansion*, 77–80; Brummett, *Ottoman Seapower and Levantine Diplomacy*, 96, 115–16, 144, 174; Casale, *Ottoman Age of Exploration*, 201–2. On forestry and Ottoman shipbuilding in the Mediterranean, see Çizakça, "Ottomans and the Mediterranean."

10. On the Ottomans in various parts of the Arabian Peninsula and Persian Gulf, see Salih Özbaran, "Bahrain in 1559: A Narrative of Turco-Portuguese Conflict in the Gulf," *Osmanlı Araştırmaları* 3 (1982): 91–104; Salih Özbaran, *Yemen'den Basra'ya Sınırdaki Osmanlı* (Istanbul: Kitap Yayınevi, 2004); Salih Özbaran, "The Ottoman Turks and the Portuguese in the Persian Gulf, 1534–1581," *Journal of Asian History* 6 (1972): 45–88; Jan E. Mandaville, "The Ottoman Province of Al-Hasâ in the Sixteenth and Seventeenth Centuries," *Journal of the American Oriental Society* 90 (1970): 486–513; Casale, *Ottoman Age of Exploration*, 63–65; Patricia Risso, "Cross-Cultural Perceptions of Piracy: Maritime Violence in the Western Indian Ocean and Persian Gulf Region during a Long Eighteenth Century," *Journal of World History* 12 (2001): 293–319; Patricia Risso, "Muslim Identity in Maritime Trade: General Observations and Some Evidence from the 18th Century Persian Gulf / Indian Ocean Region," *International Journal of Middle East Studies* 21 (1989): 381–92.

11. ʿAbd al-Muʿṭī, *al-ʿAlāqāt al-Miṣriyya al-Ḥijāziyya*.

12. Suraiya Faroqhi, *Pilgrims and Sultans: The Hajj under the Ottomans* (London: I.B. Tauris, 1994).

13. Although there is little direct evidence from the early modern period about the role of the pilgrimage in the spread of disease, there are numerous examples from the end of the nineteenth century of the diffusion of cholera resulting from the pilgrimage. One can only imagine that similar situations maintained in earlier centuries as well. For the late nineteenth-century examples, see Kuhnke, *Lives at Risk*, 95, 107–8; J.R. McNeill, *Something New under the Sun*, 196.

14. On the Ottoman provisioning of the pilgrimage from Egypt, see Mikhail, *Nature and Empire*, 113–22.

15. See, for example, BOA, MM, 3:210 (Evail Ş 1133/27 May–5 June 1721); BOA, HAT, 29/1358 (29 Z 1197/24 Nov. 1783); BOA, HAT, 28/1354 (7 Za 1198/22 Sept. 1784); BOA, HAT, 26/1256 (10 Za 1200/3 Sept. 1786). There is no internal evidence for the date of this final case. The date given is the one assigned by the BOA. TSMA, E. 3218 (n.d.); TSMA, E. 5657 (13 Ra 1204/1 Dec. 1789); TSMA, E. 664/40 (n.d.); TSMA, E. 5225/12 (Evahir S 1194/27 Feb.–7 Mar. 1780); TSMA, E. 664/51 (n.d.); TSMA, E. 2229/3 (n.d.).

16. On the provisioning of Istanbul from Egypt, see Mikhail, *Nature and Empire*, 103–13.

17. For some of this history of water management in Ottoman Egypt, see ibid., 38–81.

18. BOA, İbnülemin Umur-i Nafia, 94 (Evasıt Ra 1121/21–30 May 1709); BOA, MM, 1:116 (Evail R 1122/30 May–8 June 1710); BOA, MM, 1:167 (Evasıt S 1123/31 Mar.–9 Apr. 1711).

19. Shaw, *Financial and Administrative Organization*, 269–70. One of the most famous and lucrative of these *awqāf* in the late seventeenth century was very near Fayyum and possibly administratively connected to it. It consisted of a group of nine villages in al-Bahnasa (Beni Suef) controlled by Ḥasan Aghā Bilīfyā, a Faqari leader and commander of the Gönüllüyan military bloc. Hathaway, "Role of the Kızlar Ağası"; Hathaway, *Politics of Households*, 157–60; Hathaway, "Egypt in the Seventeenth Century," 50.

20. On shifts in trading patterns in the fifteenth century, see Hanna, *Urban History of Būlāq*, 7–32.

21. For studies of the Red Sea shipwreck site of a particular Ottoman vessel that illuminate the ship's cargo, carrying capacity, structure, and so forth, see Cheryl Ward, "The Sadana Island Shipwreck: An Eighteenth-Century AD Merchantman off the Red Sea Coast of Egypt," *World Archaeology* 32 (2001): 368–82; Cheryl Ward, "The Sadana Island Shipwreck: A Mideighteenth-Century Treasure Trove," in *A Historical Archaeology of the Ottoman Empire: Breaking New Ground*, ed. Uzi Baram and Lynda Carroll, 185–202 (New York: Kluwer Academic / Plenum, 2000); Cheryl Ward and Uzi Baram, "Global Markets, Local Practice: Ottoman-Period Clay Pipes and Smoking Paraphernalia from the Red Sea Shipwreck at Sadana Island, Egypt," *International Journal of Historical Archaeology* 10 (2006): 135–58.

22. I am relying on the following court cases from Rosetta to tell the story of the wood's movement: DWQ, Maḥkamat Rashīd, 132, p. 88, case 140 (17 Ş 1137/30 Apr. 1725); DWQ, Maḥkamat Rashīd, 132, pp. 200–201, case 311 (3 N 1137/16 May 1725); DWQ, Maḥkamat Rashīd, 132, p. 199, case 308 (16 Ş 1137/29 Apr. 1725); DWQ, Maḥkamat Rashīd, 132, p. 199, case 309 (17 Ş 1137/30 Apr. 1725). All of these cases are written in Ottoman Turkish. The recording of these imperial orders in Ottoman Turkish in the normally Arabic-language registers of the court of Rosetta shows both the imperial nature of this project and the way the empire used its courts to manage these kinds of imperial endeavors.

23. Earlier examples of attempts to gain access to wood from Anatolia for the construction of ships in Suez further highlight the strategic nature of this commodity. In 1510, for instance, eleven galleons were dispatched to the Anatolian port of Ayas at the very northeastern corner of the Mediterranean from the Egyptian port of Damietta to secure wood supplies for the construction of ships in Suez. Suspicious that this movement of ships to Ayas was part of an Ottoman-Mamluk plot against Rhodes, the leaders of this still-independent island territory attacked and destroyed the convoy. Brummett, *Ottoman Seapower and Levantine Diplomacy*, 115–16. For another example of the transport of wood from Anatolia to build ships on the Red Sea, see ibid., 174.

24. For a sketch of historic forest locations and coverage in the Middle East, see Cordova, *Millennial Landscape Change in Jordan*, 3–4.

25. There is relatively little work on the history of Ottoman forestry. For some of the current literature, see Dursun, "Forest and the State"; S. White, *Climate of Rebellion*, 16–17, 28–31, 72, 278, 289; Mikhail, *Nature and Empire*, 124–69. See also the following very general history of Turkish forestry: Yücel Çağlar, *Türkiye Ormanları ve Ormancılık* (Istanbul: İletişim

Yayınları, 1992). For useful collections of documents on Ottoman forestry, see Çevre ve Orman Bakanlığı, *Osmanlı Ormancılığı ile İlgili Belgeler*, 3 vols. (Ankara: Çevre ve Orman Bakanlığı, 1999–2003); Halil Kutluk, ed., *Türkiye Ormancılığı ile İlgili Tarihi Vesikalar, 893–1339 (1487–1923)* (Istanbul: Osmanbey Matbaası, 1948).

26. On the use of wood to construct ships in the imperial dockyards of Istanbul, see Bostan, *Tersâne-i Âmire*, 102–18.

27. On the office of the *kereste emîni*, see Çevre ve Orman Bakanlığı, *Osmanlı Ormancılığı*, 1:94–95. For a discussion of Ottoman forestry guilds in the context of the wider early modern Mediterranean world, see J. Donald Hughes, *The Mediterranean: An Environmental History* (Santa Barbara, CA: ABC-CLIO, 2005), 97–99.

28. For cases involving the organization of laborers for the harvesting of lumber from Anatolian forests, see Çevre ve Orman Bakanlığı, *Osmanlı Ormancılığı*, 1:8–9, 1:46–47, 1:48–49, 1:56–57, 1:60–61.

29. For more on these and other positions related to the harvesting of timber, see ibid., 1:XIII.

30. For an example of timber harvests on the southern Black Sea coast near the town of Sinop that were used to repair Egyptian vessels, see BOA, Cevdet Bahriye, 1413 (Evasıt R 1120/30 June–9 July 1708 and 20 Za 1124/19 Dec. 1712).

31. For a statement of the historic role of Rhodes in funneling wood to the imperial timber stores in this period, see BOA, Cevdet Nafia, 302 (23 Za 1216/28 Mar. 1802).

32. For a general discussion of Ottoman imperial forest management policies, see Çevre ve Orman Bakanlığı, *Osmanlı Ormancılığı*, 1:XI–XVI. For specific regulations, see ibid., 1:2–3, 1:6–7, 1:18–19, 1:22–23, 1:24–25, 1:26–27, 1:38–39, 1:104–5, 1:106–7, 1:110–11, 1:114–15, 1:120–21, 1:124–25, 1:150–51, 1:172–73, 2:2–3, 2:42–43, 2:46–47, 2:48–49, 3:4–5, 3:6–7, 3:8–9, 3:16–17, 3:18–19.

33. For useful comparative examples of sustainable forest management techniques in early modern Japan, Germany, and Spain, see Conrad Totman, *The Green Archipelago: Forestry in Preindustrial Japan* (Berkeley: University of California Press, 1989); Conrad Totman, *The Lumber Industry in Early Modern Japan* (Honolulu: University of Hawai'i Press, 1995); Warde, *Ecology, Economy and State Formation*; John Thomas Wing, "Roots of Empire: State Formation and the Politics of Timber Access in Early Modern Spain, 1556–1759" (PhD diss., University of Minnesota, 2009); John T. Wing, "Keeping Spain Afloat: State Forestry and Imperial Defense in the Sixteenth Century," *Environmental History* 17 (2012): 116–45.

34. DWQ, Maḥkamat Rashīd, 132, p. 88, case 140 (17 Ş 1137/30 Apr. 1725); DWQ, Maḥkamat Rashīd, 132, pp. 200–201, case 311 (3 N 1137/16 May 1725); DWQ, Maḥkamat Rashīd, 132, p. 199, case 308 (16 Ş 1137/29 Apr. 1725); DWQ, Maḥkamat Rashīd, 132, p. 199, case 309 (17 Ş 1137/30 Apr. 1725).

35. DWQ, Maḥkamat Rashīd, 132, p. 88, case 140 (17 Ş 1137/30 Apr.

1725); DWQ, Maḥkamat Rashīd, 132, p. 199, case 309 (17 Ş 1137/30 Apr. 1725).

36. Bostan, "Ottoman Base in Eastern Mediterranean"; Reimer, "Ottoman Alexandria."

37. Panzac, "International and Domestic Maritime Trade."

38. On Ottoman Egypt's many trading links, see Raymond, *Artisans et commerçants*.

39. DWQ, Maḥkamat Rashīd, 132, p. 88, case 140 (17 Ş 1137/30 Apr. 1725); DWQ, Maḥkamat Rashīd, 132, pp. 200–201, case 311 (3 N 1137/16 May 1725).

40. On pirates and corsairs in the Ottoman Mediterranean and imperial attempts to stop them, see Brummett, *Ottoman Seapower and Levantine Diplomacy*, 94–102, 135–36; İdris Bostan, *Kürekli ve Yelkenli Osmanlı Gemileri* (Istanbul: Bilge, 2005), 372, 376; Molly Greene, *Catholic Pirates and Greek Merchants: A Maritime History of the Mediterranean* (Princeton, NJ: Princeton University Press, 2010); Molly Greene, "The Ottomans in the Mediterranean," in *The Early Modern Ottomans: Remapping the Empire*, ed. Virginia H. Aksan and Daniel Goffman (Cambridge: Cambridge University Press, 2007), 113–16. For the case of a pirate attack near Rhodes on Egyptian grain ships on their way to Istanbul, see TSMA, E. 7008/12 (n.d.).

41. Bostan, "Ottoman Base in Eastern Mediterranean," 76–77.

42. DWQ, Maḥkamat Rashīd, 132, p. 88, case 140 (17 Ş 1137/30 Apr. 1725).

43. For accounts of various Ottoman plans for a Suez Canal that ultimately never materialized, see Casale, *Ottoman Age of Exploration*, 135–37, 159–70, 201–2; Colin Imber, *The Ottoman Empire, 1300–1650: The Structure of Power* (New York: Palgrave Macmillan, 2002), 62; Mustafa Bilge, "Suez Canal in the Ottoman Sources," in *Proceedings of the International Conference on Egypt during the Ottoman Era: 26–30 November 2007, Cairo, Egypt*, ed. Research Centre for Islamic History, Art and Culture, 89–113 (Istanbul: IRCICA, 2010).

44. The absence of such a waterway was recognized as a problem by various governments throughout Egypt's history. For accounts of attempts to build such a canal, see Hairy and Sennoune, "Géographie historique du canal d'Alexandrie"; Ṭūsūn, *Tārīkh Khalīj al-Iskandariyya*; Mikhail, *Nature and Empire*, 242–90.

45. DWQ, Maḥkamat Rashīd, 132, p. 88, case 140 (17 Ş 1137/30 Apr. 1725); DWQ, Maḥkamat Rashīd, 132, pp. 200–201, case 311 (3 N 1137/16 May 1725). On *cerîm* ships, see Bostan, *Osmanlı Gemileri*, 253–59. For an example of the empire's hiring of sailors of *cerîm* ships, see BOA, Cevdet Bahriye, 208 (14 Ra 1204/2 Dec. 1789).

46. DWQ, Maḥkamat Rashīd, 132, p. 88, case 140 (17 Ş 1137/30 Apr. 1725).

47. For studies of the Nile's sediment load adding to the force of its seaward flow, see Frihy et. al., "Patterns of Nearshore Sediment Transport"; Smith and Abdel-Kader, "Coastal Erosion"; Elsayed et. al., "Accretion and Erosion Patterns."

48. DWQ, Maḥkamat Rashīd, 132, p. 88, case 140 (17 Ş 1137/30 Apr. 1725); DWQ, Maḥkamat Rashīd, 132, pp. 200–201, case 311 (3 N 1137/16 May 1725). On Bulaq during the Ottoman period, see Hanna, *Urban History of Būlāq*.

49. DWQ, Maḥkamat Rashīd, 132, pp. 200–201, case 311 (3 N 1137/16 May 1725).

50. Generally on the use of camels in transport in the Middle East, see Bulliet, *Camel and the Wheel.*

51. DWQ, Maḥkamat Rashīd, 132, pp. 200–201, case 311 (3 N 1137/16 May 1725). The case lists these amounts in *niṣf faḍḍa*, but, again, according to Stanford J. Shaw, "The silver coin in common use during Mamlûk and Ottoman times in Egypt was called *nıṣf fiḍḍe* colloquially and *para* officially." Shaw, *Financial and Administrative Organization*, 65n169.

52. Shaw, *Financial and Administrative Organization*, 264–67.

53. On the importance of camels for Ottoman transportation and military ventures, see İnalcık, "'Arab' Camel Drivers"; İnalcık, "Ottoman State," 1:38–39, 1:62–63; Faroqhi, "Camels, Wagons, and the Ottoman State." In 1399, for example, Bayezid the Thunderbolt (r. 1389–1402) took ten thousand camels as booty from his conquest of the region of Antalya. İnalcık, "'Arab' Camel Drivers," 265.

54. Quataert, *Ottoman Empire*, 119.

55. For a similar example of the movement of wood for the construction of ships in Suez in 1810, see al-Jabartī, *ʿAjāʾib al-Āthār* (1994), 4:146.

56. It was of course this lack of alternatives that made the project of the Suez Canal so appealing.

57. Şevket Pamuk, *A Monetary History of the Ottoman Empire* (Cambridge: Cambridge University Press, 2000); Şevket Pamuk, "Prices in the Ottoman Empire, 1469–1914," *International Journal of Middle East Studies* 36 (2004): 451–68.

58. I of course do not mean to imply that the Ottoman Empire intervened in all economic relations in its realm or that somehow a Muslim polity would be more economically interventionist than a non-Muslim one. Obviously, most economic relationships in the empire did not involve the state in any way. The ship construction analyzed here, however, is an instance in which the state did play a central role and therefore provides an opportunity to understand an important aspect of Ottoman economic history. For a discussion of the empire's selective protectionism in the sixteenth century, see Brummett, *Ottoman Seapower and Levantine Diplomacy*, 181–82.

59. For studies pointing to some of the administrative weaknesses of the empire in Egypt and in some of its other provinces in the eighteenth century, see ʿAbd al-Raḥīm, *al-Rīf al-Miṣrī*; Albert Hourani, "Ottoman Reforms and the Politics of Notables," in *The Beginnings of Modernization in the Middle East: The Nineteenth Century*, ed. William R. Polk and Richard L. Chambers, 41–68 (Chicago: University of Chicago Press, 1968); Abdul-Karim Rafeq, "ʿAbd al-Ghani al-Nabulsi: Religious Tolerance and 'Arabness' in Ottoman Damascus," in *Transformed Landscapes: Essays on Palestine and the Middle East in Honor of Walid Khalidi*, ed. Camille Mansour and Leila Fawaz, 1–17

(Cairo: American University in Cairo Press, 2009). For a very useful review of much of this literature, see Suraiya Faroqhi, "Coping with the Central State, Coping with Local Power: Ottoman Regions and Notables from the Sixteenth to the Early Nineteenth Century," in *The Ottomans and the Balkans: A Discussion of Historiography*, ed. Fikret Adanır and Suraiya Faroqhi, 351–81 (Leiden, Netherlands: Brill, 2002). On the Ottoman army in the eighteenth century and its provisioning problems, see Aksan, *Ottoman Wars*, 83–179.

60. William Cronon, *Nature's Metropolis: Chicago and the Great West* (New York: Norton, 1991), 149. Emphasis in original.

61. Mikhail, *Nature and Empire*, 128–36.

CHAPTER 9

1. Panzac, *La peste*, 29–57, 381–407; Ibrāhīm, *al-Azamāt al-Ijtimāʿiyya fī Miṣr*; Raymond, "Les Grandes Épidémies de peste au Caire"; W. Tucker, "Natural Disasters and the Peasantry."

2. Many Muslim writers on the subject of plague commented on and struggled with the following three tenets derived from the teachings of the Prophet Muḥammad: plague is a mercy and a form of martyrdom from God for a pious Muslim (and a form of punishment for infidels), a Muslim should neither flee from nor enter a region affected by plague, and plague is not contagious because it comes directly from God. Differing opinions about these principles came to constitute the religio-medico-legal underpinnings of many ideas about plague in the Muslim world. Dols, *Black Death in the Middle East*, 23–25, 109–21; Michael W. Dols, "Ibn al-Wardī's *Risālah al-Nabaʾ ʿan al-Wabaʾ*, a Translation of a Major Source for the History of the Black Death in the Middle East," in *Near Eastern Numismatics, Iconography, Epigraphy and History: Studies in Honor of George C. Miles*, ed. Dickran K. Kouymjian (Beirut: American University of Beirut, 1974), 444–45; Jacqueline Sublet, "La peste prise aux rêts de la jurisprudence: Le Traité d'Ibn Ḥağar al-ʿAsqalānī sur la peste," *Studia Islamica* 33 (1971): 141–49.

3. Dols, "Second Plague," 169, 176. For the period from 1416 to 1514, David Neustadt (Ayalon) reports that an outbreak of plague struck Egypt once every seven years on average. Neustadt (Ayalon), "Plague and Its Effects upon the Mamlûk Army," 68. See also Dols, *Black Death in the Middle East*, 223–24; Panzac, *La peste*, 197–207.

4. One could rightly make the point that the arbitrary geographic and political division of Egypt is just as misleading in a discussion of plague as is the identification of the year 1517 as a chronological divide.

5. Works on plague in the medieval Middle East include Dols, "General Mortality of the Black Death"; Dols, *Black Death in the Middle East*; Dols, "Ibn al-Wardī's *Risālah al-Nabaʾ*"; Michael W. Dols, "al-Manbijī's 'Report of the Plague': A Treatise on the Plague of 764–765/1362–1364 in the Middle East," in *The Black Death: The Impact of the Fourteenth-Century Plague*, ed. Daniel Williman, 65–75, Papers of the Eleventh Annual Conference of the Center for Medieval and Early Renaissance Studies (Binghamton, NY: Center for Medieval and Early Renaissance Studies, 1982); Neustadt (Ayalon), "Plague and Its Effects upon the Mamlûk Army"; Borsch, *Black Death in*

Egypt and England; Justin K. Stearns, *Infectious Ideas: Contagion in Premodern Islamic and Christian Thought in the Western Mediterranean* (Baltimore: Johns Hopkins University Press, 2011). For works on earlier plague epidemics, see Lawrence I. Conrad, "The Plague in the Early Medieval Near East" (PhD diss., Princeton University, 1981); Lawrence I. Conrad, "The Biblical Tradition for the Plague of the Philistines," *Journal of the American Oriental Society* 104 (1984): 281–87; Dols, "Plague in Early Islamic History"; Josiah C. Russell, "That Earlier Plague," *Demography* 5 (1968): 174–84. For a critical reading of many of the primary sources for the history of plague, see Lawrence I. Conrad, "Arabic Plague Chronologies and Treatises: Social and Historical Factors in the Formation of a Literary Genre," *Studia Islamica* 54 (1981): 51–93. The bibliographical information provided in these major studies points to the enormity of the primary and secondary literature on plague in the Middle East.

6. Dols, "Second Plague," 164–65.

7. For a very useful summation of the recent literature on plague in the Ottoman Empire that employs some of these sources, see S. White, "Rethinking Disease in Ottoman History."

The records of Islamic courts, for example, have proved extremely useful for the study of the history of plague in the Middle East. One of the primary functions of the court was to administer the inheritance of the deceased. Thus, one regularly finds inventories of estates (*tarikāt*) in the records of local courts. By tracing vicissitudes in the number of such inventories before, during, and after known plague epidemics, one can gain an approximate idea of the scale of mortality for a given region. This is clearly not an exact science, however, as the majority of plague deaths leave no record in court registers. At present, we have no reliable way of knowing how to determine what percentage of an area's total deaths were recorded in a given court.

For references to some of the numerous manuscript sources available for the study of plague (many of which are copies or compilations of earlier sources), see the bibliographical information found in the following: Ibrāhīm, *al-Azamāt al-Ijtimāʿiyya fī Miṣr*, 316–20; Dols, *Black Death in the Middle East*, 320–39; Mohammed Melhaoui, *Peste, contagion et martyre: Histoire du fléau en Occident musulman médiéval* (Paris: Publisud, 2005), 20–57. For a compilation that includes translated excerpts of several medieval Arabic plague treatises, see John Aberth, *The Black Death: The Great Mortality of 1348–1350, a Brief History with Documents* (New York: Palgrave Macmillan, 2005).

8. For examples of studies of plague in Egypt during this period, see Ibrāhīm, *al-Azamāt al-Ijtimāʿiyya fī Miṣr*; Raymond, "Les Grandes Épidémies de peste au Caire," 203–10; Raymond, *Artisans et commerçants*; Max Meyerhof, "La peste en Égypte à la fin du XVIII siècle et le Mèdecin Enrico di Wolmar," *La Revue Médicale d'Égypte* 1 (1913): 1–13. Seventeenth- and eighteenth-century Arabic chronicles that contain information on various plague epidemics in Egypt during the period include—but are by no means limited to—the following: al-Khashshāb, *Akhbār Ahl al-Qarn al-Thānī ʿAshar*; al-Khashshāb, *Akhbār al-Amīr Murād*; al-Damurdāshī Katkhudā ʿAzabān, *Kitāb al-Durra al-Muṣāna*; Muṣṭafā ibn al-Ḥājj Ibrāhīm Tābiʿ al-Marḥūm

Ḥasan Aghā ʿAzabān al-Damardāshī, *Tārīkh Waqāʾiʿ Miṣr al-Qāhira al-Maḥrūsa Kinānat Allah fī Arḍihi*, ed. Ṣalāḥ Aḥmad Harīdī ʿAlī, 2nd ed. (Cairo: Dār al-Kutub wa al-Wathāʾiq al-Qawmiyya, 2002); al-ʿAwfī al-Ḥanbalī, *Tarājim al-Ṣawāʿiq*; al-Jabartī, *ʿAjāʾib al-Āthār* (1994); Muḥammad ibn Abī al-Surūr al-Bakrī, *al-Nuzha al-Zahiyya fī Dhikr Wulāt Miṣr wa al-Qāhira al-Muʿizziyya*, ed. ʿAbd al-Rāziq ʿAbd al-Rāziq ʿĪsā (Cairo: al-ʿArabī lil-Nashr wa al-Tawzīʿ, 1998); Aḥmad Shalabī ibn ʿAbd al-Ghanī, *Awḍaḥ al-Ishārāt fī man Tawallā Miṣr al-Qāhira min al-Wuzarāʾ wa al-Bāshāt*, ed. ʿAbd al-Raḥīm ʿAbd al-Raḥman ʿAbd al-Raḥīm (Cairo: Maktabat al-Khānjī, 1978).

9. The story of this earthquake is reported in ʿAbd al-Raḥman ibn Ḥasan al-Jabartī, *ʿAjāʾib al-Āthār fī al-Tarājim wa al-Akhbār*, ed. Ḥasan Muḥammad Jawhar, ʿAbd al-Fattāḥ al-Saranjāwī, ʿUmar al-Dasūqī, and al-Sayyid Ibrāhīm Sālim, 7 vols. (Cairo: Lajnat al-Bayān al-ʿArabī, 1958–67), 4:132.

10. For a discussion of earthquakes and their psychological impact on the population of Mamluk Egypt, see W. Tucker, "Natural Disasters and the Peasantry," 219–20, 222–23.

11. Al-Jabartī, *ʿAjāʾib al-Āthār* (1958–67), 4:132. The verse is: "*wa kam dhā bi-Miṣr min al-muḍḥikāt / wa lakinahu ḍiḥkun ka-al-bukkāʾ*."

12. Al-Khashshāb, *Akhbār Ahl al-Qarn al-Thānī ʿAshar*, 58.

13. Al-Khashshāb, *Akhbār al-Amīr Murād*, 40.

14. According to the American missionary John Antes, however, "It has been observed in Turkey, and particularly in Egypt, that persons of the age of seventy, and upwards, are not so much subject to the infection, and very old people not at all. The most vigorous and the strongest appear to be most subject to it." Antes, *Observations*, 47. Born in 1740 in Frederick Township, Pennsylvania, John Antes was the first American missionary in Egypt, and during his residence there from 13 January 1771 to 27 January 1782, he witnessed three plague epidemics. Since Antes was one of the few foreigners resident in Egypt at the end of the eighteenth century who wrote specifically on the subject of plague, his account is particularly useful as a supplement to the various Egyptian chronicles.

15. Al-Jabartī, *ʿAjāʾib al-Āthār* (1958–67), 4:132.

16. Ibid., 4:133; al-Khashshāb, *Akhbār al-Amīr Murād*, 40; al-Khashshāb, *Akhbār Ahl al-Qarn al-Thānī ʿAshar*, 58.

17. BOA, Cevdet Dahiliye, 1722 (Evasıt N 1205/15–24 May 1791).

18. *Aghā* was a title given to the head of each of the seven military blocs stationed in Egypt. Here the reference is most likely to the head of the Mustahfızan military bloc, who served as a kind of chief of police in Cairo. For more on the role of the *aghā*, see Shaw, *Nizâmnâme-i Mısır*, 10–11; Aḥmad, *al-Idāra fī Miṣr*, 176–77, 229–32.

19. On the position of the *wālī*, see Aḥmad, *al-Idāra fī Miṣr*, 233–35.

20. Al-Jabartī, *ʿAjāʾib al-Āthār* (1958–67), 4:133.

21. Al-Khashshāb, *Akhbār Ahl al-Qarn al-Thānī ʿAshar*, 58.

22. Al-Jabartī, *ʿAjāʾib al-Āthār* (1958–67), 4:138. Ṭurā is located on the east bank of the Nile south of Old Cairo in the province of Aṭfīḥ. For more on Ṭurā, see Ramzī, *al-Qāmūs al-Jughrāfī*, pt. 2, 3:15–16.

23. Al-Khashshāb, *Akhbār al-Amīr Murād*, 40.

24. Al-Jabartī, *ʿAjāʾib al-Āthār* (1958–67), 4:133.

25. Ibid.

26. Ibid., 4:140.

27. Ibid.

28. BOA, Cevdet Dahiliye, 1722 (Evasıt N 1205/15–24 May 1791).

29. Because these endemic foci of plague cover huge geographical areas of sparse human and vast rodent populations, the complete eradication of plague remains unlikely. On this point, see Dols, "Second Plague," 178; Conrad, "Plague in the Early Medieval Near East," 6–7.

30. Dols, "Plague in Early Islamic History," 381; Raymond, "Les Grandes Épidémies de peste au Caire," 208–9. For more general discussions of the relationships between the movements of peoples and goods and the spread of plague, see W. McNeill, *Plagues and Peoples*; Abu-Lughod, *Before European Hegemony*.

31. Antes, *Observations*, 39.

32. Raymond, "Les Grandes Épidémies de peste au Caire," 208–9; Dols, "Second Plague," 179–80. On the basis of the following, Dols compiles a list of plagues that came to Egypt and North Africa from Sudan and Central Africa. Sticker, *Abhandlungen aus der Seuchengeschichte und Seuchenlehre*. For more on plague in the Sudan, see Walz, *Trade between Egypt and Bilād as-Sūdān*, 200–201.

33. Raymond, "Les Grandes Épidémies de peste au Caire," 208–9.

34. On the epidemiological fact that Egypt did not house an endemic focus of plague, see Dols, "Second Plague," 183; Dols, *Black Death in the Middle East*, 35.

35. Kuhnke, *Lives at Risk*, 70; J. Worth Estes and LaVerne Kuhnke, "French Observations of Disease and Drug Use in Late Eighteenth-Century Cairo," *Journal of the History of Medicine and Allied Sciences* 39 (1984): 123. There were, of course, differing opinions on this point. John Antes writes, "I think Egypt cannot, with any truth, be called the mother of the plague." Antes, *Observations*, 41. See also ibid., 36–37. Ibn al-Wardī writes of plague beginning "in the land of darkness," which Dols (citing Alfred von Kremer) identifies as "northern Asia." Dols, "Ibn al-Wardī's *Risālah al-Nabaʾ*," 448. He then goes on to trace the disease's course from China and India, through Sind and the land of the Uzbeks, to Persia and the Crimea, and finally into Rūm, Egypt, Syria, and Palestine. Ibid., 448–53. The Mamluk chronicler al-Maqrīzī and others also place the origins of plague in a vague "East" or in parts of Mongolia. For more on these accounts, see Borsch, *Black Death in Egypt and England*, 4–5; Dols, *Black Death in the Middle East*, 35–42.

36. Al-Jabartī, *ʿAjāʾib al-Āthār* (1958–67), 4:322.

37. In Dols's words, "What we cannot judge accurately is the severity of the major plague epidemics from the late fifteenth to the late eighteenth centuries in the Middle East. . . . Therefore, we cannot propose, as we have done in the earlier period, a significant demographic effect of these epidemics." Dols, "Second Plague," 176–77.

38. For statistics on and discussions of the demographic effects of the Black Death in Egypt, see Dols, *Black Death in the Middle East*, 143–235; Dols,

"General Mortality of the Black Death"; Borsch, *Black Death in Egypt and England*, 40–54. Studies of the demographic effects of plague in nineteenth-century Egypt include Panzac, *La peste*, 231–78, 339–80; Kuhnke, *Lives at Risk*, 84–86. For a critical discussion of plague mortality statistics, see Conrad, "Plague in the Early Medieval Near East," 415–47.

39. By way of comparison to reported plague deaths reaching 1,000–2,000 a day, take, for example, the records of French doctors who tracked plague deaths during the months of the French occupation of Egypt from 1798 to 1801. They recorded plague deaths on the order of approximately 500–800 deaths per month, with the total number of deaths surpassing 1,000 for only four out of twenty-nine reported months. Panzac, *La peste*, 346.

40. For a discussion of the veracity of mortality figures as cited in Mamluk chronicler reports, see Neustadt (Ayalon), "Plague and Its Effects upon the Mamlûk Army," 68–71.

41. Raymond, "Les Grandes Épidémies de peste au Caire," 209–10.

42. Panzac, *La peste*, 361. André Raymond writes that the population of Cairo when the French expedition arrived in 1798 was 260,000. Raymond, "La population du Caire." For comparative plague mortality figures from Milan, Aleppo, Izmir, Marseille, and other cities during the seventeenth, eighteenth, and nineteenth centuries, see Panzac, *La peste*, 353–62; Panzac, *Quarantaines et lazarets*, 12.

43. Al-Jabartī, *ʿAjāʾib al-Āthār* (1958–67), 4:129–30.

44. Ibid., 4:129.

45. Ibid., 4:130. For more on the gathering points of pilgrims in Cairo, see Antes, *Observations*, 69.

46. A useful comparison to the rains of 1790 is Antes's description of the rains and flooding in Cairo in 1771. "It once happened, during my abode, in November 1771, that heavy showers of rain, accompanied with some thunder and lightning, followed one another for five successive nights, though it did not rain in the day time. . . . Some houses fell down on that occasion, and several lives were lost." Antes, *Observations*, 99.

47. Al-Jabartī, *ʿAjāʾib al-Āthār* (1958–67), 4:130.

48. For a discussion of precedents to this sort of damage from rain in Mamluk Egypt, see W. Tucker, "Natural Disasters and the Peasantry," 216–17.

49. On this point Antes writes, "Sometimes the river rises so rapidly, and to such a height, that all their [peasants'] endeavours are in vain, and all such vegetables are destroyed." Antes, *Observations*, 72.

50. For instance, in 638 or 639, plague struck Syria, killing at least twenty-five thousand soldiers and countless others. Important for our purposes here is the observation that this instance of plague was preceded by a severe famine that likely, as with the 1791 plague in Egypt, weakened the population, making them all the more vulnerable to infection. Dols, "Plague in Early Islamic History," 376. See also W. Tucker, "Natural Disasters and the Peasantry," 217–19.

51. For more on the physical attributes and abilities of rats, see the following classic study of typhus: Hans Zinsser, *Rats, Lice and History, Being a Study in Biography, Which, after Twelve Preliminary Chapters Indispensable for the*

Preparation of the Lay Reader, Deals with the Life History of Typhus Fever (London: George Routledge and Sons, 1935), 197–204.

52. Conrad, "Plague in the Early Medieval Near East," 35.

53. Antes, *Observations*, 69.

54. Dols makes a similar point about the proximity of humans and rats in the plague epidemic of 638 or 639 in Syria. Dols, "Plague in Early Islamic History," 376.

55. For more on the epidemiology, pathology, and etiology of plague, see Conrad, "Plague in the Early Medieval Near East," 4–38; Dols, *Black Death in the Middle East*, 68–83; Borsch, *Black Death in Egypt and England*, 2–8. For the most recent works in this regard, see Hugo Kupferschmidt, *Die Epidemiologie der Pest: Der Konzeptwandel in der Erforschung der Infektionsketten seit der Entdeckung des Pesterregers im Jahre 1894* (Aarau, Switzerland: Sauerländer, 1993); Graham Twigg, *The Black Death: A Biological Reappraisal* (London: Batsford, 1984). The standard works are Robert Pollitzer, *Plague* (Geneva, Switzerland: World Health Organization, 1954); L. Fabian Hirst, *The Conquest of Plague: A Study of the Evolution of Epidemiology* (Oxford: Clarendon Press, 1953).

56. The disease exists in the blood of infected rodents, and when a rodent is bitten by a flea, the flea ingests blood infected with plague bacilli. When the rat population begins to die off from plague, fleas seek out new hosts. Very commonly, especially in situations of close proximity like the ones described here, these hosts are human.

57. When blood infected with plague enters the human bloodstream, the bacilli rapidly multiply. Lymph glands filter plague bacilli from the blood, and these bacilli accumulate in the glands, where they continue to multiply, thus producing the characteristic buboes on the neck or groin associated with bubonic plague. The highly contagious pneumonic version of plague occurs when bacilli settle in the lungs rather than the lymph glands. The septicemic variety of plague, in which bacilli attack primarily the victim's blood, is the most virulent and quickest to kill, with victims often dying within hours of the onset of the disease. On septicemic plague, see Twigg, *Biological Reappraisal*, 19; Hirst, *Conquest of Plague*, 29; Pollitzer, *Plague*, 439–40. Other varieties of plague include the tonsillar and the vesicular. For a discussion of these and other types of plague, see Dols, *Black Death in the Middle East*, 73–74; Borsch, *Black Death in Egypt and England*, 4; Hirst, *Conquest of Plague*, 30.

58. Al-Jabartī, *ʿAjāʾib al-Āthār* (1958–67), 4:133.

59. Ibid., 4:140.

60. Antes, *Observations*, 42.

61. Ibid., 47.

62. Ibid.

63. Ibid., 42. Arab physicians and observers of plague used many terms to describe both individual and collective plague buboes. These included "the cucumber," "the almond," "the pustule," "grains," and "blistering." For more on the terminology used for plague buboes, see Dols, *Black Death in the Middle East*, 75, 77–79, 316–19.

64. Dols, "Second Plague," 176–77, 182–89; Kuhnke, *Lives at Risk*, 72.

Dols goes so far as to suggest that pneumonic plague disappeared from most parts of the Middle East in the second half of the fifteenth century. Dols, "Second Plague," 182.

65. LaVerne Kuhnke explains this singular anomaly of pneumonic plague in Asyūṭ as a product of late nineteenth-century Egyptian and British efforts to expand the irrigation network in the south of Egypt, which, unwittingly, created a permanent population of rats and fleas. Kuhnke, *Lives at Risk*, 72–73, 200–201n16. As I discuss here, though, evidence suggests the presence of pneumonic plague in Asyūṭ long before the end of the nineteenth century.

66. Al-Jabartī, *ʿAjāʾib al-Āthār* (1958–67), 5:241–42.

67. Ibid., 5:241. Mention of this plague is also made in Panzac, *La peste*, 284.

68. Dols, *Black Death in the Middle East*, 60n92.

69. Kuhnke herself concedes as much in her discussion of plague in Asyūṭ. Kuhnke, *Lives at Risk*, 73.

70. Al-Jabartī, *ʿAjāʾib al-Āthār* (1958–67), 4:141–42.

71. On this waterway, Antes writes, "The remaining water is horribly corrupted, by the filth thrown in from the adjoining houses, and the great number of necessaries that empty themselves into it, which occasions a most abominable stench for several months of the year, tarnishing in a short time even gold and silver in the houses near it." Antes, *Observations*, 38.

72. For a description of the function and importance of these festivities in Fatimid Cairo, see Paula Sanders, *Ritual, Politics, and the City in Fatimid Cairo* (Albany: State University of New York Press, 1994), 99–119.

73. For a discussion of the relationship between famine and plague in Mamluk Egypt, see W. Tucker, "Natural Disasters and the Peasantry," 217–19. Instructive also on this point is Elisabeth Carpentier, "Autour de la Peste Noire: Famines et épidémies dans l'histoire du XIVe siècle," *Annales* 17 (1962): 1062–92.

74. Al-Jabartī, *ʿAjāʾib al-Āthār* (1958–67), 4:197.

75. Ibid., 4:141–42.

76. Ibid., 4:199.

77. Ibid.

78. On the absence in Arabic plague treatises of any association between the pathology of plague and rodent populations, see Dols, "al-Manbijī's 'Report of the Plague,'" 71.

79. For more on the presence of rodents in the Egyptian countryside, see Antes, *Observations*, 85–86.

80. Suggestive of a similar concept is Ira M. Lapidus's use of the phrase an "economic geography of Egypt." Ira M. Lapidus, "The Grain Economy of Mamluk Egypt," *Journal of the Economic and Social History of the Orient* 12 (1969): 13.

81. For another example of the relationship between food shortages and price increases during plague epidemics, see Dols, "al-Manbijī's 'Report of the Plague,'" 71.

82. This account is related in al-Khashshāb, *Akhbār al-Amīr Murād*, 33–34.

83. Dols, *Black Death in the Middle East*, 154–69. On the movement of civilian populations during plague epidemics in the Mamluk period, see Neustadt (Ayalon), "Plague and Its Effects upon the Mamlûk Army," 72.

84. Lapidus, "Grain Economy," 8n2. Cited also in Borsch, *Black Death in Egypt and England*, 50. On the procurement, politics, and economy of grain in Mamluk Egypt, see also Boaz Shoshan, "Grain Riots and the 'Moral Economy': Cairo, 1350–1517," *Journal of Interdisciplinary History* 10 (1980): 459–78.

85. Dols, *Black Death in the Middle East*, 163; W. Tucker, "Natural Disasters and the Peasantry," 222–24; Borsch, *Black Death in Egypt and England*, 49–50.

86. Al-Jabartī, *ʿAjāʾib al-Āthār* (1958–67), 4:188.

87. Ibid., 4:192.

88. Dols, "Ibn al-Wardī's *Risālah al-Nabaʾ*," 450–51.

89. Ibid., 454.

90. Al-Jabartī, *ʿAjāʾib al-Āthār* (1958–67), 5:242.

91. Ibrāhīm, *al-Azamāt al-Ijtimāʿiyya fī Miṣr*, 72–75; Dols, "Second Plague," 181.

92. Antes, *Observations*, 94. On wind in Egypt more generally, see ibid., 93–99.

93. Al-Damurdāshī Katkhudā ʿAzabān, *Kitāb al-Durra al-Muṣāna*, 29. Aḥmad al-Damurdāshī Katkhudā ʿAzabān, moreover, goes on to write that during this plague, one would wake to find ten new victims every morning. As in other plague outbreaks, there were also shortages of corpse-washers, and given the great number of dead bodies, gravediggers were forced to work long into the night.

94. For a general discussion of the periodicity, timing, and seasonal incidence of plague in the Middle East, see Panzac, *La peste*, 195–227; Conrad, "Plague in the Early Medieval Near East," 323–27; Kuhnke, *Lives at Risk*, 72–78. For a more general treatment of these subjects, see Hirst, *Conquest of Plague*, 254–82.

95. Panzac, *La peste*, 223.

96. Kuhnke, *Lives at Risk*, 201n18.

97. Antes, *Observations*, 39.

98. Conrad, "Plague in the Early Medieval Near East," 326.

99. Panzac, *La peste*, 225; Dols, "Second Plague," 181.

100. Antes writes the following about the climate of Egypt: "There is scarcely a country on the globe, where the climate is so very regular as it is in Egypt. . . . The difference between the greatest degree of cold and the greatest, or, more properly, the most usual heat in summer, does not exceed thirty degrees, according to Fahrenheit's thermometer." Antes, *Observations*, 89, 91.

101. Ibid., 44.

102. Antes uses this evidence about the effects of heat on plague to argue against the notion suggested by some that plague was a putrid fever, since heat was thought to increase, not diminish, the severity of putrid fevers. Ibid., 44–45.

103. Ibid., 43, 67.

104. Kuhnke identifies 26 June as the date of the festival of Saint John and of "the death of the plague" in Egypt. For more on the celebrations of this day by Egyptians and Europeans, see Kuhnke, *Lives at Risk*, 73.

CHAPTER 10

1. The fissure's name is Lakagígar, or the Craters of Laki. The volcanic eruption consisted of a series of craters along the fissure exploding together. For the sake of ease, I will refer to the fissure as just "Laki" or simply "the volcano," but readers should keep in mind that this was a series of volcanoes, not just one.

2. Using a particular climate simulation model, Oman, Robock, Stenchikov, and Thordarson do make the scientific link between Laki's eruptions and Nile floods, but they do not use the Arabic and Ottoman historical materials to analyze the effects of the volcano on Egypt. Oman, Robock, Stenchikov, and Thordarson, "High-Latitude Eruptions Cast Shadow."

3. The history of climate change is obviously a vast and rapidly expanding field. Notable works include Parker, *Global Crisis*; Le Roy Ladurie, *Times of Feast, Times of Famine*; William F. Ruddiman, *Plows, Plagues, and Petroleum: How Humans Took Control of Climate* (Princeton, NJ: Princeton University Press, 2005); John L. Brooke, *Climate Change and the Course of Global History: A Rough Journey* (Cambridge: Cambridge University Press, 2014); Fagan, *Long Summer*; Linden, *Winds of Change*. For recent discussions of the state of the art, see the following sets of essays: Mark Carey, Philip Garone, Adrian Howkins, Georgina H. Endfield, Lawrence Culver, Sam White, Sherry Johnson, and James Rodger Fleming, "Forum: Climate Change and Environmental History," *Environmental History* (2014): 281–364; Morgan Kelly, Cormac Ó Gráda, Sam White, Ulf Büntgen, Lena Hellmann, and Jan de Vries, "The Little Ice Age: Climate and History Reconsidered," *Journal of Interdisciplinary History* 44 (2014): 301–77. For studies of climate change in the Middle East, see S. White, *Climate of Rebellion*; Bulliet, *Cotton, Climate, and Camels*; Murphey, "Decline of North Africa"; Ellenblum, *Collapse of the Eastern Mediterranean*; Griswold, "Climatic Change: A Possible Factor in the Social Unrest of Seventeenth Century Anatolia"; Issar and Zohar, *Climate Change*.

4. In this regard, see Paolo Squatriti, "The Floods of 589 and Climate Change at the Beginning of the Middle Ages: An Italian Microhistory," *Speculum* 85 (2010): 799–826.

5. For the climatological data, see Oman, Robock, Stenchikov, and Thordarson, "High-Latitude Eruptions Cast Shadow."

6. The latest studies of the global history of climate change include Parker's *Global Crisis* on the Little Ice Age and Brooke's *Climate Change and the Course of Global History*.

7. The closest work I have found is the following about a raid on Iceland by Barbary pirates in 1627: Bernard Lewis, "Corsairs in Iceland," *Revue de l'Occident musulman et de la Méditerranée* 15 (1973): 139–44.

8. Karen Oslund, *Iceland Imagined: Nature, Culture, and Storytelling in the North Atlantic* (Seattle: University of Washington Press, 2011), 35–36.

9. Jón Steingrímsson, *Fires of the Earth: The Laki Eruption, 1783–1784*,

trans. Keneva Kunz (Reykjavík: University of Iceland Press and Nordic Volcanological Institute, 1998), 25.

10. The largest lava flow of the last ten thousand years was also from an Icelandic volcano—Katla in 943. Guðrún Larsen, "Katla: Tephrochronology and Eruption History," *Developments in Quaternary Sciences* 13 (2010): 23–49. See also Steingrímsson, *Fires of the Earth*, 5; Thorvaldur Thordarson and Stephen Self, "Atmospheric and Environmental Effects of the 1783–1784 Laki Eruption: A Review and Reassessment," *Journal of Geophysical Research* 108 (2003): 3.

11. On Iceland's population reductions, see Thordarson and Self, "Atmospheric and Environmental Effects of Laki," 13.

12. For an intellectual and cultural history of volcanism, see Haraldur Sigurdsson, *Melting the Earth: The History of Ideas on Volcanic Eruptions* (New York: Oxford University Press, 1999).

13. For studies of Icelandic geology and glaciology, see Judith K. Maizels and Chris Caseldine, eds., *Environmental Change in Iceland: Past and Present* (Dordrecht, Netherlands: Kluwer Academic Publishers, 1991); C. Caseldine, A. Russell, J. Harðardóttir, and Ó. Knudsen, eds., *Iceland—Modern Processes and Past Environments* (Amsterdam: Elsevier, 2005); Thor Thordarson and Armann Hoskuldsson, *Iceland* (Harpenden, UK: Terra, 2002).

14. For comparative perspective, see Sigurdsson, *Melting the Earth*; Larsen, "Katla"; Thordarson and Self, "Atmospheric and Environmental Effects of Laki," 3; Oslund, *Iceland Imagined*, 34.

15. Steingrímsson, *Fires of the Earth*, 15; Oslund, *Iceland Imagined*, 44.

16. Thordarson and Self, "Atmospheric and Environmental Effects of Laki," 3.

17. Ibid., 5.

18. Ibid., 3.

19. Steingrímsson, *Fires of the Earth*, 35.

20. Ibid., 36.

21. Ibid., 45.

22. Ibid., 47.

23. Ibid., 58.

24. Ibid., 59, 63.

25. Ibid., 65.

26. Ibid., 65–70.

27. Thordarson and Self, "Atmospheric and Environmental Effects of Laki," 3.

28. On various interpretations of Laki and its eruptions by Icelanders, foreign scientists, nationalists, the Danish state, naturalists, travelers, and others, see Oslund, *Iceland Imagined*, 34–60.

29. Charles A. Wood, "Climatic Effects of the 1783 Laki Eruption," in *The Year without a Summer? World Climate in 1816*, ed. C.R. Harington (Ottawa: Canadian Museum of Nature, 1992), 60.

30. Steingrímsson, *Fires of the Earth*, 68.

31. Ibid., 70.

32. E.L. Jackson, "The Laki Eruption of 1783: Impacts on Population and Settlement in Iceland," *Geography* 67 (1982): 44.

33. Steingrímsson, *Fires of the Earth*, 76.

34. Ibid., 77–78.

35. Ibid., 84.

36. Thordarson and Self, "Atmospheric and Environmental Effects of Laki," 13.

37. Oslund, *Iceland Imagined*, 36.

38. Jackson, "Impacts on Population and Settlement in Iceland," 47. Oslund puts the number of dead at "more than 10,000." Oslund, *Iceland Imagined*, 36. On the connections between volcanic dry fogs and disease, see Richard B. Stothers, "Volcanic Dry Fogs, Climate Cooling, and Plague Pandemics in Europe and the Middle East," *Climate Change* 42 (1999): 713–23; John Grattan, Roland Rabartin, Stephen Self, and Thorvaldur Thordarson, "Volcanic Air Pollution and Mortality in France, 1783–1784," *C. R. Geoscience* 337 (2005): 641–51.

39. D.S. Stevenson, C.E. Johnson, E.J. Highwood, V. Gauci, W.J. Collins, and R.G. Derwent, "Atmospheric Impact of the 1783–1784 Laki Eruption: Part I, Chemistry Modelling," *Atmospheric Chemistry and Physics Discussions* 3 (2003): 551–96.

40. Thordarson and Self, "Atmospheric and Environmental Effects of Laki," 1.

41. Ibid., 6–7. For supporting scientific evidence, see Luke Oman, Alan Robock, Georgiy L. Stenchikov, Thorvaldur Thordarson, Dorothy Koch, Drew T. Shindell, and Chaochao Gao, "Modeling the Distribution of the Volcanic Aerosol Cloud from the 1783–1784 Laki Eruption," *Journal of Geophysical Research* 111 (2006): D12209. For a discussion of Franklin's sighting of the haze, see Wood, "Climatic Effects of the 1783 Laki Eruption."

42. Thordarson and Self, "Atmospheric and Environmental Effects of Laki," 8.

43. Ibid., 20–22.

44. Generally on Italian volcanoes and dry fogs, see D. Camuffo and S. Enzi, "Chronology of 'Dry Fogs' in Italy, 1374–1891," *Theoretical and Applied Climatology* 50 (1994): 31–33.

45. Thordarson and Self, "Atmospheric and Environmental Effects of Laki," 8; Gordon C. Jacoby, Karen W. Workman, and Rosanne D. D'Arrigo, "Laki Eruption of 1783, Tree Rings, and Disaster for Northwest Alaska Inuit," *Quaternary Science Reviews* 18 (1999): 1365–71; Rosanne D. D'Arrigo and Gordon C. Jacoby, "Northern North American Tree-Ring Evidence for Regional Temperature Changes after Major Volcanic Events," *Climate Change* 41 (1999): 1–15.

46. Thordarson and Self, "Atmospheric and Environmental Effects of Laki," 14.

47. Generally on the historical relationship between volcanoes and climate, see Atwell, "Volcanism and Short-Term Climatic Change"; Michael McCormick, Paul Edward Dutton, and Paul A. Mayewski, "Volcanoes and the Climate Forcing of Carolingian Europe, A.D. 750–950," *Speculum* 82

(2007): 865–95; Alan Robock, "Volcanic Eruptions and Climate," *Reviews of Geophysics* 38 (2000): 191–219; Drew T. Shindell and Gavin A. Schmidt, "Dynamic Winter Climate Response to Large Tropical Volcanic Eruptions since 1600," *Journal of Geophysical Research* 109 (2004): D05104; Luke Oman, Alan Robock, Georgiy Stenchikov, Gavin A. Schmidt, and Reto Ruedy, "Climatic Response to High-Latitude Volcanic Eruptions," *Journal of Geophysical Research* 110 (2005): D13103; H.H. Lamb, "Volcanic Dust in the Atmosphere; with a Chronology and Assessment of Its Meteorological Significance," *Philosophical Transactions of the Royal Society of London. Series A, Mathematical and Physical Sciences* 266 (1970): 425–533. For discussion of various global volcanic eruptions' climate effects in the Ottoman Empire, see S. White, *Climate of Rebellion*, 133–37, 142, 181–82, 212, 220.

48. Thordarson and Self, "Atmospheric and Environmental Effects of Laki," 1.

49. On the specific climatic effects of high-latitude eruptions, see Oman, Robock, Stenchikov, Schmidt, and Ruedy, "Climatic Response to High-Latitude Volcanic Eruptions."

50. Thordarson and Self, "Atmospheric and Environmental Effects of Laki," 6; E.J. Highwood and D.S. Stevenson, "Atmospheric Impact of the 1783–1784 Laki Eruption: Part II, Climatic Effect of Sulphate Aerosol," *Atmospheric Chemistry and Physics* 3 (2003): 1177–89.

51. July 1783 was exceptionally warm in southwest, west, and northwest Europe, for example. Thordarson and Self, "Atmospheric and Environmental Effects of Laki," 15.

52. Ibid., 19.

53. Ibid., 16. A minority opinion holds that this winter's extremely low temperatures were a function not of Laki but of negative North Atlantic oscillation (NAO) and El Niño southern oscillation (ENSO) events. Rosanne D'Arrigo, Richard Seager, Jason E. Smerdon, Allegra N. LeGrande, and Edward R. Cook, "The Anomalous Winter of 1783–1784: Was the Laki Eruption or an Analog of the 2009–2010 Winter to Blame?," *Geophysical Research Letters* 38 (2011): L05706. On the connections between ENSO and the Nile, see Luc Ortlieb, "Historical Chronology of ENSO and the Nile Flood Record," in *Past Climate Variability through Europe and Africa*, ed. Richard W. Battarbee, Françoise Gasse, and Catherine E. Stickley, 257–78 (Dordrecht, Netherlands: Springer, 2004).

54. Thordarson and Self, "Atmospheric and Environmental Effects of Laki," 16.

55. Ibid., 15.

56. Ibid., 16.

57. Ibid., 15.

58. The Japanese case in the summer of 1783 is slightly more complicated given the eruption of the Asama volcano in central Japan that same summer. For a description of this explosion, see Maya Yasui and Takehiro Koyaguchi, "Sequence and Eruptive Style of the 1783 Eruption of Asama Volcano, Central Japan: A Case Study of an Andesitic Explosive Eruption Generating Fountain-

Fed Lava Flow, Pumice Fall, Scoria Flow and Forming a Cone," *Bulletin of Volcanology* 66 (2004): 243–62. For an argument that Laki's climate effects were much more significant than Asama's, see G.A. Zielinski, R.J. Fiacco, P.A. Mayewski, L.D. Meeker, S. Whitlow, M.S. Twickler, M.S. Germani, K. Endo, and M. Yasui, "Climatic Impact of the A.D. 1783 Asama (Japan) Eruption Was Minimal: Evidence from the GISP2 Ice Core," *Geophysical Research Letters* 21 (1994): 2365–68.

59. Thordarson and Self, "Atmospheric and Environmental Effects of Laki," 22.

60. Jacoby, Workman, and D'Arrigo, "Laki, Tree Rings, and Disaster for Northwest Alaska Inuit."

61. Thordarson and Self, "Atmospheric and Environmental Effects of Laki," 19.

62. Ibid.

63. Oman, Robock, Stenchikov, Schmidt, and Ruedy, "Climatic Response to High-Latitude Volcanic Eruptions," 1.

64. On Indian Ocean monsoons and the Nile, see Pierre Camberlin, "Rainfall Anomalies in the Source Region of the Nile and Their Connection with the Indian Summer Monsoon," *Journal of Climate* 10 (1997): 1380–92; Marie Revel, E. Ducassou, F.E. Grousset, S.M. Bernasconi, S. Migeon, S. Revillon, J. Mascle, A. Murat, S. Zaragosi, and D. Bosch, "100,000 Years of African Monsoon Variability Recorded in Sediments of the Nile Margin," *Quaternary Science Reviews* 29 (2010): 1342–62; Barbara Bell, "The Oldest Records of the Nile Floods," *Geographical Journal* 136 (1970): 569–73.

65. Oman, Robock, Stenchikov, and Thordarson, "High-Latitude Eruptions Cast Shadow."

66. Oman, Robock, Stenchikov, Schmidt, and Ruedy, "Climatic Response to High-Latitude Volcanic Eruptions," 3.

67. H.G. Lyons, "On the Nile Flood and Its Variation," *Geographical Journal* 26 (1905): 406.

68. For a general discussion of this period, see Crecelius, "Egypt in the Eighteenth Century," 82–86.

69. For a discussion of this empire-wide phenomenon, see Yaycıoğlu, "Provincial Power-Holders and the Empire"; Engin D. Akarli, "Provincial Power Magnates in Ottoman Bilad al-Sham and Egypt, 1740–1840," in *La vie sociale dans les Provinces Arabes à l'époque Ottoman*, vol. 3, ed. Abdeljelil Temimi, 41–56 (Zaghwān, Tunisia: Markaz al-Dirāsāt wa al-Buḥūth al-ʿUthmāniyya wa al-Mūrīskiyya wa al-Tawthīq wa al-Maʿlūmāt, 1988).

70. Crecelius, *Roots of Modern Egypt.*

71. Ibid., 79–91, 159–68.

72. For discussions of some of these late eighteenth-century imperial processes and stresses, see Aksan, *Ottoman Statesman*, 100–205; Aksan, *Ottoman Wars*; Tezcan, *Second Ottoman Empire*; Kasaba, *Ottoman Empire and the World Economy.*

73. On plague in 1784, see BOA, HAT, 29/1361 (13 Ş 1198/1 July 1784); BOA, HAT, 28/1354 (7 Za 1198/22 Sept. 1784). For the plague of 1785, see

al-Khashshāb, *Akhbār al-Amīr Murād*, 24–25; al-Jabartī, *ʿAjāʾib al-Āthār* (1994), 2:157. On plague in 1787, see ibid., 2:228–29, 2:232, 2:241. In 1788, see ibid., 2:260, 2:263, 2:275, 2:280, 2:282. In 1791, see BOA, Cevdet Dahiliye, 1722 (Evasıt N 1205/15–24 May 1791); al-Jabartī, *ʿAjāʾib al-Āthār* (1994), 2:315; al-Khashshāb, *Akhbār al-Amīr Murād*, 33–34; al-Khashshāb, *Akhbār Ahl al-Qarn al-Thānī ʿAshar*, 58. On the plague of 1792, see BOA, HAT, 209/11213 (29 Z 1206/18 Aug. 1792). There is no internal evidence for the date of this case. The date given is the one assigned by the BOA. For examples of British correspondence about plague in Egypt in 1791 and 1792, see the following reports sent by the British consul in Egypt, George Baldwin, to London: TNA, FO, 24/1, 183r–185v (4 July 1791), 191r–196v (12 July 1791), 211r–212v (21 June 1792). On plague in 1799, see BOA, HAT, 245/13801A (3 Za 1213/9 Apr. 1799).

74. Lyons, "On the Nile Flood," 406, 412.

75. Al-Jabartī, *ʿAjāʾib al-Āthār* (1994), 2:123. On famine in 1783, see also Constantin-François Volney, *Travels through Egypt and Syria, in the Years 1783, 1784, and 1785. Containing the Present Natural and Political State of Those Countries, Their Productions, Arts, Manufactures, and Commerce; with Observations on the Manners, Customs, and Government of the Turks and Arabs*, trans. from the French, 2 vols. (New York: Evert Duyckinck, 1798), 1:101.

76. BOA, HAT, 28/1354 (7 Za 1198/22 Sept. 1784).

77. Al-Jabartī, *ʿAjāʾib al-Āthār* (1994), 2:138.

78. Shaw, *Financial and Administrative Organization*, 355–57.

79. Al-Jabartī, *ʿAjāʾib al-Āthār* (1994), 2:138.

80. Ibid., 2:139–40.

81. Volney, *Travels through Egypt and Syria*, 1:122.

82. Al-Jabartī, *ʿAjāʾib al-Āthār* (1994), 2:155.

83. Volney, *Travels through Egypt and Syria*, 1:123.

84. Ibid., 1:122.

85. Ibid., 1:121.

86. BOA, HAT, 29/1361 (13 Ş 1198/1 July 1784); BOA, HAT, 28/1354 (7 Za 1198/22 Sept. 1784).

87. Al-Khashshāb, *Akhbār al-Amīr Murād*, 24–25; al-Jabartī, *ʿAjāʾib al-Āthār* (1994), 2:157.

88. Volney, *Travels through Egypt and Syria*, 1:122.

89. For a detailed analysis of an earlier period of political crisis in the Ottoman Empire induced in part by climate change, see S. White, *Climate of Rebellion*.

90. Al-Jabartī, *ʿAjāʾib al-Āthār* (1994), 2:133.

91. On these various predations, see ibid., 2:123, 2:133–34, 2:140, 2:173, 2:175, 2:178, 2:189; al-Khashshāb, *Akhbār al-Amīr Murād*, 25.

92. Al-Jabartī, *ʿAjāʾib al-Āthār* (1994), 2:138–39.

93. Crecelius, "Egypt in the Eighteenth Century," 82–86.

94. There was a parallel political phenomenon in Iceland. In the same way that Egyptian local elites took the environmental stresses brought on by Laki as an occasion to act against the Ottoman Empire, some Icelanders pointed

to the catastrophe of the volcano as an example of the failures of the Danish kingdom to put in place policies that would allow the island to overcome such crises. Both the Egyptian and Icelandic arguments were thus that imperial states had failed to help mitigate local environmental, social, and economic crises and that these larger political powers were therefore inadequate. As with anti-Ottoman rhetoric in Egypt, critiques of Danish power would later inform Icelandic nationalist sentiment in the first half of the nineteenth century. Oslund, *Iceland Imagined*, 45–48.

95. This report was published and translated as Shaw, *Nizâmnâme-i Mısır*.

96. On the Russo-Ottoman wars of the second half of the eighteenth century, see Aksan, *Ottoman Statesman*, 100–205; Aksan, *Ottoman Wars*.

97. For discussions of this historiography, see Gran, *Islamic Roots of Capitalism*; Fahmy, *All the Pasha's Men*, 1–37; Fahmy, *Mehmed Ali*, 112–27; Toledano, "Mehmet Ali Paşa or Muhammad Ali Basha?"; Mikhail, "Unleashing the Beast," 319–21.

CONCLUSION

1. "Economy," as William Cronon usefully reminds us, and to which we might also attach politics, society, and imperial sovereignty, is "a subset of ecology." Cronon, *Changes in the Land*, xv–xvi.

2. Of course, not everything in Egypt or elsewhere in the empire was an outcome of Ottoman rule. As is true in all polities, whether ancient or modern, imperial or not, sources from the Ottoman Empire show that multiple histories of Egypt from the period of Ottoman rule bear no recognizable trace of the empire or any of its laws or institutions. For example, whole portions of al-Shirbīnī's seventeenth-century satirical chronicle of the Egyptian countryside *Hazz al-Quḥūf* seemingly have nothing to do with anything one would identify as connected to the Ottoman political presence in Egypt at the time.

3. See, for example, Beshara Doumani, *Rediscovering Palestine: Merchants and Peasants in Jabal Nablus, 1700–1900* (Berkeley: University of California Press, 1995); Amy Singer, *Palestinian Peasants and Ottoman Officials: Rural Administration around Sixteenth-Century Jerusalem* (Cambridge: Cambridge University Press, 1994); Peirce, *Morality Tales*; Dror Ze'evi, *An Ottoman Century: The District of Jerusalem in the 1600s* (Albany: State University of New York Press, 1996).

4. Timothy Mitchell, "Are Environmental Imaginaries Culturally Constructed?," in *Environmental Imaginaries of the Middle East and North Africa*, ed. Diana K. Davis and Edmund Burke III (Athens: Ohio University Press, 2011), 266.

Bibliography

ARCHIVAL SOURCES

Başbakanlık Osmanlı Arşivi, Istanbul

Cevdet Bahriye

208, 1413, 5701

Cevdet Dahiliye

1722

Cevdet Nafia

120, 302, 458, 644, 2570

Hatt-ı Hümayun

26/1256, 28/1354, 29/1358, 29/1361, 86/3520, 88/3601, 130/5404, 131/5411, 209/11213, 240/13451, 245/13801A, 342/19546, 593/29055, 656/32064, 795/36893, 1399/56283, 1412/57500

İbnülemin Umur-i Nafia

94

Mühimme-i Mısır

1, 3, 4, 5, 6, 8, 9

Dār al-Wathāʾiq al-Qawmiyya, Cairo

al-Jusūr al-Sulṭāniyya

784, 785, 786, 787, 788

Maḍābiṭ al-Daqahliyya

19, 20, 34

Maḥkamat Asyūṭ

1, 2, 4, 5, 8

Maḥkamat al-Baḥayra

5, 7, 8, 10, 11, 14, 16, 21, 22, 23, 24, 25, 26, 37, 38

Maḥkamat Isnā

3, 6, 8

Maḥkamat Manfalūṭ

1, 2, 3

Maḥkamat al-Manṣūra

1, 2, 3, 4, 7, 9, 12, 14, 15, 16, 17, 18, 19, 22, 24, 51

Maḥkamat Rashīd

122, 123, 124, 125, 130, 132, 134, 139, 142, 144, 145, 146, 148, 151, 154, 157

al-Rūznāma, Daftar Irtifāʿ al-Miyāh bi-Baḥr Sayyidnā Yūsuf lihi al-Ṣalāh wa al-Salām ʿan al-Qabḍa al-Yūsufiyya Tābiʿ Wilāyat al-Fayyūm

Raqam al-Ḥifẓ al-Nauʿī 1, ʿAyn 59, Makhzin Turkī 1, Musalsal 4557

The National Archives of the United Kingdom, Kew

Foreign Office

24/1

Privy Council

1/19/24

Topkapı Sarayı Müzesi Arşivi, Istanbul

Evrak

664/4, 664/40, 664/51, 664/52, 664/55, 664/63, 664/64, 664/66, 2229/3, 2444/107, 3218, 3522, 4675/2, 5207/49, 5207/57, 5207/58, 5207/62, 5225/12, 5657, 7008/12, 7016/95

PUBLISHED PRIMARY SOURCES

ʿAbd al-Raḥīm, ʿAbd al-Raḥīm ʿAbd al-Raḥman. *Wathāʾiq al-Maḥākim al-Sharʿiyya al-Miṣriyya ʿan al-Jāliya al-Maghāribiyya ibbāna al-ʿAṣr al-ʿUthmānī*. Edited and introduced by ʿAbd al-Jalīl al-Tamīmī. Zaghwān, Tunisia: Markaz al-Dirāsāt wa al-Buḥūth al-ʿUthmāniyya wa al-Mūrīskiyya wa al-Tawthīq wa al-Maʿlūmāt, 1992.

Antes, John. *Observations on the Manners and Customs of the Egyptians, the Overflowing of the Nile and Its Effects; with Remarks on the Plague and Other Subjects. Written during a Residence of Twelve Years in Cairo and Its Vicinity*. London: printed for J. Stockdale, 1800.

al-ʿAwfī al-Ḥanbalī, Ibrāhīm ibn Abī Bakr al-Ṣawāliḥī. *Tarājim al-Ṣawāʿiq fī Wāqiʿat al-Ṣanājiq*. Edited by ʿAbd al-Raḥīm ʿAbd al-Raḥman ʿAbd al-Raḥīm. Cairo: Institut français d'archéologie orientale, 1986.

Barkan, Ömer Lûtfi. *Kanunlar*. Vol. 1 of *XV ve XVIinci asırlarda Osmanlı İmparatorluğunda Ziraî Ekonominin Hukukî ve Malî Esasları*. İstanbul Üniversitesi Yayınlarından 256. Istanbul: Bürhaneddin Matbaası, 1943.

Commission des sciences et arts d'Egypte. *Description de l'Égypte, ou, recueil de observations et des recherches qui ont été faites en Égypte pendant l'éxpédition de l'armée française, publié par les ordres de Sa Majesté l'empereur Napoléon le Grand.* 9 vols. in 3 pts. Paris: Imprimerie impériale, 1809–28.

al-Damardāshī, Muṣṭafā ibn al-Ḥājj Ibrāhīm Tābiʿ al-Marḥūm Ḥasan Aghā ʿAzabān. *Tārīkh Waqāʾiʿ Miṣr al-Qāhira al-Maḥrūsa Kinānat Allah fī Arḍihi.* Edited by Ṣalāḥ Aḥmad Harīdī ʿAlī. 2nd ed. Cairo: Dār al-Kutub wa al-Wathāʾiq al-Qawmiyya, 2002.

al-Damurdāshī Katkhudā ʿAzabān, Aḥmad. *Kitāb al-Durra al-Muṣāna fī Akhbār al-Kināna.* Edited by ʿAbd al-Raḥīm ʿAbd al-Raḥman ʿAbd al-Raḥīm. Cairo: Institut français d'archéologie orientale, 1989.

Herodotus. *The History.* Translated by David Grene. Chicago: University of Chicago Press, 1987.

Heywood, Colin. "A Red Sea Shipping Register of the 1670s for the Supply of Foodstuffs from Egyptian *Wakf* Sources to Mecca and Medina (Turkish Documents from the Archive of ʿAbdurrahman 'ʿAbdi' Pasha of Buda, I)." *Anatolia Moderna* 6 (1996): 111–74.

Ibn ʿAbd al-Ghanī, Aḥmad Shalabī. *Awḍaḥ al-Ishārāt fī man Tawallā Miṣr al-Qāhira min al-Wuzarāʾ wa al-Bāshāt.* Edited by ʿAbd al-Raḥīm ʿAbd al-Raḥman ʿAbd al-Raḥīm. Cairo: Maktabat al-Khānjī, 1978.

Ibn Abī al-Surūr al-Bakrī, Muḥammad. *al-Nuzha al-Zahiyya fī Dhikr Wulāt Miṣr wa al-Qāhira al-Muʿizziyya.* Edited by ʿAbd al-Rāziq ʿAbd al-Rāziq ʿĪsā. Cairo: al-ʿArabī lil-Nashr wa al-Tawzīʿ, 1998.

al-Jabartī, ʿAbd al-Raḥman ibn Ḥasan. *ʿAbd al-Raḥman al-Jabartī's History of Egypt: ʿAjāʾib al-Āthār fī al-Tarājim wa al-Akhbār.* Edited by Thomas Philipp and Moshe Perlmann. 4 vols. Stuttgart: Franz Steiner Verlag, 1994.

———. *ʿAjāʾib al-Āthār fī al-Tarājim wa al-Akhbār.* Edited by Ḥasan Muḥammad Jawhar, ʿAbd al-Fattāḥ al-Saranjāwī, ʿUmar al-Dasūqī, and al-Sayyid Ibrāhīm Sālim. 7 vols. Cairo: Lajnat al-Bayān al-ʿArabī, 1958–67.

———. *ʿAjāʾib al-Āthār fī al-Tarājim wa al-Akhbār.* Edited by ʿAbd al-Raḥīm ʿAbd al-Raḥman ʿAbd al-Raḥīm. 4 vols. Cairo: Maṭbaʿat Dār al-Kutub al-Miṣriyya, 1998.

al-Jirjāwī, Muḥammad ibn Muḥammad Ḥāmid al-Marāghī. *Tārīkh Wilāyat al-Ṣaʿīd fī al-ʿAṣrayn al-Mamlūkī wa al-ʿUthmānī: al-Musammā bi-"Nūr al-ʿUyūn fī Dhikr Jirjā min ʿAhd Thalāthat Qurūn."* Edited by Aḥmad Ḥusayn al-Namakī. Cairo: Maktabat al-Nahḍa al-Miṣriyya, 1998.

al-Khashshāb, Ismāʿīl ibn Saʿd. *Akhbār Ahl al-Qarn al-Thānī ʿAshar: Tārīkh al-Mamālīk fī al-Qāhira.* Edited by ʿAbd al-ʿAzīz Jamāl al-Dīn and ʿImād Abū Ghāzī. Cairo: al-ʿArabī lil-Nashr wa al-Tawzīʿ, 1990.

———. *Khulāṣat mā Yurād min Akhbār al-Amīr Murād.* Edited and translated by Hamza ʿAbd al-ʿAzīz Badr and Daniel Crecelius. Cairo: al-ʿArabī lil-Nashr wa al-Tawzīʿ, 1992.

Kutluk, Halil, ed. *Türkiye Ormancılığı ile İlgili Tarihi Vesikalar, 893–1339 (1487–1923).* Istanbul: Osmanbey Matbaası, 1948.

al-Maqrīzī, Aḥmad ibn ʿAlī. *al-Mawāʿiẓ wa al-Iʿtibār bi-Dhikr al-Khiṭaṭ wa al-Āthār.* 2 vols. Būlāq, Egypt: Dār al-Ṭibāʿa al-Miṣriyya, 1853.

Mutawallī, Aḥmad Fu'ād, trans. and intro. *Qānūn Nāmah Miṣr, alladhī Aṣdarahu al-Sulṭān al-Qānūnī li-Ḥukm Miṣr*. Cairo: Maktabat al-Anjlū al-Miṣriyya, 1986.

Norden, Frederik Ludvig. *Voyage d'Égypte et de Nubie, par Frederic Louis Norden, ouvrage enrichi de cartes & de figures dessinées sur les lieux, par l'auteur même*. 2 vols. Copenhagen: Imprimerie de la Maison Royale, 1755.

al-Ṣafadī al-Shāfi'ī, Abū 'Uthmān al-Nābulusī. *Tārīkh al-Fayyūm wa Bilādihi*. Cairo: al-Maṭba'a al-Ahliyya, 1898.

———. *Tārīkh al-Fayyūm wa Bilādihi*. Beirut: Dār al-Jīl, 1974.

Sāmī, Amīn. *Taqwīm al-Nīl*. 5 vols. in 3 pts. Cairo: Dār al-Kutub wa al-Wathā'iq al-Qawmiyya, 2003.

Sharqāwī, 'Abd Allāh. *Tuḥfat al-Nāẓirīn fī man Waliya Miṣr min al-Mulūk wa al-Salāṭīn*. Edited by Riḥāb 'Abd al-Ḥamīd al-Qārī. Cairo: Maktabat Madbūlī, 1996.

Shaw, Stanford J. *The Budget of Ottoman Egypt, 1005–1006/1596–1597*. The Hague: Mouton, 1968.

———, ed. and trans. *Ottoman Egypt in the Eighteenth Century: The Niẓâmnâme-i Mısır of Cezzâr Aḥmed Pasha*. Cambridge, MA: Center for Middle Eastern Studies of Harvard University, 1964.

al-Shirbīnī, Yūsuf ibn Muḥammad. *Kitāb Hazz al-Quḥūf bi-Sharḥ Qaṣīd Abī Shādūf*. Edited and translated by Humphrey Davies. 2 vols. Leuven, Belgium: Peeters, 2005–7.

Steingrímsson, Jón. *Fires of the Earth: The Laki Eruption, 1783–1784*. Translated by Keneva Kunz. Reykjavík: University of Iceland Press and Nordic Volcanological Institute, 1998.

al-Suyūṭī, Jalāl al-Dīn. *Kawkab al-Rauḍa*. Edited by Muḥammad al-Shashtāwī. Cairo: Dār al-Āfāq al-'Arabiyya, 2002.

Tietze, Andreas. *Muṣṭafā 'Ālī's Description of Cairo of 1599: Text, Transliteration, Translation, Notes*. Vienna: Verlag der Österreichischen Akademie der Wissenschaften, 1975.

Volney, Constantin-François. *Travels through Egypt and Syria, in the Years 1783, 1784, and 1785. Containing the Present Natural and Political State of Those Countries, Their Productions, Arts, Manufactures, and Commerce; with Observations on the Manners, Customs, and Government of the Turks and Arabs*. Translated from the French. 2 vols. New York: Evert Duyckinck, 1798.

Warner, Nicholas. *The True Description of Cairo: A Sixteenth-Century Venetian View*. 3 vols. Oxford: Arcadian Library in association with Oxford University Press, 2006.

SECONDARY SOURCES

Abbas, Raouf, and Assem El-Dessouky. *The Large Landowning Class and the Peasantry in Egypt, 1837–1952*. Translated by Amer Mohsen with Mona Zikri. Edited by Peter Gran. Syracuse, NY: Syracuse University Press, 2011.

'Abd al-Fattāḥ, Muḥammad Ḥusām al-Dīn Ismā'īl, and Suhayr Ṣāliḥ. "A Wikāla of Sulṭān Mu'ayyid: Wikālat 'Ūda Pasha." *Annales Islamologiques* 28 (1994): 71–96.

ʿAbd al-Karīm, Aḥmad ʿIzzat. *Tārīkh al-Taʿlīm fī ʿAṣr Muḥammad ʿAlī*. Cairo: Maktabat al-Nahḍa al-Miṣriyya, 1938.

ʿAbd al-Mutajallī, Naṣra. "al-Muqāwama bil-Tasaḥḥub fī Rīf Miṣr al-ʿUthmāniyya." In *al-Rafḍ wa al-Iḥtijāj fī al-Mujtamaʿal-Miṣrī fī al-ʿAṣr al-ʿUthmānī*, edited by Nāṣir Ibrāhīm and Raʾūf ʿAbbās, 127–36. Cairo: Markaz al-Buḥūth wa al-Dirāsāt al-Ijtimāʿiyya, 2004.

ʿAbd al-Mutajallī Ibrāhīm ʿAlī, Nāṣira. "al-Daqahliyya fī al-ʿAṣr al-ʿUthmānī." MA thesis, ʿAyn Shams University, 2005.

ʿAbd al-Muʿṭī, Ḥusām Muḥammad. *al-ʿAlāqāt al-Miṣriyya al-Ḥijāziyya fī al-Qarn al-Thāmin ʿAshar*. Cairo: al-Hayʾa al-Miṣriyya al-ʿĀmma lil-Kitāb, 1999.

———. "al-Buyūt al-Tijāriyya al-Maghribiyya fī Miṣr fī al-ʿAṣr al-ʿUthmānī." PhD diss., Mansura University, 2002.

———. "Riwāq al-Maghāriba fī al-Jāmiʿ al-Azhar fī al-ʿAṣr al-ʿUthmānī." *al-Rūznāma: al-Ḥauliyya al-Miṣriyya lil-Wathāʾiq* 3 (2005): 165–204.

ʿAbd al-Raḥīm, ʿAbd al-Raḥīm ʿAbd al-Raḥman. *al-Rīf al-Miṣrī fī al-Qarn al-Thāmin ʿAshar*. Cairo: Maktabat Madbūlī, 1986.

Aberth, John. *The Black Death: The Great Mortality of 1348–1350, a Brief History with Documents*. New York: Palgrave Macmillan, 2005.

Abou-El-Haj, Rifaʿat ʿAli. *Formation of the Modern State: The Ottoman Empire, Sixteenth to Eighteenth Centuries*. 2nd ed. Syracuse, NY: Syracuse University Press, 2005.

Abu-Lughod, Janet L. *Before European Hegemony: The World System A.D. 1250–1350*. New York: Oxford University Press, 1989.

Adams, Robert McC. *Land behind Baghdad: A History of Settlement on the Diyala Plains*. Chicago: University of Chicago Press, 1965.

Adas, Michael. "From Avoidance to Confrontation: Peasant Protest in Precolonial and Colonial Southeast Asia." *Comparative Studies in Society and History* 23 (1981): 217–47.

Ágoston, Gábor. *Guns for the Sultan: Military Power and the Weapons Industry in the Ottoman Empire*. Cambridge: Cambridge University Press, 2005.

Aḥmad, Laylā ʿAbd al-Laṭīf. *al-Idāra fī Miṣr fī al-ʿAṣr al-ʿUthmānī*. Cairo: Maṭbaʿat Jāmiʿat ʿAyn Shams, 1978.

———. *al-Mujtamaʿ al-Miṣrī fī al-ʿAṣr al-ʿUthmānī*. Cairo: Dār al-Kitāb al-Jāmiʿī, 1987.

———. *al-Ṣaʿīd fī ʿAhd Shaykh al-ʿArab Hammām*. Cairo: al-Hayʾa al-Miṣriyya al-ʿAmma lil-Kitāb, 1987.

———. *Tārīkh wa Muʾarrikhī Miṣr wa al-Shām ibbāna al-ʿAṣr al-ʿUthmānī*. Cairo: Maktabat al-Khānjī, 1980.

Akarli, Engin D. "Provincial Power Magnates in Ottoman Bilad al-Sham and Egypt, 1740–1840." In *La vie sociale dans les Provinces Arabes à l'époque Ottoman*, vol. 3, edited by Abdeljelil Temimi, 41–56. Zaghwān, Tunisia: Markaz al-Dirāsāt wa al-Buḥūth al-ʿUthmāniyya wa al-Mūrīskiyya wa al-Tawthīq wa al-Maʿlūmāt, 1988.

Aksan, Virginia H. *An Ottoman Statesman in War and Peace: Ahmed Resmi Efendi, 1700–1783*. Leiden, Netherlands: Brill, 1995.

———. *Ottoman Wars, 1700–1870: An Empire Besieged*. Harlow, UK: Routledge, 2007.

Alam, Muzaffar, and Sanjay Subrahmanyam. *Indo-Persian Travels in the Age of Discoveries, 1400–1800*. Cambridge: Cambridge University Press, 2007.

Albert, Jeff, Magnus Bernhardsson, and Roger Kenna, eds. *Transformations of Middle Eastern Natural Environments: Legacies and Lessons*. Bulletin Series, no. 103. New Haven, CT: Yale School of Forestry and Environmental Sciences, 1998.

Alexander, Jennifer Karns. *The Mantra of Efficiency: From Waterwheel to Social Control*. Baltimore: Johns Hopkins University Press, 2008.

ʿAlī, Ṣalāḥ Aḥmad Harīdī. "al-Ḥayāh al-Iqtiṣādiyya wa al-Ijtimāʿiyya fī Madīnat Rashid fī al-ʿAṣr al-ʿUthmānī, Dirāsa Wathāʾiqiyya." *Egyptian Historical Review* 30–31 (1983–84): 327–78.

Allsen, Thomas T. *The Royal Hunt in Eurasian History*. Philadelphia: University of Pennsylvania Press, 2006.

Ambraseys, Nicholas. *Earthquakes in the Eastern Mediterranean and Middle East: A Multidisciplinary Study of Seismicity up to 1900*. Cambridge: Cambridge University Press, 2009.

Ambraseys, N.N., and C.F. Finkel. *The Seismicity of Turkey and Adjacent Areas: A Historical Review, 1500–1800*. Istanbul: Eren, 1995.

Ambraseys, N.N., and C.P. Melville. *A History of Persian Earthquakes*. Cambridge: Cambridge University Press, 1982.

Ambraseys, N.N., C.P. Melville, and R.D. Adams. *The Seismicity of Egypt, Arabia and the Red Sea: A Historical Review*. Cambridge: Cambridge University Press, 1994.

Amer, M.H., and N.A. de Ridder, eds. *Land Drainage in Egypt*. Cairo: Drainage Research Institute, 1989.

Anderson, Virginia DeJohn. *Creatures of Empire: How Domestic Animals Transformed Early America*. New York: Oxford University Press, 2004.

Appadurai, Arjun. "Wells in Western India: Irrigation and Cooperation in an Agricultural Society." *Expedition* 26 (1984): 3–14.

Appuhn, Karl. "Ecologies of Beef: Eighteenth-Century Epizootics and the Environmental History of Early Modern Europe." *Environmental History* 15 (2010): 268–87.

———. *A Forest on the Sea: Environmental Expertise in Renaissance Venice*. Baltimore: Johns Hopkins University Press, 2009.

Arnold, David. *Colonizing the Body: State Medicine and Epidemic Disease in Nineteenth-Century India*. Berkeley: University of California Press, 1993.

Asmar, Basel N. "The Science and Politics of the Dead Sea: Red Sea Canal or Pipeline." *Journal of Environment and Development* 12 (2003): 325–39.

Aston, T.H., and C.H.E. Philpin, eds. *The Brenner Debate: Agrarian Class Structure and Economic Development in Pre-industrial Europe*. Cambridge: Cambridge University Press, 1985.

Atwell, William S. "Volcanism and Short-Term Climatic Change in East Asian and World History, c. 1200–1699." *Journal of World History* 12 (2001): 29–98.

Ayalon, Yaron. *Natural Disasters in the Ottoman Empire: Plague, Famine, and Other Misfortunes*. Cambridge: Cambridge University Press, 2015.

ʿAzabāwī, ʿAbd Allah Muḥammad. "al-ʿAlāqāt al-ʿUthmāniyya-al-Maghribiyya fī ʿAhd Kullin min Maulāya Muḥammad (1757–1790) wa Ibnihi Yazīd (1790–1792)." *Egyptian Historical Review* 30–31 (1983–84): 379–413.

Bacqué-Grammont, Jean-Louis, and Anne Kroell. *Mamlouks, ottomans et portugais en Mer Rouge: L'affaire de Djedda en 1517*. Cairo: Institut français d'archéologie orientale, 1988.

Bacqué-Grammont, Jean-Louis, Joséphine Lesur-Gebremariam, and Catherine Mayeur-Jaouen. "Quelques aspects de la faune nilotique dans la relation d'Evliyâ Çelebî, voyageur ottoman." *Journal Asiatique* 296 (2008): 331–74.

Badawi, El-Said, and Martin Hinds. *A Dictionary of Egyptian Arabic*. Beirut: Librairie du Liban, 1986.

Baer, Gabriel. "The Dissolution of the Egyptian Village Community." *Die Welt des Islams* 6 (1959): 56–70.

———. *A History of Landownership in Modern Egypt, 1800–1950*. London: Oxford University Press, 1962.

Bagis, Ali Ihsan. "Turkey's Hydropolitics of the Euphrates-Tigris Basin." *International Journal of Water Resources Development* 13 (1997): 567–82.

Bagnall, Roger S. "The Camel, the Wagon, and the Donkey in Later Roman Egypt." *Bulletin of the American Society of Papyrologists* 22 (1985): 1–6.

———. *Egypt in Late Antiquity*. Princeton, NJ: Princeton University Press, 1993.

Baldwin, James Edward. "Islamic Law in an Ottoman Context: Resolving Disputes in Late Seventeenth and Early Eighteenth-Century Cairo." PhD diss., New York University, 2010.

Barkan, Ömer Lütfi. *Süleymaniye Cami ve İmareti İnşaatı (1550–1557)*. 2 vols. Ankara: Türk Tarih Kurumu, 1972–79.

Barkey, Karen. *Bandits and Bureaucrats: The Ottoman Route to State Centralization*. Ithaca, NY: Cornell University Press, 1994.

———. *Empire of Difference: The Ottomans in Comparative Perspective*. Cambridge: Cambridge University Press, 2008.

Beaumont, Peter. "Water Resource Development in Iran." *Geographical Journal* 140 (1974): 418–31.

Behrens-Abouseif, Doris. *Egypt's Adjustment to Ottoman Rule: Institutions, Waqf, and Architecture in Cairo, 16th and 17th Centuries*. Leiden, Netherlands: Brill, 1994.

Beinin, Joel. *Workers and Peasants in the Modern Middle East*. Cambridge: Cambridge University Press, 2001.

Beinin, Joel, and Zachary Lockman. *Workers on the Nile: Nationalism, Communism, Islam, and the Egyptian Working Class, 1882–1954*. Princeton, NJ: Princeton University Press, 1987.

Bell, Barbara. "The Oldest Records of the Nile Floods." *Geographical Journal* 136 (1970): 569–73.

Benkheira, Mohamed Hocine, Catherine Mayeur-Jaouen, and Jacqueline Sublet. *L'animal en islam*. Paris: Indes savantes, 2005.

Bernasconi, Maria Pia, Daniel Jean Stanley, and Italo Di Geronimo. "Molluscan Faunas and Paleobathymetry of Holocene Sequences in the Northeastern Nile Delta, Egypt." *Marine Geology* 99 (1991): 29–43.

Bilge, Mustafa. "Suez Canal in the Ottoman Sources." In *Proceedings of the International Conference on Egypt during the Ottoman Era: 26–30 November 2007, Cairo, Egypt*, edited by Research Centre for Islamic History, Art and Culture, 89–113. Istanbul: IRCICA, 2010.

Binder, Leonard. *In a Moment of Enthusiasm: Political Power and the Second Stratum in Egypt*. Chicago: University of Chicago Press, 1978.

Bishop, Elizabeth. "Talking Shop: Egyptian Engineers and Soviet Specialists at the Aswan High Dam." PhD diss., University of Chicago, 1997.

Boak, A.E.R. "Irrigation and Population in Faiyum, the Garden of Egypt." *Geographical Review* 16 (1926): 353–64.

Borsch, Stuart J. *The Black Death in Egypt and England: A Comparative Study*. Austin: University of Texas Press, 2005.

———. "Environment and Population: The Collapse of Large Irrigation Systems Reconsidered." *Comparative Studies in Society and History* 46 (2004): 451–68.

Bostan, İdris. *Kürekli ve Yelkenli Osmanlı Gemileri*. Istanbul: Bilge, 2005.

———. *Osmanlı Bahriye Teşkilâtı: XVII. Yüzyılda Tersâne-i Âmire*. Ankara: Türk Tarih Kurumu Basımevi, 1992.

———. "An Ottoman Base in Eastern Mediterranean: Alexandria of Egypt in the 18th Century." In *Proceedings of the International Conference on Egypt during the Ottoman Era: 26–30 November 2007, Cairo, Egypt*, edited by Research Centre for Islamic History, Art and Culture, 63–77. Istanbul: IRCICA, 2010.

Braudel, Fernand. *The Wheels of Commerce*. Translated by Siân Reynolds. Vol. 2 of *Civilization and Capitalism, 15th–18th Century*. London: Collins, 1982.

Brice, William C., ed. *The Environmental History of the Near and Middle East since the Last Ice Age*. London: Academic Press, 1978.

Brooke, John L. *Climate Change and the Course of Global History: A Rough Journey*. Cambridge: Cambridge University Press, 2014.

Brookfield, Michael. "The Desertification of the Egyptian Sahara during the Holocene (the Last 10,000 Years) and Its Influence on the Rise of Egyptian Civilization." In *Landscapes and Societies: Selected Cases*, edited by I. Peter Martini and Ward Chesworth, 91–108. Dordrecht, Netherlands: Springer, 2010.

Brown, Karen, and Daniel Gilfoyle, eds. *Healing the Herds: Disease, Livestock Economies, and the Globalization of Veterinary Medicine*. Athens: Ohio University Press, 2010.

Brown, Nathan J. *Peasant Politics in Modern Egypt: The Struggle against the State*. New Haven, CT: Yale University Press, 1990.

Brummett, Palmira. *Ottoman Seapower and Levantine Diplomacy in the Age of Discovery*. Albany: State University of New York Press, 1994.

Bulliet, Richard W. "The Camel and the Watermill." *International Journal of Middle East Studies* 42 (2010): 666–68.

———. *The Camel and the Wheel*. New York: Columbia University Press, 1990.

———. *Cotton, Climate, and Camels in Early Islamic Iran: A Moment in World History*. New York: Columbia University Press, 2009.

———. "History and Animal Energy in the Arid Zone." In *Water on Sand: Environmental Histories of the Middle East and North Africa*, edited by Alan Mikhail, 51–69. New York: Oxford University Press, 2013.

———. *Hunters, Herders, and Hamburgers: The Past and Future of Human-Animal Relationships*. New York: Columbia University Press, 2005.

Bulmuş, Birsen. *Plague, Quarantines and Geopolitics in the Ottoman Empire*. Edinburgh: Edinburgh University Press, 2012.

Burke, Edmund, III. "The Big Story: Human History, Energy Regimes, and the Environment." In *The Environment and World History*, edited by Edmund Burke III and Kenneth Pomeranz, 33–53. Berkeley: University of California Press, 2009.

———. "The Transformation of the Middle Eastern Environment, 1500 B.C.E.–2000 C.E." In *The Environment and World History*, edited by Edmund Burke III and Kenneth Pomeranz, 81–117. Berkeley: University of California Press, 2009.

Butzer, Karl W. *Early Hydraulic Civilization in Egypt: A Study in Cultural Ecology*. Chicago: University of Chicago Press, 1976.

Çağlar, Yücel. *Türkiye Ormanları ve Ormancılık*. Istanbul: İletişim Yayınları, 1992.

Camberlin, Pierre. "Rainfall Anomalies in the Source Region of the Nile and Their Connection with the Indian Summer Monsoon." *Journal of Climate* 10 (1997): 1380–92.

Campopiano, Michele. "Rural Communities, Land Clearance and Water Management in the Po Valley in the Central and Late Middle Ages." *Journal of Medieval History* 39 (2013): 377–93.

Camuffo, D., and S. Enzi. "Chronology of 'Dry Fogs' in Italy, 1374–1891." *Theoretical and Applied Climatology* 50 (1994): 31–33.

Carey, Mark, Philip Garone, Adrian Howkins, Georgina H. Endfield, Lawrence Culver, Sam White, Sherry Johnson, and James Rodger Fleming. "Forum: Climate Change and Environmental History." *Environmental History* (2014): 281–364.

Çarkoğlu, Ali, and Mine Eder. "Development *alla Turca*: The Southeastern Anatolia Development Project (GAP)." In *Environmentalism in Turkey: Between Democracy and Development?*, edited by Fikret Adaman and Murat Arsel, 167–84. Aldershot, UK: Ashgate, 2005.

———. "Domestic Concerns and the Water Conflict over the Euphrates-Tigris River Basin." *Middle Eastern Studies* 37 (2001): 41–71.

Carpentier, Elisabeth. "Autour de la Peste Noire: Famines et épidémies dans l'histoire du XIVe siècle." *Annales* 17 (1962): 1062–92.

Casale, Giancarlo. "The Ottoman Administration of the Spice Trade in the Sixteenth-Century Red Sea and Persian Gulf." *Journal of the Economic and Social History of the Orient* 49 (2006): 170–98.

———. *The Ottoman Age of Exploration*. New York: Oxford University Press, 2010.

Casalilla, Bartolomé Yun. *Las redes del imperio: Élites sociales en la articulación de la Monarquía Hispánica, 1492–1714*. Madrid: Marcial Pons; Seville: Universidad Pablo de Olavide, 2009.

Caseldine, C., A. Russell, J. Harðardóttir, and Ó. Knudsen, eds. *Iceland—Modern Processes and Past Environments*. Amsterdam: Elsevier, 2005.

Caton-Thompson, Gertrude, and E.W. Gardner. "Recent Work on the Problem of Lake Moeris." *Geographical Journal* 73 (1929): 20–58.

Çevre ve Orman Bakanlığı. *Osmanlı Ormancılığı ile İlgili Belgeler*. 3 vols. Ankara: Çevre ve Orman Bakanlığı, 1999–2003.

Chaiklin, Martha. "Exotic-Bird Collecting in Early-Modern Japan." In *JAPANimals: History and Culture in Japan's Animal Life*, edited by Gregory M. Pflugfelder and Brett L. Walker, 125–61. Ann Arbor: Center for Japanese Studies at the University of Michigan, 2005.

Christensen, Peter. *The Decline of Iranshahr: Irrigation and Environments in the History of the Middle East, 500 B.C. to A.D. 1500*. Copenhagen: Museum Tusculanum Press, 1993.

Çizakça, Murat. "Ottomans and the Mediterranean: An Analysis of the Ottoman Shipbuilding Industry as Reflected by the Arsenal Registers of Istanbul, 1529–1650." In *Le genti del Mare Mediterraneo*, vol. 2, edited by Rosalba Ragosta, 773–89. Naples: Lucio Pironti, 1981.

Clapp, Gordon R. "Iran: A TVA for the Khuzestan Region." *Middle East Journal* 11 (1957): 1–11.

Clarence-Smith, William Gervase, and Steven Topik, eds. *The Global Coffee Economy in Africa, Asia, and Latin America, 1500–1989*. Cambridge: Cambridge University Press, 2003.

Cohen, Ammon. *The Guilds of Ottoman Jerusalem*. Leiden, Netherlands: Brill, 2001.

Collins, Robert O. *The Nile*. New Haven, CT: Yale University Press, 2002.

Conrad, Lawrence I. "Arabic Plague Chronologies and Treatises: Social and Historical Factors in the Formation of a Literary Genre." *Studia Islamica* 54 (1981): 51–93.

———. "The Biblical Tradition for the Plague of the Philistines." *Journal of the American Oriental Society* 104 (1984): 281–87.

———. "The Plague in the Early Medieval Near East." PhD diss., Princeton University, 1981.

Cooper, David E., and Simon P. James. *Buddhism, Virtue and Environment*. Aldershot, UK: Ashgate, 2005.

Cordova, Carlos E. *Millennial Landscape Change in Jordan: Geoarchaeology and Cultural Ecology*. Tucson: University of Arizona Press, 2007.

Creager, Angela N.H., and William Chester Jordan, eds. *The Animal/Human Boundary: Historical Perspectives*. Rochester, NY: University of Rochester Press, 2002.

Crecelius, Daniel. "Egypt in the Eighteenth Century." In *Modern Egypt, from 1517 to the End of the Twentieth Century*. Vol. 2 of *The Cambridge His-*

tory of Egypt, edited by M.W. Daly, 59–86. Cambridge: Cambridge University Press, 1998.

———. "The Importance of Qusayr in the Late Eighteenth Century." *Journal of the American Research Center in Egypt* 24 (1987): 53–60.

———. *The Roots of Modern Egypt: A Study of the Regimes of ʿAli Bey al-Kabir and Muhammad Bey Abu al-Dhahab, 1760–1775*. Minneapolis: Bibliotheca Islamica, 1981.

Crecelius, Daniel, and Hamza ʿAbd al-ʿAziz Badr. "French Ships and Their Cargoes Sailing between Damiette and Ottoman Ports, 1777–1781." *Journal of the Economic and Social History of the Orient* 37 (1994): 251–86.

Cronon, William. *Changes in the Land: Indians, Colonists, and the Ecology of New England*. 1st rev. ed. New York: Hill and Wang, 2003.

———. *Nature's Metropolis: Chicago and the Great West*. New York: Norton, 1991.

———. "The Trouble with Wilderness; or, Getting Back to the Wrong Nature." *Environmental History* 1 (1996): 7–28.

Crosby, Alfred W. *Ecological Imperialism: The Biological Expansion of Europe, 900–1900*. Cambridge: Cambridge University Press, 2004.

Cuno, Kenneth M. "Commercial Relations between Town and Village in Eighteenth and Early Nineteenth-Century Egypt." *Annales Islamologiques* 24 (1988): 111–35.

———. *The Pasha's Peasants: Land, Society, and Economy in Lower Egypt, 1740–1858*. Cambridge: Cambridge University Press, 1992.

Curth, Louise Hill. *The Care of Brute Beasts: A Social and Cultural Study of Veterinary Medicine in Early Modern England*. Leiden, Netherlands: Brill, 2010.

D'Arrigo, Rosanne D., and Gordon C. Jacoby. "Northern North American Tree-Ring Evidence for Regional Temperature Changes after Major Volcanic Events." *Climate Change* 41 (1999): 1–15.

D'Arrigo, Rosanne, Richard Seager, Jason E. Smerdon, Allegra N. LeGrande, and Edward R. Cook. "The Anomalous Winter of 1783–1784: Was the Laki Eruption or an Analog of the 2009–2010 Winter to Blame?" *Geophysical Research Letters* 38 (2011): L05706.

Davis, Diana K. "Brutes, Beasts, and Empire: Veterinary Medicine and Environmental Policy in French North Africa and British India." *Journal of Historical Geography* 34 (2008): 242–67.

———. "Potential Forests: Degradation Narratives, Science, and Environmental Policy in Protectorate Morocco, 1912–1956." *Environmental History* 10 (2005): 211–38.

———. "Power, Knowledge, and Environmental History in the Middle East and North Africa." *International Journal of Middle East Studies* 42 (2010): 657–59.

———. *Resurrecting the Granary of Rome: Environmental History and French Colonial Expansion in North Africa*. Athens: Ohio University Press, 2007.

Davis, Diana K., and Edmund Burke III, eds. *Environmental Imaginaries of the Middle East and North Africa*. Athens: Ohio University Press, 2011.

Davis, Mike. *Late Victorian Holocausts: El Niño Famines and the Making of the Third World*. London: Verso, 2001.

Decker, Michael. "Plants and Progress: Rethinking the Islamic Agricultural Revolution." *Journal of World History* 20 (2009): 187–206.

Delaporte, François. *Disease and Civilization: The Cholera in Paris, 1832*. Translated by Arthur Goldhammer. Cambridge: Massachusetts Institute of Technology Press, 1986.

Denevan, William M. "The Pristine Myth: The Landscape of the Americas in 1492." *Annals of the Association of American Geographers* 82 (1992): 369–85.

Derr, Jennifer Leslee. "Cultivating the State: Cash Crop Agriculture, Irrigation, and the Geography of Authority in Colonial Southern Egypt, 1868–1931." PhD diss., Stanford University, 2009.

de Vries, Jan. *The Dutch Rural Economy in the Golden Age, 1500–1700*. New Haven, CT: Yale University Press, 1974.

Dewald, Jonathan, Geoffrey Parker, Michael Marmé, and J.B. Shank. "*AHR* Forum: The General Crisis of the Seventeenth Century Revisited." *American Historical Review* 113 (2008): 1029–99.

Dols, Michael W. *The Black Death in the Middle East*. Princeton, NJ: Princeton University Press, 1977.

———. "The General Mortality of the Black Death in the Mamluk Empire." In *The Islamic Middle East, 700–1900: Studies in Social and Economic History*, edited by Abraham L. Udovitch, 397–428. Princeton, NJ: Darwin Press, 1981.

———. "Ibn al-Wardī's *Risālah al-Naba' 'an al-Waba'*, a Translation of a Major Source for the History of the Black Death in the Middle East." In *Near Eastern Numismatics, Iconography, Epigraphy and History: Studies in Honor of George C. Miles*, edited by Dickran K. Kouymjian, 443–55. Beirut: American University of Beirut, 1974.

———. "al-Manbijī's 'Report of the Plague': A Treatise on the Plague of 764–765/1362–1364 in the Middle East." In *The Black Death: The Impact of the Fourteenth-Century Plague*, edited by Daniel Williman, 65–75. Papers of the Eleventh Annual Conference of the Center for Medieval and Early Renaissance Studies. Binghamton, NY: Center for Medieval and Early Renaissance Studies, 1982.

———. "Plague in Early Islamic History." *Journal of the American Oriental Society* 94 (1974): 371–83.

———. "The Second Plague Pandemic and Its Recurrences in the Middle East: 1347–1894." *Journal of the Economic and Social History of the Orient* 22 (1979): 162–89.

Dominik, Janusz, and Daniel Jean Stanley. "Boron, Beryllium and Sulfur in Holocene Sediments and Peats of the Nile Delta, Egypt: Their Use as Indicators of Salinity and Climate." *Chemical Geology* 104 (1993): 203–16.

Doumani, Beshara. *Rediscovering Palestine: Merchants and Peasants in Jabal Nablus, 1700–1900*. Berkeley: University of California Press, 1995.

Dursun, Selçuk. "Forest and the State: History of Forestry and Forest Administration in the Ottoman Empire." PhD diss., Sabancı University, 2007.

Eaton, Richard M. "Islamic History as Global History." In *Islamic and European Expansion: The Forging of a Global Order*, edited by Michael Adas, 1–36. Philadelphia: Temple University Press, 1993.

Ellenblum, Ronnie. *The Collapse of the Eastern Mediterranean: Climate Change and the Decline of the East, 950–1072*. Cambridge: Cambridge University Press, 2012.

Elmusa, Sharif S., ed. *Culture and the Natural Environment: Ancient and Modern Middle Eastern Texts*. Vol. 26, no. 1, *Cairo Papers in Social Science*. Cairo: American University in Cairo Press, 2003.

Elsayed, Mohamed A.K., Nazeih A. Younan, Alfy M. Fanos, and Khalid H. Baghdady. "Accretion and Erosion Patterns along Rosetta Promontory, Nile Delta Coast." *Journal of Coastal Research* 21 (2005): 412–20.

Elvin, Mark. *The Retreat of the Elephants: An Environmental History of China*. New Haven, CT: Yale University Press, 2004.

Elvin, Mark, and Liu Ts'ui-Jung, eds. *Sediments of Time: Environment and Society in Chinese History*. Cambridge: Cambridge University Press, 1998.

Ergene, Boğaç A. *Local Court, Provincial Society and Justice in the Ottoman Empire: Legal Practice and Dispute Resolution in Çankırı and Kastamonu (1652–1744)*. Leiden, Netherlands: Brill, 2003.

Erler, Mehmet Yavuz. *Osmanlı Devleti'nde Kuraklık ve Kıtlık Olayları, 1800–1880*. Istanbul: Libra Kitap, 2010.

Erlich, Haggai. *The Cross and the River: Ethiopia, Egypt, and the Nile*. Boulder, CO: Lynne Rienner, 2002.

Erlich, Haggai, and Israel Gershoni, eds. *The Nile: Histories, Cultures, Myths*. Boulder, CO: Lynne Rienner, 2000.

Estes, J. Worth, and LaVerne Kuhnke. "French Observations of Disease and Drug Use in Late Eighteenth-Century Cairo." *Journal of the History of Medicine and Allied Sciences* 39 (1984): 121–52.

Evans, J.A.S. "Herodotus and the Problem of the 'Lake of Moeris.'" *Classical World* 56 (1963): 275–77.

Evans, Richard J. *Death in Hamburg: Society and Politics in the Cholera Years, 1830–1910*. New York: Oxford University Press, 1987.

Fagan, Brian. *The Long Summer: How Climate Changed Civilization*. New York: Basic Books, 2004.

Fahim, Hussein M. *Dams, People and Development: The Aswan High Dam Case*. New York: Pergamon Press, 1981.

Fahmy, Khaled. *All the Pasha's Men: Mehmed Ali, His Army and the Making of Modern Egypt*. Cambridge: Cambridge University Press, 1997.

———. "The Era of Muhammad ʿAli Pasha, 1805–1848." In *Modern Egypt, from 1517 to the End of the Twentieth Century*. Vol. 2 of *The Cambridge History of Egypt*, edited by M.W. Daly, 139–79. Cambridge: Cambridge University Press, 1998.

———. *al-Jasad wa al-Ḥadātha: al-Ṭibb wa al-Qānūn fī Miṣr al-Ḥadītha*.

Translated by Sharīf Yūnis. Cairo: Dār al-Kutub wa al-Wathā'iq al-Qawmiyya, 2004.

———. *Mehmed Ali: From Ottoman Governor to Ruler of Egypt*. Oxford: Oneworld Publications, 2009.

Farnie, D.A. *East and West of Suez: The Suez Canal in History, 1854–1956*. Oxford: Clarendon Press, 1969.

Faroqhi, Suraiya. "Agriculture and Rural Life in the Ottoman Empire (ca 1500–1878) (A Report on Scholarly Literature Published between 1970 and 1985)." *New Perspectives on Turkey* 1 (1987): 3–34.

———, ed. *Animals and People in the Ottoman Empire*. Istanbul: Eren, 2010.

———. *Artisans of Empire: Crafts and Craftspeople under the Ottomans*. London: I.B. Tauris, 2009.

———. "Camels, Wagons, and the Ottoman State in the Sixteenth and Seventeenth Centuries." *International Journal of Middle East Studies* 14 (1982): 523–39.

———. "Coffee and Spices: Official Ottoman Reactions to Egyptian Trade in the Later Sixteenth Century." *Wiener Zeitschrift für die Kunde des Morgenlandes* 76 (1986): 87–93.

———. "Coping with the Central State, Coping with Local Power: Ottoman Regions and Notables from the Sixteenth to the Early Nineteenth Century." In *The Ottomans and the Balkans: A Discussion of Historiography*, edited by Fikret Adanır and Suraiya Faroqhi, 351–81. Leiden, Netherlands: Brill, 2002.

———. "Ottoman Peasants and Rural Life: The Historiography of the Twentieth Century." *Archivum Ottomanicum* 18 (2000): 153–82.

———. "The Peasants of Saideli in the Late Sixteenth Century." *Archivum Ottomanicum* 8 (1983): 215–50.

———. *Pilgrims and Sultans: The Hajj under the Ottomans*. London: I.B. Tauris, 1994.

———. "Red Sea Trade and Communications as Observed by Evliya Çelebi (1671–72)." *New Perspectives on Turkey* 5–6 (1991): 87–105.

———. "Rural Society in Anatolia and the Balkans during the Sixteenth Century, I." *Turcica* 9 (1977): 161–95.

———. "Rural Society in Anatolia and the Balkans during the Sixteenth Century, II." *Turcica* 11 (1979): 103–53.

———. *Towns and Townsmen in Ottoman Anatolia: Trade, Crafts, and Food Production in an Urban Setting, 1520–1650*. Cambridge: Cambridge University Press, 1984.

———. "Trade Controls, Provisioning Policies, and Donations: The Egypt-Hijaz Connection during the Second Half of the Sixteenth Century." In *Süleymân the Second and His Time*, edited by Halil İnalcık and Cemal Kafadar, 131–43. Istanbul: Isis Press, 1993.

Faroqhi, Suraiya, and Randi Deguilhem, eds. *Crafts and Craftsmen of the Middle East: Fashioning the Individual in the Muslim Mediterranean*. London: I.B. Tauris, 2005.

Faroqhi, Suraiya, and Christoph K. Neumann, eds. *Ottoman Costumes: From Textile to Identity*. Istanbul: Eren, 2004.

Foltz, Richard C. *Animals in Islamic Tradition and Muslim Cultures*. Oxford: Oneworld, 2006.

———, ed. *Environmentalism in the Muslim World*. New York: Nova Science Publishers, 2005.

———. "Is There an Islamic Environmentalism?" *Environmental Ethics* 22 (2000): 63–72.

Foltz, Richard C., Frederick M. Denny, and Azizan Baharuddin, eds. *Islam and Ecology: A Bestowed Trust*. Cambridge, MA: Harvard University Press, 2003.

Fournier, Patrick, and Sandrine Lavaud, eds. *Eaux et conflits dans l'Europe médiévale et moderne*. Toulouse: Presses universitaires du Mirail, 2012.

Frihy, Omran E., Alfy M. Fanos, Ahmed A. Khafagy, and Paul D. Komar. "Patterns of Nearshore Sediment Transport along the Nile Delta, Egypt." *Coastal Engineering* 15 (1991): 409–29.

Fudge, Erica. *Brutal Reasoning: Animals, Rationality, and Humanity in Early Modern England*. Ithaca, NY: Cornell University Press, 2006.

———. *Perceiving Animals: Humans and Beasts in Early Modern English Culture*. New York: St. Martin's Press, 2000.

———. *Pets*. Stocksfield, UK: Acumen, 2008.

———, ed. *Renaissance Beasts: Of Animals, Humans, and Other Wonderful Creatures*. Urbana: University of Illinois Press, 2004.

Gadgil, Madhav, and Ramachandra Guha. *This Fissured Land: An Ecological History of India*. Berkeley: University of California Press, 1993.

Geertz, Clifford. "The Wet and the Dry: Traditional Irrigation in Bali and Morocco." *Human Ecology* 1 (1972): 23–39.

Glacken, Clarence J. *Traces on the Rhodian Shore: Nature and Culture in Western Thought from Ancient Times to the End of the Eighteenth Century*. Berkeley: University of California Press, 1967.

Glick, Thomas F. *Irrigation and Hydraulic Technology: Medieval Spain and Its Legacy*. Brookfield, VT: Variorum, 1996.

———. *Irrigation and Society in Medieval Valencia*. Cambridge, MA: Harvard University Press, 1970.

Goldberg, Ellis Jay, ed. *The Social History of Labor in the Middle East*. Boulder, CO: Westview Press, 1996.

———. *Tinker, Tailor, and Textile Worker: Class and Politics in Egypt, 1930–1952*. Berkeley: University of California Press, 1986.

Goubert, Jean-Pierre. *The Conquest of Water: The Advent of Health in the Industrial Age*. Translated by Andrew Wilson. Princeton, NJ: Princeton University Press, 1989.

Gould, Andrew Gordon. "Pashas and Brigands: Ottoman Provincial Reform and Its Impact on the Nomadic Tribes of Southern Anatolia, 1840–1885." PhD diss., University of California, Los Angeles, 1973.

Gran, Peter. *Islamic Roots of Capitalism: Egypt, 1760–1840*. Austin: University of Texas Press, 1979.

Grattan, John, Roland Rabartin, Stephen Self, and Thorvaldur Thordarson. "Volcanic Air Pollution and Mortality in France, 1783–1784." *C. R. Geoscience* 337 (2005): 641–51.

Greene, Ann Norton. *Horses at Work: Harnessing Power in Industrial America*. Cambridge, MA: Harvard University Press, 2008.

Greene, Molly. *Catholic Pirates and Greek Merchants: A Maritime History of the Mediterranean*. Princeton, NJ: Princeton University Press, 2010.

———. "The Ottomans in the Mediterranean." In *The Early Modern Ottomans: Remapping the Empire*, edited by Virginia H. Aksan and Daniel Goffman, 104–16. Cambridge: Cambridge University Press, 2007.

Greenwood, Antony. "Istanbul's Meat Provisioning: A Study of the *Celepkeşan* System." PhD diss., University of Chicago, 1988.

Grehan, James. *Everyday Life & Consumer Culture in 18th-Century Damascus*. Seattle: University of Washington Press, 2007.

Grier, Katherine C. *Pets in America: A History*. Chapel Hill: University of North Carolina Press, 2006.

Grigsby, Darcy Grimaldo. *Colossal: Engineering the Suez Canal, Statue of Liberty, Eiffel Tower, and Panama Canal: Transcontinental Ambition in France and the United States during the Long Nineteenth Century*. Pittsburgh: Periscope, 2012.

Griswold, William. "Climatic Change: A Possible Factor in the Social Unrest of Seventeenth Century Anatolia." In *Humanist and Scholar: Essays in Honor of Andreas Tietze*, edited by Heath W. Lowry and Donald Quataert, 37–57. Istanbul: Isis Press, 1993.

———. *The Great Anatolian Rebellion, 1000–1020/1591–1611*. Berlin: Klaus Schwarz Verlag, 1983.

Grove, Richard H., Vinita Damodaran, and Satpal Sangwan, eds. *Nature and the Orient: The Environmental History of South and Southeast Asia*. Delhi: Oxford University Press, 1998.

Guerrini, Anita. *Experimenting with Humans and Animals: From Galen to Animal Rights*. Baltimore: Johns Hopkins University Press, 2003.

Guha, Ramachandra, and J. Martinez-Alier. *Varieties of Environmentalism: Essays North and South*. London: Earthscan Publications, 1997.

Guillerme, André E. *The Age of Water: The Urban Environment in the North of France, A.D. 300–1800*. College Station: Texas A&M University Press, 1988.

Hairy, Isabelle, and Oueded Sennoune. "Géographie historique du canal d'Alexandrie." *Annales Islamologiques* 40 (2006): 247–78.

Haleem, Harfiyah Abdel, ed. *Islam and the Environment*. London: Ta-Ha Publishers, 1998.

Hanioğlu, M. Şükrü. *A Brief History of the Late Ottoman Empire*. Princeton, NJ: Princeton University Press, 2008.

Hanna, Nelly. *Construction Work in Ottoman Cairo (1517–1798)*. Cairo: Institut français d'archéologie orientale, 1984.

———. *In Praise of Books: A Cultural History of Cairo's Middle Class, Sixteenth to the Eighteenth Century*. Syracuse, NY: Syracuse University Press, 2003.

———. *Making Big Money in 1600: The Life and Times of Isma'il Abu Taqiyya, Egyptian Merchant*. Syracuse, NY: Syracuse University Press, 1998.

———. *An Urban History of Būlāq in the Mamluk and Ottoman Periods.* Cairo: Institut français d'archéologie orientale, 1983.
Haraway, Donna. "Teddy Bear Patriarchy: Taxidermy in the Garden of Eden, New York City, 1908–1936." *Social Text* 11 (1984): 20–64.
———. *When Species Meet.* Minneapolis: University of Minnesota Press, 2008.
Harris, Leila. "Postcolonialism, Postdevelopment, and Ambivalent Spaces of Difference in Southeastern Turkey." *Geoforum* 39 (2008): 1698–1708.
———. "Water and Conflict Geographies of the Southeastern Anatolia Project." *Society and Natural Resources* 15 (2002): 743–59.
Hathaway, Jane. "Egypt in the Seventeenth Century." In *Modern Egypt, from 1517 to the End of the Twentieth Century.* Vol. 2 of *The Cambridge History of Egypt,* edited by M.W. Daly, 34–58. Cambridge: Cambridge University Press, 1998.
———. *The Politics of Households in Ottoman Egypt: The Rise of the Qazdağlıs.* Cambridge: Cambridge University Press, 1997.
———. "The Role of the Kızlar Ağası in 17th–18th Century Ottoman Egypt." *Studia Islamica* 75 (1992): 141–58.
———. *A Tale of Two Factions: Myth, Memory, and Identity in Ottoman Egypt and Yemen.* Albany: State University of New York Press, 2003.
Hattox, Ralph. *Coffee and Coffeehouses: The Origins of a Social Beverage in the Medieval Near East.* Seattle: University of Washington Press, 1985.
Henninger-Voss, Mary J., ed. *Animals in Human Histories: The Mirror of Nature and Culture.* Rochester, NY: University of Rochester Press, 2002.
Hess, Andrew C. "The Ottoman Conquest of Egypt (1517) and the Beginning of the Sixteenth-Century World War." *International Journal of Middle East Studies* 4 (1973): 55–76.
Hewison, R. Neil. *The Fayoum: A Practical Guide.* Cairo: American University in Cairo Press, 1984.
Highwood, E.J., and D.S. Stevenson. "Atmospheric Impact of the 1783–1784 Laki Eruption: Part II, Climatic Effect of Sulphate Aerosol." *Atmospheric Chemistry and Physics* 3 (2003): 1177–89.
Hinz, Walther. *Islamische Masse und Gewichte umgerechnet ins metrische System.* Leiden, Netherlands: Brill, 1955.
Hirst, L. Fabian. *The Conquest of Plague: A Study of the Evolution of Epidemiology.* Oxford: Clarendon Press, 1953.
Hoage, R.J., and William A. Deiss, eds. *New Worlds, New Animals: From Menagerie to Zoological Park in the Nineteenth Century.* Baltimore: Johns Hopkins University Press, 1996.
Hobsbawm, E.J. "The Crisis of the 17th Century—II." *Past and Present* 6 (1954): 44–64.
———. "The General Crisis of the European Economy in the 17th Century." *Past and Present* 5 (1954): 33–53.
Holt, P.M. "The Beylicate in Ottoman Egypt during the Seventeenth Century." *Bulletin of the School of Oriental and African Studies* 24 (1961): 214–48.
———. *Egypt and the Fertile Crescent, 1516–1922: A Political History.* Ithaca, NY: Cornell University Press, 1966.

———. "al-Fayyūm." *Encyclopaedia of Islam*. 2nd ed. Leiden, Netherlands: Brill, 2006.

Hourani, Albert. "Ottoman Reforms and the Politics of Notables." In *The Beginnings of Modernization in the Middle East: The Nineteenth Century*, edited by William R. Polk and Richard L. Chambers, 41–68. Chicago: University of Chicago Press, 1968.

Huber, Valeska. *Channelling Mobilities: Migration and Globalisation in the Suez Canal Region and Beyond, 1869–1914*. Cambridge: Cambridge University Press, 2013.

Hughes, J. Donald. *The Mediterranean: An Environmental History*. Santa Barbara, CA: ABC-CLIO, 2005.

Husain, Faisal. "In the Bellies of the Marshes: Water and Power in the Countryside of Ottoman Baghdad." *Environmental History* 19 (2014): 638–64.

Hütteroth, Wolf-Dieter. "Ecology of the Ottoman Lands." In *The Later Ottoman Empire, 1603–1839*. Vol. 3 of *The Cambridge History of Turkey*, edited by Suraiya N. Faroqhi, 18–43. Cambridge: Cambridge University Press, 2006.

Ibrāhīm, Nāṣir Aḥmad. *al-Azamāt al-Ijtimāʿiyya fī Miṣr fī al-Qarn al-Sābiʿ ʿAshar*. Cairo: Dār al-Āfāq al-ʿArabiyya, 1998.

Imber, Colin. "The Navy of Süleiman the Magnificent." *Archivum Ottomanicum* 6 (1980): 211–82.

———. "The Ottoman Dynastic Myth." *Turcica* 19 (1987): 7–27.

———. *The Ottoman Empire, 1300–1650: The Structure of Power*. New York: Palgrave Macmillan, 2002.

İnalcık, Halil. "'Arab' Camel Drivers in Western Anatolia in the Fifteenth Century." *Revue d'Histoire Maghrebine* 10 (1983): 256–70.

———. "Capital Formation in the Ottoman Empire." *Journal of Economic History* 29 (1969): 97–140.

———. "Centralization and Decentralization in Ottoman Administration." In *Studies in Eighteenth Century Islamic History*, edited by Thomas Naff and Roger Owen, 27–52. Carbondale: Southern Illinois University Press, 1977.

———. "Māʾ. 8. Irrigation in the Ottoman Empire." *Encyclopaedia of Islam*. 2nd ed. Leiden, Netherlands: Brill, 2006.

———. "The Origins of the Ottoman-Russian Rivalry and the Don-Volga Canal, 1569." *Les annales de l'Université d'Ankara* 1 (1946–47): 47–106.

———. *The Ottoman Empire: The Classical Age, 1300–1600*. Translated by Norman Itzkowitz and Colin Imber. New York: Praeger Publishers, 1973.

———. "Bursa and the Silk Trade." In *An Economic and Social History of the Ottoman Empire*, 2 vols., edited by Halil İnalcık with Donald Quataert, 1:218–55. Cambridge: Cambridge University Press, 1994.

———. "Şikâyet Hakkı: *ʿArż-i Ḥâl ve ʿArż-i Maḥzar*'lar." In *Osmanlı'da Devlet, Hukuk, Adâlet*, 49–71. Istanbul: Eren Yayıncılık, 2000.

Isenberg, Andrew C. *The Destruction of the Bison: An Environmental History, 1750–1920*. Cambridge: Cambridge University Press, 2000.

İslamoğlu-İnan, Huri, ed. *The Ottoman Empire and the World-Economy*. Cambridge: Cambridge University Press, 1987.

———. *State and Peasant in the Ottoman Empire: Agrarian Power Relations and Regional Economic Development in Ottoman Anatolia during the Sixteenth Century*. Leiden, Netherlands: Brill, 1994.

Issar, Arie S. *Water Shall Flow from the Rock: Hydrogeology and Climate in the Lands of the Bible*. Berlin: Springer-Verlag, 1990.

Issar, Arie S., and Mattanyah Zohar. *Climate Change—Environment and Civilization in the Middle East*. Berlin: Springer, 2004.

Izzi Dien, Mawil. *The Environmental Dimensions of Islam*. Cambridge: Lutterworth Press, 2000.

———. "Islam and the Environment: Theory and Practice." *Journal of Beliefs and Values* 18 (1997): 47–57.

Jackson, E.L. "The Laki Eruption of 1783: Impacts on Population and Settlement in Iceland." *Geography* 67 (1982): 42–50.

Jacobs, Nancy J. "The Great Bophuthatswana Donkey Massacre: Discourse on the Ass and the Politics of Class and Grass." *American Historical Review* 106 (2001): 485–507.

Jacobsen, Thorkild. *Salinity and Irrigation Agriculture in Antiquity: Diyala Basin Archaeological Report on Essential Results, 1957–58*. Bibliotheca Mesopotamica, vol. 14. Malibu, CA: Undena Publications, 1982.

Jacobsen, Thorkild, and Robert M. Adams, "Salt and Silt in Ancient Mesopotamian Agriculture." *Science* 128 (1958): 1251–58.

Jacoby, Gordon C., Karen W. Workman, and Rosanne D. D'Arrigo. "Laki Eruption of 1783, Tree Rings, and Disaster for Northwest Alaska Inuit." *Quaternary Science Reviews* 18 (1999): 1365–71.

Jirjis, Majdī. "Manhaj al-Dirāsāt al-Wathāʾiqiyya wa Wāqiʿ al-Baḥth fī Miṣr." *al-Rūznāma: al-Ḥauliyya al-Miṣriyya lil-Wathāʾiq* 2 (2004): 237–87.

Jones, Susan D. *Death in a Small Package: A Short History of Anthrax*. Baltimore: Johns Hopkins University Press, 2010.

———. *Valuing Animals: Veterinarians and Their Patients in Modern America*. Baltimore: Johns Hopkins University Press, 2002.

Jones, Toby Craig. *Desert Kingdom: How Oil and Water Forged Modern Saudi Arabia*. Cambridge: Harvard University Press, 2010.

Jordan, William Chester. *The Great Famine: Northern Europe in the Early Fourteenth Century*. Princeton, NJ: Princeton University Press, 1996.

Karabell, Zachary. *Parting the Desert: The Creation of the Suez Canal*. New York: Vintage, 2003.

Kasaba, Reşat. *A Moveable Empire: Ottoman Nomads, Migrants, and Refugees*. Seattle: University of Washington Press, 2009.

———. *The Ottoman Empire and the World Economy: The Nineteenth Century*. Albany: State University of New York Press, 1988.

Kashef, Abdel-Aziz I. "Salt-Water Intrusion in the Nile Delta." *Ground Water* 21 (1983): 160–67.

Kassler, P. "The Structural and Geomorphic Evolution of the Persian Gulf." In *The Persian Gulf: Holocene Carbonate Sedimentation and Diagenesis in a Shallow Epicontinental Sea*, edited by B.H. Purser, 11–32. Berlin: Springer-Verlag, 1973.

Keenan, G. "Fayyum Agriculture at the End of the Ayyubid Era: Nabulsi's *Survey*." In *Agriculture in Egypt: From Pharaonic to Modern Times*, edited by

Alan K. Bowman and Eugene Rogan, 287–99. Oxford: Oxford University Press for the British Academy, 1999.
Kelly, Morgan, Cormac Ó Gráda, Sam White, Ulf Büntgen, Lena Hellmann, and Jan de Vries. "The Little Ice Age: Climate and History Reconsidered." *Journal of Interdisciplinary History* 44 (2014): 301–77.
Kete, Kathleen. *The Beast in the Boudoir: Petkeeping in Nineteenth-Century Paris*. Berkeley: University of California Press, 1994.
Keyder, Çağlar, and Faruk Tabak, eds. *Landholding and Commercial Agriculture in the Middle East*. Albany: State University of New York Press, 1991.
Khalid, Fazlun M., with Joanne O'Brien, eds. *Islam and Ecology*. New York: Cassell, 1992.
Khazeni, Arash. *Tribes and Empire on the Margins of Nineteenth-Century Iran*. Seattle: University of Washington Press, 2009.
Khoury, Dina Rizk. "The Ottoman Centre versus Provincial Power-Holders: An Analysis of the Historiography." In *The Later Ottoman Empire, 1603–1839*. Vol. 3 of *The Cambridge History of Turkey*, edited by Suraiya N. Faroqhi, 135–56. Cambridge: Cambridge University Press, 2006.
———. *State and Provincial Society in the Ottoman Empire: Mosul, 1540–1834*. Cambridge: Cambridge University Press, 1997.
Kudlick, Catherine J. *Cholera in Post-Revolutionary Paris: A Cultural History*. Berkeley: University of California Press, 1996.
Kuhnke, LaVerne. *Lives at Risk: Public Health in Nineteenth-Century Egypt*. Berkeley: University of California Press, 1990.
Kupferschmidt, Hugo. *Die Epidemiologie der Pest: Der Konzeptwandel in der Erforschung der Infektionsketten seit der Entdeckung des Pesterregers im Jahre 1894*. Aarau, Switzerland: Sauerländer, 1993.
Kurat, A.N. "The Turkish Expedition to Astrakhan and the Problem of the Don-Volga Canal." *Slavonic and East European Review* 40 (1961): 7–23.
LaCapra, Dominick. *History and Its Limits: Human, Animal, Violence*. Ithaca, NY: Cornell University Press, 2009.
Lamb, H.H. "Volcanic Dust in the Atmosphere; with a Chronology and Assessment of Its Meteorological Significance." *Philosophical Transactions of the Royal Society of London. Series A, Mathematical and Physical Sciences* 266 (1970): 425–533.
Lane, Edward William. *An Arabic-English Lexicon*. 8 vols. Beirut: Librairie du Liban, 1968.
Lapidus, Ira M. "The Grain Economy of Mamluk Egypt." *Journal of the Economic and Social History of the Orient* 12 (1969): 1–15.
Larsen, Guðrún. "Katla: Tephrochronology and Eruption History." *Developments in Quaternary Sciences* 13 (2010): 23–49.
Lawson, Fred H. "Rural Revolt and Provincial Society in Egypt, 1820–1824." *International Journal of Middle East Studies* 13 (1981): 131–53.
Le Roy Ladurie, Emmanuel. *Times of Feast, Times of Famine: A History of Climate since the Year 1000*. Translated by Barbara Bray. Garden City, NY: Doubleday, 1971.
Lewis, Bernard. "Corsairs in Iceland." *Revue de l'Occident musulman et de la Méditerranée* 15 (1973): 139–44.

Linant de Bellefonds, M.A. *Mémoires sur les principaux travaux d'utilité publiqué exécutés en Egypte depuis la plus haute antiquité jusqu'à nos jours: Accompagné d'un atlas renfermant neuf planches grand in-folio imprimées en couleur*. Paris: Arthus Bertrand, 1872–73.

Linden, Eugene. *The Winds of Change: Climate, Weather, and the Destruction of Civilizations*. New York: Simon and Schuster, 2006.

Lisān al-'Arab. 4 vols. Beirut: Dār Lisān al-'Arab, 1970.

Little, Tom. *High Dam at Aswan: The Subjugation of the Nile*. New York: Methuen, 1965.

Lockman, Zachary, ed. *Workers and Working Classes in the Middle East: Struggles, Histories, Historiographies*. Albany: State University of New York Press, 1994.

Low, Michael Christopher. "Empire and the Hajj: Pilgrims, Plagues, and Pan-Islam under British Surveillance, 1865–1908." *International Journal of Middle East Studies* 40 (2008): 269–90.

Lydon, Ghislaine. *On Trans-Saharan Trails: Islamic Law, Trade Networks, and Cross-Cultural Exchange in Nineteenth-Century Western Africa*. Cambridge: Cambridge University Press, 2009.

———. "Writing Trans-Saharan History: Methods, Sources and Interpretations across the African Divide." *Journal of North African Studies* 10 (2005): 293–324.

Lyons, H.G. "On the Nile Flood and Its Variation." *Geographical Journal* 26 (1905): 249–72, 395–415.

Magnusson, Roberta J. *Water Technology in the Middle Ages: Cities, Monasteries, and Waterworks after the Roman Empire*. Baltimore: Johns Hopkins University Press, 2001.

Maḥārīq, Yāsir 'Abd al-Min'am. *al-Minūfiyya fī al-Qarn al-Thāmin 'Ashar*. Cairo: al-Hay'a al-Miṣriyya al-'Amma lil-Kitāb, 2000.

Maizels, Judith K., and Chris Caseldine, eds. *Environmental Change in Iceland: Past and Present*. Dordrecht, Netherlands: Kluwer Academic Publishers, 1991.

Malamud, Randy. *Reading Zoos: Representations of Animals and Captivity*. New York: New York University Press, 1998.

Mandaville, Jan E. "The Ottoman Province of Al-Hasâ in the Sixteenth and Seventeenth Centuries." *Journal of the American Oriental Society* 90 (1970): 486–513.

Mardin, Şerif. "Center-Periphery Relations: A Key to Turkish Politics." *Daedalus* 102 (1973): 169–91.

Marlowe, John. *World Ditch: The Making of the Suez Canal*. New York: Macmillan, 1964.

Marsot, Afaf Lutfi al-Sayyid. *Egypt in the Reign of Muhammad Ali*. Cambridge: Cambridge University Press, 1984.

Martin, John. "Sheep and Enclosure in Sixteenth-Century Northamptonshire." *Agricultural History Review* 36 (1988): 39–54.

Masri, Basheer Ahmad. *Animals in Islam*. Petersfield, UK: Athene Trust, 1989.

———. *Animal Welfare in Islam*. Markfield, Leicestershire, UK: Islamic Foundation, 2007.

Mayeur-Jaouen, Catherine. "Badawi and His Camel: An Animal as the Attribute of a Muslim Saint in Mamluk and Ottoman Egypt." Translated by Suraiya Faroqhi. In *Animals and People in the Ottoman Empire*, edited by Suraiya Faroqhi, 113–28. Istanbul: Eren, 2010.

Mazzaoui, Michel M. "Global Policies of Sultan Selim, 1512–1520." In *Essays on Islamic Civilization: Presented to Niyazi Berkes*, edited by Donald P. Little, 224–43. Leiden, Netherlands: Brill, 1976.

McCann, James C. *Maize and Grace: Africa's Encounter with a New World Crop, 1500–2000*. Cambridge, MA: Harvard University Press, 2007.

McCarthy, Justin A. "Nineteenth-Century Egyptian Population." *Middle Eastern Studies* 12 (1976): 1–39.

McCormick, Michael, Paul Edward Dutton, and Paul A. Mayewski. "Volcanoes and the Climate Forcing of Carolingian Europe, A.D. 750–950." *Speculum* 82 (2007): 865–95.

McGowan, Bruce. "Peasants and Pastoralists." In *An Economic and Social History of the Ottoman Empire*, 2 vols., edited by Halil İnalcık with Donald Quataert, 2:680–94. Cambridge: Cambridge University Press, 1994.

McNeill, J.R. "The First Hundred Thousand Years." In *The Turning Points of Environmental History*, edited by Frank Uekoetter, 13–28. Pittsburgh: University of Pittsburgh Press, 2010.

———. *The Mountains of the Mediterranean World: An Environmental History*. Cambridge: Cambridge University Press, 1992.

———. *Something New under the Sun: An Environmental History of the Twentieth-Century World*. New York: Norton, 2000.

McNeill, William H. *Plagues and Peoples*. Garden City, NY: Anchor Press / Doubleday, 1976.

McShane, Clay. *Down the Asphalt Path: The Automobile and the American City*. New York: Columbia University Press, 1994.

Meiggs, Russell. *Trees and Timber in the Ancient Mediterranean World*. Oxford: Clarendon Press, 1982.

Melhaoui, Mohammed. *Peste, contagion et martyre: Histoire du fléau en Occident musulman medieval*. Paris: Publisud, 2005.

Melville, Elinor G.K. *A Plague of Sheep: Environmental Consequences of the Conquest of Mexico*. Cambridge: Cambridge University Press, 1997.

Meyerhof, Max. "La peste en Égypte à la fin du XVIII siècle et le Mèdecin Enrico di Wolmar." *La Revue Médicale d'Égypte* 1 (1913): 1–13.

Michel, Nicolas. "Les Dafātir al-ǧusūr, source pour l'histoire du réseau hydraulique de l'Égypte ottomane." *Annales Islamologiques* 29 (1995): 151–68.

———. "Migrations de paysans dans le Delta du Nil au début de l'époque ottoman." *Annales Islamologiques* 35 (2001): 241–90.

———. "Villages désertés, terres en friche et reconstruction rurale en Égypte au début de l'époque ottoman." *Annales Islamologiques* 36 (2002): 197–251.

Mikhail, Alan. "Anatolian Timber and Egyptian Grain: Things That Made the Ottoman Empire." In *Early Modern Things: Objects and Their Histories, 1500–1800*, edited by Paula Findlen, 274–93. New York: Routledge, 2013.

———. *The Animal in Ottoman Egypt*. New York: Oxford University Press, 2014.
———. "Animals as Property in Early Modern Ottoman Egypt." *Journal of the Economic and Social History of the Orient* 53 (2010): 621–52.
———. *Nature and Empire in Ottoman Egypt: An Environmental History*. Cambridge: Cambridge University Press, 2011.
———. "Tārīkh Dirāsāt al-Tābiʿ wa Naẓariyyatayn ʿan al-Sulṭa." In *Thaqāfat al-Nukhba wa Thaqāfat al-ʿĀmma fī Miṣr fī al-ʿAṣr al-ʿUthmānī*, edited by Nāṣir Aḥmad Ibrāhīm, 349–60. Cairo: Markaz al-Buḥūth wa al-Dirāsāt al-Ijtimāʿiyya, 2008.
———. "Unleashing the Beast: Animals, Energy, and the Economy of Labor in Ottoman Egypt." *American Historical Review* 118 (2013): 317–48.
———, ed. *Water on Sand: Environmental Histories of the Middle East and North Africa*. New York: Oxford University Press, 2013.
Miller, Ian Jared. *The Nature of the Beasts: Empire and Exhibition at the Tokyo Imperial Zoo*. Berkeley: University of California Press, 2013.
Mitchell, Timothy. "Are Environmental Imaginaries Culturally Constructed?" In *Environmental Imaginaries of the Middle East and North Africa*, edited by Diana K. Davis and Edmund Burke III, 265–73. Athens: Ohio University Press, 2011.
———. "Can the Mosquito Speak?" In *Rule of Experts: Egypt, Techno-Politics, Modernity*, 19–53. Berkeley: University of California Press, 2002.
———. "Carbon Democracy." *Economy and Society* 38 (2009): 399–432.
———. "Dreamland." In *Rule of Experts: Egypt, Techno-Politics, Modernity*, 272–303. Berkeley: University of California Press, 2002.
Muḥammad, ʿIrāqī Yūsuf. *al-Wujūd al-ʿUthmānī al-Mamlūkī fī Miṣr fī al-Qarn al-Thāmin ʿAshar wa Awāʾil al-Qarn al-Tāsiʿ ʿAshar*. Cairo: Dār al-Maʿārif, 1985.
———. *al-Wujūd al-ʿUthmānī fī Miṣr fī al-Qarnayn al-Sādis ʿAshar wa al-Sābiʿ ʿAshar (Dirāsa Wathāʾiqiyya)*. Cairo: Markaz Kliyūbātrā lil-Kumbiyūtar, 1996.
Mukerji, Chandra. *Impossible Engineering: Technology and Territoriality on the Canal du Midi*. Princeton, NJ: Princeton University Press, 2009.
Murphey, Rhoads. "The Decline of North Africa since the Roman Occupation: Climatic or Human?" *Annals of the Association of American Geographers* 41 (1951): 116–32.
———. "The Ottoman Centuries in Iraq: Legacy or Aftermath? A Survey Study of Mesopotamian Hydrology and Ottoman Irrigation Projects." *Journal of Turkish Studies* 11 (1987): 17–29.
Muslu, Emire Cihan. "Ottoman-Mamluk Relations: Diplomacy and Perceptions." PhD diss., Harvard University, 2007.
El-Nahal, Galal H. *The Judicial Administration of Ottoman Egypt in the Seventeenth Century*. Minneapolis: Bibliotheca Islamica, 1979.
Najm, Zayn al-ʿĀbidīn Shams al-Dīn. "Tasaḥḥub al-Fallāḥīn fī ʿAṣr Muḥammad ʿAlī, Asbābuhu wa Natāʾijuhu." *Egyptian Historical Review* 36 (1989): 259–316.
Neustadt (Ayalon), David. "The Plague and Its Effects upon the Mamlûk

Army." *Journal of the Royal Asiatic Society of Great Britain and Ireland* (1946): 67–73.

Newfield, Timothy P. "A Cattle Panzootic in Early Fourteenth-Century Europe." *Agricultural History Review* 57 (2009): 155–90.

Nijland, H.J., ed. *Drainage along the River Nile*. Egypt: Egyptian Public Authority for Drainage Projects; Netherlands: Directorate-General of Public Works and Water Management, 2000.

Oman, Luke, Alan Robock, Georgiy Stenchikov, Gavin A. Schmidt, and Reto Ruedy. "Climatic Response to High-Latitude Volcanic Eruptions." *Journal of Geophysical Research* 110 (2005): D13103.

Oman, Luke, Alan Robock, Georgiy L. Stenchikov, and Thorvaldur Thordarson. "High-Latitude Eruptions Cast Shadow over the African Monsoon and the Flow of the Nile." *Geophysical Research Letters* 33 (2006): L18711.

Oman, Luke, Alan Robock, Georgiy L. Stenchikov, Thorvaldur Thordarson, Dorothy Koch, Drew T. Shindell, and Chaochao Gao. "Modeling the Distribution of the Volcanic Aerosol Cloud from the 1783–1784 Laki Eruption." *Journal of Geophysical Research* 111 (2006): D12209.

Omar, Omar Abdel-Aziz. "Anglo-Egyptian Relations and the Construction of the Alexandria-Cairo-Suez Railway (1833–1858)." DPhil thesis, University of London, 1966.

Ortlieb, Luc. "Historical Chronology of ENSO and the Nile Flood Record." In *Past Climate Variability through Europe and Africa*, edited by Richard W. Battarbee, Françoise Gasse, and Catherine E. Stickley, 257–78. Dordrecht, Netherlands: Springer, 2004.

Oslund, Karen. *Iceland Imagined: Nature, Culture, and Storytelling in the North Atlantic*. Seattle: University of Washington Press, 2011.

Özbaran, Salih. "Bahrain in 1559: A Narrative of Turco-Portuguese Conflict in the Gulf." *Osmanlı Araştırmaları* 3 (1982): 91–104.

———. *Ottoman Expansion towards the Indian Ocean in the 16th Century*. Istanbul: Bilgi University Press, 2009.

———. "Ottoman Naval Power in the Indian Ocean in the 16th Century." In *The Kapudan Pasha, His Office and His Domain: Halcyon Days in Crete IV*, edited by Elizabeth Zachariadou, 109–17. Rethymnon: Crete University Press, 2002.

———. *The Ottoman Response to European Expansion: Studies on Ottoman-Portuguese Relations in the Indian Ocean and Ottoman Administration in the Arab Lands during the Sixteenth Century*. Istanbul: Isis Press, 1994.

———. "The Ottoman Turks and the Portuguese in the Persian Gulf, 1534–1581." *Journal of Asian History* 6 (1972): 45–88.

———. "A Turkish Report on the Red Sea and the Portuguese in the Indian Ocean (1525)." *Arabian Studies* 4 (1978): 81–88.

———. *Yemen'den Basra'ya Sınırdaki Osmanlı*. Istanbul: Kitap Yayınevi, 2004.

Pamuk, Şevket. *A Monetary History of the Ottoman Empire*. Cambridge: Cambridge University Press, 2000.

———. "Prices in the Ottoman Empire, 1469–1914." *International Journal of Middle East Studies* 36 (2004): 451–68.

Panzac, Daniel. "Alexandrie: Évolution d'une ville cosmopolite au XIXe siècle." In *Population et santé dans l'Empire ottoman (XVIIIe–XXe siècles)*, 141–59. Istanbul: Isis, 1996.

———. "Alexandrie: Peste et croissance urbaine (XVIIe–XIXe siècles)." In *Population et santé dans l'Empire ottoman (XVIIIe–XXe siècles)*, 45–55. Istanbul: Isis, 1996.

———. *La caravane maritime: Marins européens et marchands ottomans en Méditerranée (1680–1830)*. Paris: CNRS éditions, 2004.

———. "International and Domestic Maritime Trade in the Ottoman Empire during the 18th Century." *International Journal of Middle East Studies* 24 (1992): 189–206.

———. *La peste dans l'Empire Ottoman, 1700–1850*. Leuven, Belgium: Association pour le Développement des Études Turques, 1985.

———. *Quarantaines et lazarets: l'Europe et la peste d'Orient (XVIIe–XXe siècles)*. Aix-en-Provence, France: Édisud, 1986.

Parker, Geoffrey. *Global Crisis: War, Climate Change and Catastrophe in the Seventeenth Century*. New Haven, CT: Yale University Press, 2013.

Pehlivan, Zozan. "Beyond 'The Desert and the Sown': Peasants, Pastoralists, and Climate Crises in Ottoman Diyarbekir, 1840–1890." PhD diss., Queen's University, 2016.

Peirce, Leslie. *Morality Tales: Law and Gender in the Ottoman Court of Aintab*. Berkeley: University of California Press, 2003.

Perlin, John. *A Forest Journey: The Story of Wood and Civilization*. Woodstock, VT: Countryman Press, 2005.

Pflugfelder, Gregory M., and Brett L. Walker, eds. *JAPANimals: History and Culture in Japan's Animal Life*. Ann Arbor: Center for Japanese Studies at the University of Michigan, 2005.

Philliou, Christine M. *Biography of an Empire: Governing Ottomans in an Age of Revolution*. Berkeley: University of California Press, 2011.

Podobnik, Bruce. "Toward a Sustainable Energy Regime: A Long-Wave Interpretation of Global Energy Shifts." *Technological Forecasting and Social Change* 62 (1999): 155–72.

Pollitzer, Robert. *Plague*. Geneva, Switzerland: World Health Organization, 1954.

Pomeranz, Kenneth. *The Great Divergence: China, Europe, and the Making of the Modern World Economy*. Princeton, NJ: Princeton University Press, 2000.

Popper, William. *The Cairo Nilometer: Studies in Ibn Taghrî Birdî's Chronicles of Egypt, I*. Berkeley: University of California Press, 1951.

Posusney, Marsha Pripstein. *Labor and the State in Egypt: Workers, Unions, and Economic Restructuring*. New York: Columbia University Press, 1997.

Pyne, Stephen J. *Vestal Fire: An Environmental History, Told through Fire, of Europe and Europe's Encounter with the World*. Seattle: University of Washington Press, 1997.

———. *World Fire: The Culture of Fire on Earth*. Seattle: University of Washington Press, 1997.

al-Qaraḍāwī, Yūsuf. *Riʿāyat al-Bīʾah fī Sharīʿat al-Islām*. Cairo: Dār al-Shurūq, 2001.

Quataert, Donald, ed. *Consumption Studies and the History of the Ottoman Empire, 1550–1922: An Introduction*. Albany: State University of New York Press, 2000.

———. *The Ottoman Empire, 1700–1922*. 2nd ed. Cambridge: Cambridge University Press, 2005.

Rafeq, Abdul-Karim. "ʿAbd al-Ghani al-Nabulsi: Religious Tolerance and 'Arabness' in Ottoman Damascus." In *Transformed Landscapes: Essays on Palestine and the Middle East in Honor of Walid Khalidi*, edited by Camille Mansour and Leila Fawaz, 1–17. Cairo: American University in Cairo Press, 2009.

al-Rāfʿī, ʿAbd al-Raḥman. *ʿAṣr Muḥammad ʿAlī*. Cairo: Dār al-Maʿārif, 1989.

Ramzī, Muḥammad. *al-Qāmūs al-Jughrāfī lil-Bilād al-Miṣriyya min ʿAhd Qudamāʾ al-Miṣriyyīn ilā Sanat 1945*. 6 vols. in 2 pts. Cairo: al-Hayʾa al-Miṣriyya al-ʿĀmma lil-Kitāb, 1994.

Rapoport, Yossef. "Invisible Peasants, Marauding Nomads: Taxation, Tribalism, and Rebellion in Mamluk Egypt." *Mamlūk Studies Review* 8 (2004): 1–22.

Raymond, André. *Artisans et commerçants au Caire au XVIII*e *siècle*. 2 vols. Damascus: Institut français de Damas, 1973–74.

———. "A Divided Sea: The Cairo Coffee Trade in the Red Sea Area during the Seventeenth and Eighteenth Centuries." In *Modernity and Culture: From the Mediterranean to the Indian Ocean*, edited by Leila Tarazi Fawaz and C.A. Bayly, 46–57. New York: Columbia University Press, 2002.

———. "Une famille de grands négociants en café au Caire dans la première moitié du XVIIIe siècle: Les Sharāybī." In *Le commerce du café avant l'ère des plantations coloniales: Espaces, réseaux, sociétés (XVe–XIXe siècle)*, edited by Michel Tuchscherer, 111–24. Cairo: Institut français d'archéologie orientale, 2001.

———. "Les Grandes Épidémies de peste au Caire aux XVIIe and XVIIIe siècles." *Bulletin d'Études Orientales* 25 (1973): 203–10.

———. "La population du Caire et de l'Égypte à l'époque ottomane et sous Muḥammad ʿAlî." In *Mémorial Ömer Lûtfi Barkan*, 169–78. Paris: Librairie d'Amérique et d'Orient Adrien Maisonneuve, 1980.

———. "Une 'révolution' au Caire sous les Mamelouks: La crise de 1123/1711." *Annales Islamologiques* 6 (1966): 95–120.

Reid, Anthony. "Sixteenth-Century Turkish Influence in Western Indonesia." *Journal of South East Asian History* 10 (1969): 395–414.

Reimer, Michael J. "Ottoman Alexandria: The Paradox of Decline and the Reconfiguration of Power in Eighteenth-Century Arab Provinces." *Journal of the Economic and Social History of the Orient* 37 (1994): 107–46.

Revel, Marie, E. Ducassou, F.E. Grousset, S.M. Bernasconi, S. Migeon, S. Revillon, J. Mascle, A. Murat, S. Zaragosi, and D. Bosch. "100,000 Years of African Monsoon Variability Recorded in Sediments of the Nile Margin." *Quaternary Science Reviews* 29 (2010): 1342–62.

Reynolds, Nancy Y. "Building the Past: Rockscapes and the Aswan High Dam in Egypt." In *Water on Sand: Environmental Histories of the Middle East and North Africa*, edited by Alan Mikhail, 181–205. New York: Oxford University Press, 2013.

Richards, Alan R. "Primitive Accumulation in Egypt, 1798–1882." In *The Ottoman Empire and the World-Economy*, edited by Huri İslamoğlu-İnan, 203–43. Cambridge: Cambridge University Press, 1987.

Richards, John F. "Toward a Global System of Property Rights in Land." In *The Environment and World History*, edited by Edmund Burke III and Kenneth Pomeranz, 54–78. Berkeley: University of California Press, 2009.

———. *The Unending Frontier: An Environmental History of the Early Modern World*. Berkeley: University of California Press, 2003.

Risso, Patricia. "Cross-Cultural Perceptions of Piracy: Maritime Violence in the Western Indian Ocean and Persian Gulf Region during a Long Eighteenth Century." *Journal of World History* 12 (2001): 293–319.

———. "Muslim Identity in Maritime Trade: General Observations and Some Evidence from the 18th Century Persian Gulf / Indian Ocean Region." *International Journal of Middle East Studies* 21 (1989): 381–92.

Ritvo, Harriet. *The Animal Estate: The English and Other Creatures in the Victorian Age*. Cambridge, MA: Harvard University Press, 1987.

———. *The Platypus and the Mermaid and Other Figments of the Classifying Imagination*. Cambridge, MA: Harvard University Press, 1997.

Rivlin, Helen Anne B. *The Agricultural Policy of Muḥammad ʿAlī in Egypt*. Cambridge, MA: Harvard University Press, 1961.

Robbins, Louise E. *Elephant Slaves and Pampered Parrots: Exotic Animals in Eighteenth-Century Paris*. Baltimore: Johns Hopkins University Press, 2002.

Robock, Alan. "Volcanic Eruptions and Climate." *Reviews of Geophysics* 38 (2000): 191–219.

Rosenberg, Charles E. *The Cholera Years: The United States in 1832, 1849, and 1866*. Chicago: University of Chicago Press, 1987.

Rosenthal, Jean-Laurent, and R. Bin Wong. *Before and Beyond Divergence: The Politics of Economic Change in China and Europe*. Cambridge, MA: Harvard University Press, 2011.

Rothfels, Nigel, ed. *Representing Animals*. Bloomington: Indiana University Press, 2002.

———. *Savages and Beasts: The Birth of the Modern Zoo*. Baltimore: Johns Hopkins University Press, 2002.

el-Rouayheb, Khaled. "Opening the Gate of Verification: The Forgotten Arab-Islamic Florescence of the 17th Century." *International Journal of Middle East Studies* 38 (2006): 263–81.

———. "Sunni Muslim Scholars on the Status of Logic, 1500–1800." *Islamic Law and Society* 11 (2004): 213–32.

———. "Was There a Revival of Logical Studies in Eighteenth-Century Egypt?" *Die Welt des Islams* 45 (2005): 1–19.

Ruddiman, William F. *Plows, Plagues, and Petroleum: How Humans Took Control of Climate*. Princeton, NJ: Princeton University Press, 2005.

Russell, Josiah C. "That Earlier Plague." *Demography* 5 (1968): 174–84.

Said, Rushdi. *The Geological Evolution of the River Nile*. New York: Springer-Verlag, 1981.

———. *The Geology of Egypt*. New York: Elsevier Publishing, 1962.

Sajdi, Dana, ed. *Ottoman Tulips, Ottoman Coffee: Leisure and Lifestyle in the Eighteenth Century*. London: I.B. Tauris, 2007.

Salisbury, Joyce E. *The Beast Within: Animals in the Middle Ages*. New York: Routledge, 1994.

Sanders, Paula. *Ritual, Politics, and the City in Fatimid Cairo*. Albany: State University of New York Press, 1994.

Sardar, Ziauddin, ed. *An Early Crescent: The Future of Knowledge and the Environment in Islam*. London: Mansell, 1989.

Sbeinati, Mohamed Reda, Ryad Darawcheh, and Mikhail Mouty. "The Historical Earthquakes of Syria: An Analysis of Large and Moderate Earthquakes from 1365 B.C. to 1900 A.D." *Annals of Geophysics* 48 (2005): 347–435.

Schiebinger, Londa. "Why Mammals Are Called Mammals: Gender Politics in Eighteenth-Century Natural History." *American Historical Review* 98 (1993): 382–411.

Schimmel, Annemarie. *Islam and the Wonders of Creation: The Animal Kingdom*. London: al-Furqān Islamic Heritage Foundation, 2003.

Schoenfeld, Stuart, ed. *Palestinian and Israeli Environmental Narratives: Proceedings of a Conference Held in Association with the Middle East Environmental Futures Project*. Toronto: York University, 2005.

Scott, James C. *Seeing Like a State: How Certain Schemes to Improve the Human Condition Have Failed*. New Haven, CT: Yale University Press, 1998.

Sezgin, Fuat, Mazen Amawi, Carl Ehrig-Eggert, and Eckhard Neubauer, eds. *Studies of the Faiyūm Together with* Tārīḫ al-Faiyūm wa-Bilādihī *by Abū ʿUṯmān an-Nābulusī (d. 1261)*. Islamic Geography, vol. 54. Frankfurt am Main: Institute for the History of Arabic-Islamic Science at the Johann Wolfgang Goethe University, 1992.

Shafei Bey, Ali. "Fayoum Irrigation as Described by Nabulsi in 1245 A.D. with a Description of the Present System of Irrigation and a Note on Lake Moeris." In *Studies of the Faiyūm Together with* Tārīḫ al-Faiyūm wa-Bilādihī *by Abū ʿUṯmān an-Nābulusī (d. 1261)*, edited by Fuat Sezgin, Mazen Amawi, Carl Ehrig-Eggert, and Eckhard Neubauer, 103–55. Islamic Geography, vol. 54. Frankfurt am Main: Institute for the History of Arabic-Islamic Science at the Johann Wolfgang Goethe University, 1992.

Shaw, Stanford J. *The Financial and Administrative Organization and Development of Ottoman Egypt, 1517–1798*. Princeton, NJ: Princeton University Press, 1962.

———. "Landholding and Land-Tax Revenues in Ottoman Egypt." In *Political and Social Change in Modern Egypt: Historical Studies from the Ottoman Conquest to the United Arab Republic*, edited by P.M. Holt, 91–103. London: Oxford University Press, 1968.

———. "The Ottoman Archives as a Source for Egyptian History." *Journal of the American Oriental Society* 83 (1962): 447–52.

Shehada, Housni Alkhateeb. *Mamluks and Animals: Veterinary Medicine in Medieval Islam*. Leiden, Netherlands: Brill, 2013.

Shibl, Yusuf A. *The Aswan High Dam*. Beirut: Arab Institute for Research and Publishing, 1971.
Shindell, Drew T., and Gavin A. Schmidt. "Dynamic Winter Climate Response to Large Tropical Volcanic Eruptions since 1600." *Journal of Geophysical Research* 109 (2004): D05104.
Shokr, Ahmad. "Watering a Revolution: The Aswan High Dam and the Politics of Expertise in Mid-century Egypt." MA thesis, New York University, May 2008.
Shoshan, Boaz. "Grain Riots and the 'Moral Economy': Cairo, 1350–1517." *Journal of Interdisciplinary History* 10 (1980): 459–78.
Sigurdsson, Haraldur. *Melting the Earth: The History of Ideas on Volcanic Eruptions*. New York: Oxford University Press, 1999.
Singer, Amy. *Palestinian Peasants and Ottoman Officials: Rural Administration around Sixteenth-Century Jerusalem*. Cambridge: Cambridge University Press, 1994.
———. "Peasant Migration: Law and Practice in Early Ottoman Palestine." *New Perspectives on Turkey* 8 (1992): 49–65.
———, ed. *Starting with Food: Culinary Approaches to Ottoman History*. Princeton, NJ: Markus Wiener Publishers, 2011.
Slavin, Philip. "The Great Bovine Pestilence and Its Economic and Environmental Consequences in England and Wales, 1318–50." *Economic History Review* 65 (2012): 1239–66.
Smil, Vaclav. *Energy in Nature and Society: General Energetics of Complex Systems*. Cambridge: Massachusetts Institute of Technology Press, 2008.
———. *Energy in World History*. Boulder, CO: Westview Press, 1994.
Smith, Scot E., and Adel Abdel-Kader. "Coastal Erosion along the Egyptian Delta." *Journal of Coastal Research* 4 (1988): 245–55.
Sood, Gagan D.S. "Pluralism, Hegemony and Custom in Cosmopolitan Islamic Eurasia, ca. 1720–90, with Particular Reference to the Mercantile Arena." PhD diss., Yale University, 2008.
Soucek, Svat. "Certain Types of Ships in Ottoman-Turkish Terminology." *Turcica* 7 (1975): 233–49.
Sowers, Jeannie L. *Environmental Politics in Egypt: Activists, Experts, and the State*. London: Routledge, 2013.
———. "Remapping the Nation, Critiquing the State: Environmental Narratives and Desert Land Reclamation in Egypt." In *Environmental Imaginaries of the Middle East and North Africa*, edited by Diana K. Davis and Edmund Burke III, 158–91. Athens: Ohio University Press, 2011.
Squatriti, Paolo. "The Floods of 589 and Climate Change at the Beginning of the Middle Ages: An Italian Microhistory." *Speculum* 85 (2010): 799–826.
———. *Water and Society in Early Medieval Italy, AD 400–1000*. Cambridge: Cambridge University Press, 1998.
———, ed. *Working with Water in Medieval Europe: Technology and Resource-Use*. Leiden, Netherlands: Brill, 2000.
Stearns, Justin K. *Infectious Ideas: Contagion in Premodern Islamic and Christian Thought in the Western Mediterranean*. Baltimore: Johns Hopkins University Press, 2011.

Steeves, H. Peter, ed. *Animal Others: On Ethics, Ontology, and Animal Life.* Albany: State University of New York Press, 1999.

Sterchx, Roel. *The Animal and the Daemon in Early China.* Albany: State University of New York Press, 2002.

Stevenson, D.S., C.E. Johnson, E.J. Highwood, V. Gauci, W.J. Collins, and R.G. Derwent. "Atmospheric Impact of the 1783–1784 Laki Eruption: Part I, Chemistry Modelling." *Atmospheric Chemistry and Physics Discussions* 3 (2003): 551–96.

Sticker, Georg. *Abhandlungen aus der Seuchengeschichte und Seuchenlehre.* Giessen, Germany: A. Töpelmann, 1908–12.

Stiner, Mary C., and Gillian Feeley-Harnik. "Energy and Ecosystems." In *Deep History: The Architecture of Past and Present*, edited by Andrew Shryock and Daniel Lord Smail, 78–102. Berkeley: University of California Press, 2011.

Stothers, Richard B. "Volcanic Dry Fogs, Climate Cooling, and Plague Pandemics in Europe and the Middle East." *Climate Change* 42 (1999): 713–23.

Sublet, Jacqueline. "La peste prise aux rêts de la jurisprudence: Le Traité d'Ibn Ḥağar al-ʿAsqalānī sur la peste." *Studia Islamica* 33 (1971): 141–49.

Sufian, Sandra M. *Healing the Land and the Nation: Malaria and the Zionist Project in Palestine, 1920–1947.* Chicago: University of Chicago Press, 2007.

Sulaymān, ʿAbd al-Ḥamīd. "al-Sukhra fī Miṣr fī al-Qarnayn al-Sābiʿ ʿAshar wa al-Thāmin ʿAshar, Dirāsa fī al-Asbāb wa al-Natāʾij." In *al-Rafḍ wa al-Iḥtijāj fī al-Mujtamaʿ al-Miṣrī fī al-ʿAṣr al-ʿUthmānī*, edited by Nāṣir Ibrāhīm and Raʾūf ʿAbbās, 89–126. Cairo: Markaz al-Buḥūth wa al-Dirāsāt al-Ijtimāʿiyya, 2004.

Swabe, Joanna. *Animals, Disease, and Human Society: Human-Animal Relations and the Rise of Veterinary Medicine.* London: Routledge, 1999.

Tabak, Faruk. *The Waning of the Mediterranean, 1550–1870: A Geohistorical Approach.* Baltimore: Johns Hopkins University Press, 2008.

Taylor, Bron R., ed. *The Encyclopedia of Religion and Nature.* 2 vols. London: Thoemmes Continuum, 2005.

Taylor, Joseph E., III. *Making Salmon: An Environmental History of the Northwest Fisheries Crisis.* Seattle: University of Washington Press, 1999.

Tester, Keith. *Animals and Society: The Humanity of Animal Rights.* London: Routledge, 1991.

Tezcan, Baki. *The Second Ottoman Empire: Political and Social Transformation in the Early Modern World.* Cambridge: Cambridge University Press, 2010.

Thirgood, J.V. *Man and the Mediterranean Forest: A History of Resource Depletion.* London: Academic Press, 1981.

Thordarson, Thor, and Armann Hoskuldsson. *Iceland.* Harpenden, UK: Terra, 2002.

Thordarson, Thorvaldur, and Stephen Self. "Atmospheric and Environmental Effects of the 1783–1784 Laki Eruption: A Review and Reassessment." *Journal of Geophysical Research* 108 (2003): 1–29.

Tlili, Sarra. *Animals in the Qur'an*. Cambridge: Cambridge University Press, 2012.

Toledano, Ehud R. "Mehmet Ali Paşa or Muhammad Ali Basha? An Historiographic Appraisal in the Wake of a Recent Book." *Middle Eastern Studies* 21 (1985): 141–59.

Totman, Conrad. *The Green Archipelago: Forestry in Preindustrial Japan*. Berkeley: University of California Press, 1989.

———. *The Lumber Industry in Early Modern Japan*. Honolulu: University of Hawai'i Press, 1995.

Trawick, Paul B. "Successfully Governing the Commons: Principles of Social Organization in an Andean Irrigation System." *Human Ecology* 29 (2001): 1–25.

Tsugitaka, Sato. *State and Rural Society in Medieval Islam: Sultans, Muqta's and Fallahun*. Leiden, Netherlands: Brill, 1997.

Tuan, Yi-Fu. *Dominance and Affection: The Making of Pets*. New Haven, CT: Yale University Press, 1984.

Tuchscherer, Michel, ed. *Le commerce du café avant l'ère des plantations coloniales: Espaces, réseaux, sociétés (XVe–XIXe siècle)*. Cairo: Institut français d'archéologie orientale, 2001.

———. "Commerce et production du café en Mer Rouge au XVIe siècle." In *Le commerce du café avant l'ère des plantations coloniales: Espaces, réseaux, sociétés (XVe–XIXe siècle)*, edited by Michel Tuchscherer, 69–90. Cairo: Institut français d'archéologie orientale, 2001.

———. "La flotte impériale de Suez de 1694 à 1719." *Turcica* 29 (1997): 47–69.

———. "Some Reflections on the Place of the Camel in the Economy and Society of Ottoman Egypt." Translated by Suraiya Faroqhi. In *Animals and People in the Ottoman Empire*, edited by Suraiya Faroqhi, 171–85. Istanbul: Eren, 2010.

Tucker, Judith E. *Women in Nineteenth-Century Egypt*. Cambridge: Cambridge University Press, 1985.

Tucker, William F. "Natural Disasters and the Peasantry in Mamlūk Egypt." *Journal of the Economic and Social History of the Orient* 24 (1981): 215–24.

Ṭūsūn, ʿUmar. *Tārīkh Khalīj al-Iskandariyya al-Qadīm wa Turʿat al-Maḥmūdiyya*. Alexandria: Maṭbaʿat al-ʿAdl, 1942.

Tvedt, Terje. *The River Nile in the Age of the British: Political Ecology and the Quest for Economic Power*. London: I.B. Tauris, 2004.

Twigg, Graham. *The Black Death: A Biological Reappraisal*. London: Batsford, 1984.

Uchupi, Elazar, S.A. Swift, and D.A. Ross. "Gas Venting and Late Quaternary Sedimentation in the Persian (Arabian) Gulf." *Marine Geology* 129 (1996): 237–69.

ʿUthmān, Nāṣir. "Maḥkamat Rashid ka-Maṣdar li-Dirāsat Tijārat al-Nasīj fī Madīnat al-Iskandariyya fī al-ʿAṣr al-ʿUthmānī." *al-Rūznāma: al-Ḥauliyya al-Miṣriyya lil-Wathāʾiq* 3 (2005): 355–85.

Varlık, Nükhet. *Plague and Empire in the Early Modern Mediterranean World:*

The Ottoman Experience, 1347–1600. Cambridge: Cambridge University Press, 2015.

Wagstaff, J.M. *The Evolution of Middle Eastern Landscapes: An Outline to A.D. 1840*. London: Croon Helm, 1985.

Walker, Brett L. *The Lost Wolves of Japan*. Seattle: University of Washington Press, 2005.

Walz, Terence. *Trade between Egypt and Bilād as-Sūdān, 1700–1820*. Cairo: Institut français d'archéologie orientale, 1978.

Ward, Cheryl. "The Sadana Island Shipwreck: An Eighteenth-Century AD Merchantman off the Red Sea Coast of Egypt." *World Archaeology* 32 (2001): 368–82.

———. "The Sadana Island Shipwreck: A Mideighteenth-Century Treasure Trove." In *A Historical Archaeology of the Ottoman Empire: Breaking New Ground*, edited by Uzi Baram and Lynda Carroll, 185–202. New York: Kluwer Academic / Plenum, 2000.

Ward, Cheryl, and Uzi Baram. "Global Markets, Local Practice: Ottoman-Period Clay Pipes and Smoking Paraphernalia from the Red Sea Shipwreck at Sadana Island, Egypt." *International Journal of Historical Archaeology* 10 (2006): 135–58.

Warde, Paul. *Ecology, Economy and State Formation in Early Modern Germany*. Cambridge: Cambridge University Press, 2006.

Waterbury, John. *Hydropolitics of the Nile Valley*. Syracuse, NY: Syracuse University Press, 1979.

Watson, Andrew. *Agricultural Innovation in the Early Islamic World: The Diffusion of Crops and Farming Techniques, 700–1100*. Cambridge: Cambridge University Press, 1983.

White, Gilbert F. "The Environmental Effects of the High Dam at Aswan." *Environment* 30 (1988): 4–40.

White, Sam. *The Climate of Rebellion in the Early Modern Ottoman Empire*. Cambridge: Cambridge University Press, 2011.

———. "Rethinking Disease in Ottoman History." *International Journal of Middle East Studies* 42 (2010): 549–67.

Willcocks, W., and J.I. Craig. *Egyptian Irrigation*. 2 vols. London: E. & F.N. Spon, 1913.

Wing, John T. "Keeping Spain Afloat: State Forestry and Imperial Defense in the Sixteenth Century." *Environmental History* 17 (2012): 116–45.

———. "Roots of Empire: State Formation and the Politics of Timber Access in Early Modern Spain, 1556–1759." PhD diss., University of Minnesota, 2009.

Winter, Michael. *Egyptian Society under Ottoman Rule, 1517–1798*. London: Routledge, 1992.

Wittfogel, Karl A. "The Hydraulic Civilizations." In *Man's Role in Changing the Face of the Earth*, edited by William L. Thomas Jr., 152–64. Chicago: University of Chicago Press, 1956.

———. *Oriental Despotism: A Comparative Study of Total Power*. New Haven, CT: Yale University Press, 1957.

Wolfe, Cary. *What Is Posthumanism?* Minneapolis: University of Minnesota Press, 2010.

Wood, Charles A. "Climatic Effects of the 1783 Laki Eruption." In *The Year without a Summer? World Climate in 1816*, edited by C.R. Harington, 58–77. Ottawa: Canadian Museum of Nature, 1992.

Yaffe, Martin D., ed. *Judaism and Environmental Ethics: A Reader*. Lanham, MD: Lexington Books, 2001.

Yasui, Maya, and Takehiro Koyaguchi. "Sequence and Eruptive Style of the 1783 Eruption of Asama Volcano, Central Japan: A Case Study of an Andesitic Explosive Eruption Generating Fountain-Fed Lava Flow, Pumice Fall, Scoria Flow and Forming a Cone." *Bulletin of Volcanology* 66 (2004): 243–62.

Yaycıoğlu, Ali. "Provincial Power-Holders and the Empire in the Late Ottoman World: Conflict or Partnership?" In *The Ottoman World*, edited by Christine Woodhead, 436–52. New York: Routledge, 2012.

Yi, Eunjeong. *Guild Dynamics in Seventeenth-Century Istanbul: Fluidity and Leverage*. Leiden, Netherlands: Brill, 2004.

Zachariadou, Elizabeth, ed. *Natural Disasters in the Ottoman Empire*. Rethymnon, Greece: Crete University Press, 1999.

Ze'evi, Dror. *An Ottoman Century: The District of Jerusalem in the 1600s*. Albany: State University of New York Press, 1996.

Zielinski, G.A., R.J. Fiacco, P.A. Mayewski, L.D. Meeker, S. Whitlow, M.S. Twickler, M.S. Germani, K. Endo, and M. Yasui. "Climatic Impact of the A.D. 1783 Asama (Japan) Eruption Was Minimal: Evidence from the GISP2 Ice Core." *Geophysical Research Letters* 21 (1994): 2365–68.

Zinsser, Hans. *Rats, Lice and History, Being a Study in Biography, Which, after Twelve Preliminary Chapters Indispensable for the Preparation of the Lay Reader, Deals with the Life History of Typhus Fever*. London: George Routledge and Sons, 1935.

Index

Page numbers in italics refer to figures.

al-Sayyid ʿAbd Allāh ibn Sālim Qarqūr, 128–29
ʿAbd al-Nasser, Gamal, 35
Abdülhamid I, 195
acid rain. *See* sulfur dioxide
Adams, Robert McC., 6
Adas, Michael, 225n34
agriculture. *See* food supplies; irrigation; taxation; water
Aḥmad ibn Aḥmad, 119
Aḥmad Jalabī ibn al-Marḥūm al-Amīr Muṣṭafā Bey, 29–31
Ahmet III, 27, 61, 102
Aleppo, 55–56, 153, 180
Alexandria, 13, 39–41, 57, 87–90, 121, 132–33, 158–60, 162, 228n26
Algeria, 55
ʿAlī ʿAbd al-Qādir al-Ḥamdani, 120
ʿAlī Bey al-Kabīr, 75, *75*, 140, 143, 192
ʿAlī ibn ʿAlī al-Ḥamd, 125
ʿAlī ibn Dāwūd al-Khawlī, 125
ʿAlī ibn al-Marḥūm ʿUthmān Abū Bakr, 128
al-Shābb ʿAlī ibn Sīdī Aḥmad al-Muzayyin, 111–12, 250n2
ʿAlī Muṣṭafā ʿĪsāwī, 117–18, 229n33
ʿAlī ʿUmar, 128–29
Anatolia, 2–3, 8–9, 54, 136, 154–58, 162, 164–66, 199–200, 273n23
The Animal in Ottoman Egypt (Mikhail), xii
animals: as commodities, 112–15, 118–19, 123–30, 133, 148–49, 162, 250n2; corvée's supplanting of, 144–50; disease's devastation and, 133–44, 149, 179, 199–200, 265n45; as energy sources, 11, 111–15, 121–23; as food products, 120–21; as historical agents, xii, 129–30, 133–34, 150, 202, 251n5; historiography of,

animals (*continued*) 251n5; humans' relation to, 131–32, 266n55, 267n76; labor and, 131–32, 161, 243n21; legal disputes over, 111–12, 115, 118–19, 123–26, 250n2, 258n72; mobility of, 2, 126–30; value of, 115–19, 253n7, 255n27, 256n37. *See also specific animals*
Antes, John, 34–35, 173, 175–77, 180–82, 279n14, 281n46, 281n49, 283n71, 284n100, 284n102
Asama volcano, 288n58
Aṣfūn, 120
Ashrafiyya Canal, 39–41
astrologers, 170–71
Aswan Dam, 90, 220n2
Aswan High Dam, 35–36, 90–91, 227n9
Asyūṭ, 26, 177–78, 283n65
al-Muʿallim ʿAṭāʾ Allah, 95–97, 99, 108
al-ʿAṭṭār, Ḥasan, 177–78
Awqāf al-Ḥaramayn, 57
Aydab, 57, 156
ʿAzīza bint Ḥasan Abū Aḥmad, 127

al-Baḥayra, 39, 48, 125, 127, 230n42, 258n72
al-Bahnasa, 235n40, 273n19
Baḥr al-Fuḍālī, 98
al-Baḥr al-Ṣaghīr, 24–26, 29, 38–39, 78–79, 98–102, 122
Baḥr Yūsuf, 59–64, 102, 238n54
Banī Kalb, 85
Banu Khalid, 148
Barkey, Karen, 54
Bedouins, 116, 120–21, 124, 259n74
Bektaş Çelebi, 28
Beni Suef, 51, 59, 62, 231n4, 235n40, 273n19
beylicate, 231n8. *See also* Mamluks; *specific Beys*
Big Chill, 8
Bilbays, 138
Birkat Qārūn, 59
Black Sea littoral, 54, 157–58, 165–66
Brenner debate, 131–34, 260n2
British Empire, 53, 133, 248n77, 283n65
bubonic plague, 177, 282n57, 282n63. *See also* disease; plague
buffalo cows. See *jāmūsa*
Bulaq, 57, 161–62, 172
Bulliet, Richard W., 8, 266n55
Butzer, Karl, 5–6

Cairo, 23, 41, 44, 53, 84, 121, 132–33, 153, 161–62, 170–81, 229n36, 269n102
caloric energy, 11, 53–55, 70, 82, 113–15, 134–38, 147, 157, 167, 262n14
camels, 115–18, 122, 124, 139, 161–62, 165–67, 202
canals: classifications of, 101–2, 221n8, 237n51; Fayyum and, 59–61; forced labor and, 74–92, 133–34, 144–50; images of, 25, *87*; political power and, 24–33, 37–38, 77–81; repairs and, 42–43, 52–55, 61–66, 77–79, 85–86, 94–97, 100–102, 122, 147–48, 156, 178, 222n15, 227n19, 235n39, 238n63, 243n21; silt and dredging of, 25–27, 38–44, 98–102, 146, 227n19, 228n29; surveys of, 20–24, 222n12
Celali peasant revolts, 9
center-and-periphery model of empire, 53–55, 64–66
cerîm ships, 160–61
China, 1–2, 4, 131–32, 173, 186, 190–91, 280n35
Chios, 159
cholera, 13. *See also* disease
Christensen, Peter, 5–6
chronicles and chroniclers, 4–5
climate, 2, 8–10, 184–97, 201. *See also* environmental history; politics
Copts, 106, 182

Cordova, Carlos, 6
corruption, 62–64, 66–70
corvée, 79–81, 88, 133–34, 144–50, 267n76. *See also* labor
cows, 124–26, 128–29, 140, 255n28, 265n45. *See also* animals
Crete, 55, 159
Cronon, William, 8, 166, 291n1
crops. *See* famines; food supplies; taxation
Cuno, Kenneth M., 143

Damietta, 95, 100, 105, 121, 160, 273n23
dams, 24, 35, 52, 59–61, 63–70, 102–3, 122, 156, 178, 240n81, 241n82. *See also* irrigation; water
al-Damurdāshī Kathudā ʿAzabān, Aḥmad, 180–81, 284n93
al-Daqahliyya, 24, 41–42, 46, 94–97, 99–102, 122–23
Darwin, Charles, 251n5
delta (Nile), 34–36, 39–40, 57, 226n1, 226n3. *See also* Nile
Denmark, 188–91, 290n94
dikes, 59–61, 63–70, 78–79, 102–3, 156. *See also* Fayyum; irrigation; water
disease: animal reduction and, 133–44, 149, 179; archival sources about, 5; climate change and, 187, 193, 263n19; court records and, 278n7; as ecological phenomenon, 11–15, 169–70, 174–78, 180–83, 199–202; famines and, 10, 178–80, 187–88; *ḥajj* and, 167, 175, 271n13; strongman politics and, 140–44; trade routes and, 2, 12, 56, 169–70, 173–74, 199–200, 217n57. *See also* plague
Dols, Michael W., 217n57, 280n32, 280n37
donkeys, 115–18, 124, 127, 139, 202, 259n83
Don-Volga project, 7
dredging, 21–22, 24, 30, 37–44, 50, 79, 146, 227n19
drought, 12, 133–36, 163–64, 178–79, 193. *See also* famines; food supplies
Dubrovnik, 53

earthquakes, 7, 170–71, 186–87
Egypt: agricultural shifts in, 132–33; archival records of, 4; British colonization of, 3, 133, 248n77; climate change and, 191–97; disease's political destabilizing of, 171–74, 217n57; French invasion of, 74, 137–38, 173–74, 196; *ḥajj* and, 12–14; Herodotus and, 34–36; Hijaz and, 154–56; importance of, to Ottoman Empire, xii, 102–3, 153–55, 164–65, 167–68, 239n70; irrigation works in, xiii, 19–20, 24–33, 38–44, 59–61, 199–200, 221n8, 237nn50–51, 268n85; local expertise and, 20–33, 38–44, 56, 76; maps of, *ix*, *58*; massive infrastructural projects of, 80–92, 98–99; Ottoman management of, 59, 61, 73–74; resource circulation and, 56–59; strongman politics in, 74–75, 140–44, 192–96, 247n62; studies of, 5–6; taxation and, 19–20, 195, 224n26
Egyptian National Archives, 51, 231nn3–4
Egyptian School of Engineering, 89
enclosure. *See* land
energy: caloric, 11, 53–55, 70, 134–38, 147–49, 157, 167, 262n14; energy regimes and, 261n8; fossil fuels and, 11, 15, 134, 150; Ottoman consumption of, 165–68
engineers, 100–102, 104–8. *See also* expertise; irrigation; labor; water
England, 188–90, 265n45, 266n64
environmental history, 165–68; climate change and, 8–10, 184–97; crops and, 1–2; disease and,

environmental history (*continued*) 11–15, 169–70, 180–83, 263n19; energy and, 10–11, 165–68; labor and, 73–81; political impacts of, xi–xii, 9, 192–97, 199–203, 241n2, 290n94; pristine nature idea and, 93–94, 107; religious interpretations of, 186, 277n2; studies of, 2, 4–8, 218n66, 227n11
epizootics, 135–38, 193, 262n16
Erlich, Haggai, 234n32
eruption-people, 188
Ethiopia, 3, 56, 202, 220n2
The Evolution of Middle Eastern Landscapes (Wagstaff), 212n22
expertise, 14–15, 27; labor's commodification and, 82–85, 93–108; local autonomy and, 20–24, 59, 64–67, 103–4, 164–65, 167–68; local knowledge and, 38–44; maintenance work and, 61–64; reputations and, 101–2; workers' classifications and, 52–53

famines, 2, 10, 12, 62–64, 136–37, 169, 175, 178–80, 193–94, 266n56
Famm Ẓāfir, 29–31, 225n31
Fayyum, 51–70, 102–3, 156, 231n4, 235n39, 236n42, 237n50, 238n55, 238n63, 273n19
fishing, 120–21
fleas, 12, 173, 175–76, 181
flooding (of the Nile), 2, 63, 121, 135, 169, 174–78, 191–94, 196–97, 202–3, 220nn3–4, 238n55, 281n46
food supplies: animal labor and, 121–23; animals as providers of, 120–21; animals as transporters of, 113–15; archival records of, 6–7; circulation of, 55–70; Egypt as central provider of, 73–74, 102–3; famines and, 178–80, 193–94; *ḥajj* and, 154–56; irrigation works and, 26, 61–66; logistical management of, 153–56, 167–68; milk and, 117; Nile flooding and, 136–37, 193–94, 196–97, 202–3, 220nn3–4, 238n55; Ottoman administration of, 19–20, 163–65; prices of, 113, 163–64, 179; rim relationships of empire and, 55–59; seasonal description of, 220n3; storage spaces of, 82–84; taxes on, 19–20, 97, 103–4, 194–95, 224n26, 246n51, 257n58; technology and, 111–15
fossil fuels, 11, 15, 134, 150
France, 74
Franklin, Benjamin, 189
fraud, 123–26
Friars de Propaganda Fide, 177

Gâzî Ḥasan Paşa, 195
Gershoni, Israel, 234n32
al-Muʿallim Ghālī, 106
al-Gharaq, 59–63, 65–70, 102–3, 240n81, 241n82
al-Gharbiyya, 79, 122, 138, 147
Giza, 44, 46, 172
Glacken, Clarence, 218n66
Goa, 56
Gönüllüyan military bloc, 235n40
grain. *See* food supplies
Gran, Peter, 56
Great Lakes, 166
Greenland, 186, 189
Guha, Ramachandra, 218n66

al-Hajārsa, 41
ḥajj, 12–14, 57, 63–64, 154–56, 167, 175, 271n13
Hanna, Nelly, 56
al-Muʿallim Ḥasan, 98–99, 108
Ḥasan Abu ʿĀliyya, 128
Ḥasan Aghā Bilīfyā, 235n40, 273n19
al-Ḥajj Ḥasan ibn al-Marḥūm al-Sayyid ʿAlī al-Sharīf, 115
Ḥasan Paşa, 121
Ḥasan Shukr, 125–26
Hathaway, Jane, 231n5, 231n8
Haze Hardships, 187–88
Hazz al-Quḥūf (al-Shirbīnī), 291n2

heat, 181–82, 284n100
Hijaz, 3, 13, 55–59, 61, 63–64, 70, 97, 154–57, 162, 164, 167
historiography: animals as agents and, xiii, 129–34, 150, 251n5; center-and-periphery model of empire and, 53–55, 61–66, 154–55, 276n58; environmental history and, xii–xiii, 4–8, 184–97; labor's continuity and, 93–108; material culture and, 270n3; sources and, 4–8, 69–70, 170, 185, 270n3, 278n7
Holland, 189–91
horses, 116, 139
hub-and-spoke metaphor for empire, 54–55, 64–66
Humām ibn Yūsuf ibn Aḥmad ibn Muḥammad ibn Humām ibn Ṣubayḥ ibn Sībīh al-Hawwārī, 116, 121–22
Ḥusayn Çelebi ʿAjūwa, 105
Hyderabad, 155

Ibn al-Wardī, 180, 280n35
Ibrahim Ağa, 28
Ibrāhīm Aghā, 83
al-Shaykh Ibrāhīm ibn al-Shaykh Ramaḍān al-Shayūnī al-Khaḍrāwī, 115
Ibrahim Paşa, 106
Iceland, 2–3, 9–10, 184–97, 286n10, 290n94
Imperial Dockyards, 158, 162–63
India, 1–3, 173, 199–200, 218n66, 248n77
Indian Ocean, 3, 10, 56, 154–55, 191, 197
Iran, 6–9, 11
irrigation: animal labor and, 111–12; corruption and, 62–64, 66–70; Egyptian politics and, 19–20, 94–97, 199–200; Fayyum and, 59–61, 102–3; food supplies and, 19–20, 56, 61–64, 69–70; forced labor and, 74–92, 133–34, 144–50; local expertise and, 14–15, 19–20, 24–33, 38–44, 51–55, 64–66, 77–81; maintenance of, 23–24, 59–70, 78–79, 94–97, 122, 146–48, 156, 222n15, 227n19, 235n39, 238n63, 243n21; plague's relation to, 283n65; surveys of, 20–24, 222n12; taxes and, 26. *See also* canals; water
Islam, 14–15, 211n15, 255n28, 256n42, 276n58
Islamic green revolution, 2
islands, 44–50, 229n33
Ismaʿil Abu Taqiyya, 56
Ismāʿīl Bey, 171
Isnā, 117–18, 120–21
Istanbul, xi–xii, 2–3, 14–15, 51–55, 61, 136, 154–59, 162, 167–68, 171, 182
Izmir, 55, 136, 182

al-Jabartī, ʿAbd al-Raḥman, 62, 88, 104–5, 135, 137, 141–42, 172–80, 193–95
jāmūsa, 115–19, 122, 124, 127–28, 162, 199–200, 202, 255n27, 258n72
Japan, 191, 288n58
al-Jazīra Bākhir, 41
Jedda, 57
Jerusalem, 153
Jisr al-Ẓafar, 30, 225n31
Jordan, 6
Jordan, William Chester, 266n56
al-Jusūr al-Sulṭāniyya, 20–24, 220n5, 237n51

Kafr Ghannām, 41
Kafr al-Gharīb, 48
Kano, 56
Kasaba, Reşat, 266n58
al-Kashūfiyya, 29–31
Katla, 286n10
al-Khalīj (canal), 178, 269n102
khamāsīn, 180–81
al-Khashshāb, Ismāʿīl ibn Saʿd, 172
Kuhnke, LaVerne, 283n65
Kurdistan, 173

labor: animals as, 11, 111–15, 121–23, 131–32, 161, 165, 243n21, 258n68; circulation of, 83–85, 100–102; commodification of, 123, 132, 142–44; elite land seizures and, 81–92, 141–44, 179, 192–96, 247n62; energy studies and, 10–11, 242n11; environment's relation to, 73–74, 77–81; expertise and, 14–15, 52–53, 61–67, 93–108; forced, 2, 74, 77, 79–92, 114–15, 133–34, 144–50; humans as replacements for animals and, 133–50, 165–66, 266n58, 267n76; land ownership and, 143–44; local nature of, 77–81; politics of, 80–92. *See also* animals; corvée; expertise; irrigation; politics
al-Lāhūn, 59–62, 65–70, 102–3
Lake Azbakiyya, 171, *171*
Lake of al-Fīl, 171
Lake Maʿdiyya, 121
Lake of al-Manzala, 99, *99*
Lake of Maṭariyya, 121
Lake Nasser, 36, 91
Lake Timsāḥ, 147
Laki fissure, 9–10, 184–97, 199–200, 285n1
land: animals and, 131–32, 148–49; elite seizures of, 76, 81–92, 141–44, 179, 192–93, 195–96, 247n62, 266n64, 267n73; legal disputes over, 44–50, 230n42; Nile's effects on, 44–50. *See also* food supplies; irrigation; law; politics
law: animal ownership and, 111–12, 115, 118–19, 123–30, 250n2, 258n72; archival records and, 5, 51–52; expertise and, 94–97; inheritance and, 115, 118–19, 123–26, 278n7; land usage rights and, 44–49, 230n42; water rights and responsibilities and, 27–30, 37–38, 41–44, 46, 95–102, 223n17, 227n19, 228n29. *See also specific cases and courts*
Lebanon, 55, 155, 191
Little Ice Age, 2, 9, 201

al-Maḥalla al-Kubrā, 138, 147
Mahmud II, 88, 90
Maḥmūdiyya Canal, 87, *87*, 88–91, 148, 243n23
Malta, 159
Mamluks, 9, 51–53, 76, 153, 170, 231n8, 239n70
Manfalūṭ, 27, 85–87, 103–4, 124–27
al-Manṣūra, 24–26, 38–39, 41, 43, 46, 78, 94–102, 223nn17–18, 227n19
al-Manzala, 24, 30, 38
al-Maqrīzī, 280n35
McNeill, J.R., 6, 213n24, 261n8
Mecca, 53, 57, 61, 63, 154–56, 167–68
Medina, 53, 57, 61, 63, 154–56, 167–68
Mediterranean Sea, 1–3, 6, 13, 34–35, 54–56, 87–90, 159–63, 173, 181, 190, 213n24
Mehmed Paşa al-Yedekçi, 67–70, 240n78, 240n81
Mehmed Raghib Paşa, 68, 240n81
el-Hâc Mehmet, 103–4, 108
Mehmet ʿAli, 35, 74, 82, 86–88, 90, 104–8, 138–49, 241n2, 246n51, 247n65, 250n24, 268n85
Meiggs, Russell, 6
merchant galleys, 158–59
miasma theory of disease, 173–74, 284n102
Middle East. *See* Egypt; historiography; Ottoman Empire; *specific places*
milk, 117, 120–21, 256n42
al-Minūfiyya, 22–23
Minyyat Ḥaḍr, 122
Minyyat Ḥiṭṭiyya, 40
Minyyat Ṭalkhā, 122
Mīt ʿĀfiyya, 79

Mitchell, Timothy, 203
Móðuharðindin, 188
Mongolia, 189, 191, 280n35
monsoon rains, 9–10, 54, 191–92, 197, 202–3
Morea, 136
Morocco, 6, 11, 55
Mountains of the Mediterranean (McNeill), 213n24
Mughal Empire, 9
Muḥammad Aghā, 98–99
Muḥammad Bey Abū Dhahab, 76, 140, 143, 192
Muḥammad ibn ʿAbd al-Raḥman ibn Muḥammad Sulīkar, 116
Muḥammad ibn Baghdād, 22
Muḥammad Khalīf, 258n72
Murād Bey, 172, 179
Muṣṭafā ʿAbd Rabbu, 115
Muṣṭafā Aghā, 148
Muṣṭafā al-Faqīr ibn Muḥammad ʿAqaṣ, 118–19
al-Ḥajj Muṣṭafā ibn al-Marḥūm al-Ḥājj Darwīsh ʿIzzat ibn ʿUbayd, 111–12, 250n2
Mustafa Paşa, 40
Mustahfızan military bloc, 279n18
Mýrdalssandur, 187

al-Nābulusī al-Ṣafadī, Abū ʿUthmān, 236n42, 238n54, 238n63
Napoleon Bonaparte, 138, 196
natural disasters, 5, 7
Nature and Empire in Ottoman Egypt (Mikhail), xiii
Nile: archival records concerning, 4; delta of, 34, 39–40, 57, 91, 226n1, 226n3; Fayyum and, 59–64; flooding of, 2, 48, 63, 121, 135–37, 169, 178–80, 191–94, 196–97, 202–3, 220nn3–4, 238n55, 281n46; images of, *95*; irrigation networks and, 36–37; islands in, 44–50; Ottoman management of, 19–20; Rosetta branch of, 39–40, 160; silting of, 34–38, 226n1, 226n3; transnational importance of, 200–201. *See also* Cairo; disease; wood supplies; *specific ports and storage facilities*
The Nile (Erlich and Gershoni), 234n32
Nilometer, 46, 229n33
Nitmā, 48
Norway, 188–89
Nūb Ṭarīf, 43–44
Nuuk, 189

oil, 2, 11, 15. *See also* fossil fuels
Oriental despotism, 7, 14, 203
Orientalism, 203
Osman's tree, xi–xii
Ottoman Empire: archives of, 4–8, 69–70; bureaucracy of, 42, 44, 46–47, 54–56, 62–64, 66–70, 123, 153–55, 158–65, 223n18, 239n70, 276n58; climate change and, 191–97; decentralization in, 9, 74, 76, 140–41, 164, 192–93; as ecosystem, 199–203; Egypt's importance to, xii, 73–74, 102–3, 153–55, 164–65, 167–68, 239n70, 276n58; environmental historiography and, xii–xiii, 43–50, 165–68, 199–203; labor regimes in, 93–108; local expertise and, 14–15, 19–20, 24–33, 38–44, 51–70, 79, 96–97, 99–108, 164–65, 167–68, 230n40; maps of, *vii*, *157*; material culture of, 270n3; military expeditions of, 159; origin stories of, xi–xii, 3; resource circulation in, 55–69, 153–66; rim relationships of empire and, 54–70; Russia's wars with, 74, 195; strongman politics and, 74–75, 140–41, 192–96; taxes and, 26–30, 76, 79–80, 86–87, 97, 103–4, 153–55, 194–95, 224n26; trade and, 1–3, 12–14, 154–55, 158–60; water

Ottoman Empire (*continued*)
management and, 19–33, 37–44, 51–70, 77–81, 199–200, 222n12, 228nn29–30, 230n42, 235n39, 237nn50–51, 268n85; wood supplies and, 154–63
oxen, 80, 115, 117, 122, 202, 255n27

Panzac, Daniel, 174, 181
The Pasha's Peasants (Cuno), 143
piracy, 159
plague, 12–13, 169–73; of 1785, 140; of 1791, 170–81, 264n26; of 1801, 177–78; animals' deaths and, 133–38; carriers of, 173–74, 282n56; commodity circulation and, 56; flooding and, 174–78, 281n46; irrigation projects and, 283n65; religious understandings of, 277n2; sources on, 280n35, 289n73; translations and, 262n16. *See also* disease; *specific varieties*
pneumonic plague, 176–77, 282n57, 283n65
politics: disease and, 140–44, 171–74; environmental changes' effects on, 9, 192–96, 199–203, 241n2, 290n94; of labor, 74–77, 80–92; land ownership and, 76, 81–92, 141–44, 179, 192–96, 247n62, 266n64, 267n73; local power and, 24–33, 51–70, 96–108, 164–68; water management and, 24–33, 36–38. *See also* Egypt; Ottoman Empire; rim relationships of empire; water
Pomeranz, Kenneth, 132–33, 260n2
Portugal, 155
precedent (conceptual power of), 46–50
price ceilings, 163–64, 179
prices (of grain), 113, 163–64, 179
Prime Ministry's Ottoman Archive, 52
property: animals as, 111–12, 114–15, 118–19, 123–30, 133–34, 162, 202, 250n2; archival records about, 5; land usage rights and, 44–50, 141–44, 148–49, 179, 202, 266n64, 267n73

Qabr al-Umarāʾ, 128
quarantines, 13
Qūjindīma, 46–47
Qusayr, 57, 156, 235n38

al-Raḥmāniyya, 40
Raʾs al-Wādī Canal, 147
rats, 12, 173–79, *176*, 199–200
al-Rauḍa, 44, *45*, 46, 229n33
Raymond, André, 174, 281n42
Red Sea, 3, 54–56, 82, 153–55
reputation, 101–2
rice, 105, 111, 115–16, 119, 220n3, 233n15, 235n37. *See also* food supplies; irrigation
Richards, John F., 131–33, 210n4
Riḍwān Bey, 180
rim relationships of empire, 54–55, 61–66, 154–55, 276n58
Rivlin, Helen Anne B., 243n23
Rosetta (city), 35, 56, 82–86, 90, 105, 114, 119–21, 160, 162, 173
Rosetta branch (of the Nile), 39–40, 88–90, 160, *160*, 161–62. *See also* food supplies; wood supplies
el-Rouayheb, Khaled, 56
Rushdan ʿAbd al-Rasūl, 127
Russo-Ottoman wars, 74, 195
Ruzzāfa, 128

Safavid Empire, 7–9
Sālima bint Ḥasan ʿĀmir, 258n72
Salonica, 55
Scott, James C., 245n49
Seeing Like a State (Scott), 245n49
al-Ḥājj Shāhīn, 100–101, 104, 106, 108
Shārimsāḥ, 94–97
al-Sharqiyya, 41, 138

Shaw, Stanford J., 276n51
sheep, 117, 120, 124, 188, 254n8, 258n69, 266n64
Shibīn Canal, 147, 268n85
ships, 155–65, 273n23
al-Shirbīnī, Yūsuf, 80–81, 267n76, 291n2
Shūbar, 122
Síða, 187
silt, 24, 34, 36, 38–44, 146, 202, 226n1, 226n3
slavery, 2, 114–15, 144–50. *See also* corvée; labor
smallpox, 188–89
Sofia, 155
status symbols, 120–23, 125–26. *See also* animals; land; politics; property
steers, 125–26, 258n69. *See also* cows
Steingrímsson, Jón, 186–88
Sticker, Georg, 217n57
strongmen. *See* Egypt; Ottoman Empire; politics
Subaltern Studies, 244n48
Sudan, the, 3, 12, 56, 173, 248n77
Suez, 57, 154–57, 160–65, 273n23
Suez Canal, 7, 13, 15, 90–91, 160, 275n43
Sulaymān Jurbajī Murād, 124
sulfur dioxide, 14–15, 186–93, 201
Sunṭīs, 128
surveys (irrigation), 20–24, 222n12
Syria, 15, 54, 157, 165–66, 192–93, 281n50

Tahākān, 30
Ṭalkhā al-Gharbī, 46–47
Tanzimat, 10
taxation, 19–20, 26, 97, 103–4, 143–44, 153–55, 194–95, 257n58
tax farmers, 28–29, 44–50, 76, 79–80, 86–87, 98–99, 144–50, 224n26, 246n51
Tehran, 155
theft (of animals), 124–26
Thirgood, J.V., 6
Topkapı Palace Museum Archive, 52
Toshka scheme, 90–91
Traces on the Rhodian Shore (Glacken), 218n66
trade: disease and, 2, 12, 56, 169–70, 173–74; Ottoman Empire's designs and, 157–60; rim relationships of empire and, 54, 57; Suez and, 7, 154–56, 160–64, 273n23
Tripoli (Lebanon), 55, 189
Tripoli (Libya), 55, 136
Ṭummāy, 43–44
Ṭunāmil, 98–99
Tunis, 55
Ṭurā, 172
Turkey, 6–7, 15

ʿulamāʾ, 179
The Unending Frontier (Richards), 210n4
United States, 190–91
urbanization, 76, 132–33, 179–80
ūsya lands, 79–81, 268n77
Uzbek Empire, 9

Venice, 56, 136
volcanos, 2, 9–10, 184–97, 199–200, 286n10, 288n58
Volney, C.F., 194

Wagstaff, J.M., 6, 212n22
al-Waḥdāniyya, 96–97
Waḥīshāt, 27, 103–4
waqf, 21, 82–83, 120–21, 156, 235n39
water: animal labor and, 111–12; corruption and, 62–64; as energy source, 10–11, 261nn8–9; legal claims and, 27–30, 37–38, 41–44, 46, 94–102, 223n17, 227n19, 228n29; local autonomy and, 14–15, 24–33, 38–44, 51–70, 199–200; Nile floods and, 2, 34–38, 63, 121, 135, 169, 174–78, 191–97, 202–3; shortages of, 44–50; surveys of, 20–24, 222n12;

water (*continued*)
taxation and, 26. *See also* canals; food supplies; irrigation; Ottoman Empire; trade
water buffalo. See *jāmūsa*
waterwheels, 111–12, *112*, 114, 122–23, 138, *139*, 139–40
weather, 55, 134, 136–38, 174–75, 202. *See also* climate; environmental history; flooding (of the Nile); heat; monsoon rains
weirs, 27–28, 62–64, 103–4
wergeld, 256n37
wind, 10–11, 180–81, 202
Wittfogel, Karl, 7
wood supplies, 6; Anatolian forests and, 54, 157–58, 162, 164, 199–200, 273n23; Egypt's lack of, 154–55; logistics involved with, 156–57, 165–68; shipbuilding and, 156–63; storage of, 158–59

Yemen, 2, 55, 155
al-Sayyid Yūnī, 119
Yūsuf ibn Ḥasan ʿAṭīya, 125

al-Zababani, 180
Zaʿfarānī Canal, 147–48
al-Zaqāzīq, 147
Zāwiyat Naʿīm, 128